Direct Sum Decompositions of Torsion-Free Finite Rank Groups

PURE AND APPLIED MATHEMATICS

A Program of Monographs, Textbooks, and Lecture Notes

MONOGRAPHS AND TEXTBOOKS IN PURE AND APPLIED MATHEMATICS

Recent Titles

W. J. Wickless, A First Graduate Course in Abstract Algebra (2004)

R. P. Agarwal, M. Bohner, and W-T Li, Nonoscillation and Oscillation Theory for Functional Differential Equations (2004)

J. Galambos and I. Simonelli, Products of Random Variables: Applications to Problems of Physics and to Arithmetical Functions (2004)

Walter Ferrer and Alvaro Rittatore, Actions and Invariants of Algebraic Groups (2005)

Christof Eck, Jiri Jarusek, and Miroslav Krbec, Unilateral Contact Problems: Variational Methods and Existence Theorems (2005)

M. M. Rao, Conditional Measures and Applications, Second Edition (2005)

A. B. Kharazishvili, Strange Functions in Real Analysis, Second Edition (2006)

Vincenzo Ancona and Bernard Gaveau, Differential Forms on Singular Varieties: De Rham and Hodge Theory Simplified (2005)

Santiago Alves Tavares, Generation of Multivariate Hermite Interpolating Polynomials (2005)

Sergio Macías, Topics on Continua (2005)

Mircea Sofonea, Weimin Han, and Meir Shillor, Analysis and Approximation of Contact Problems with Adhesion or Damage (2006)

Marwan Moubachir and Jean-Paul Zolésio, Moving Shape Analysis and Control: Applications to Fluid Structure Interactions (2006)

Alfred Geroldinger and Franz Halter-Koch, Non-Unique Factorizations: Algebraic, Combinatorial and Analytic Theory (2006)

Kevin J. Hastings, Introduction to the Mathematics of Operations Research with *Mathematica®*, Second Edition (2006)

Robert Carlson, A Concrete Introduction to Real Analysis (2006)

John Dauns and Yiqiang Zhou, Classes of Modules (2006)

N. K. Govil, H. N. Mhaskar, Ram N. Mohapatra, Zuhair Nashed, and J. Szabados, Frontiers in Interpolation and Approximation (2006)

Luca Lorenzi and Marcello Bertoldi, Analytical Methods for Markov Semigroups (2006)

M. A. Al-Gwaiz and S. A. Elsanousi, Elements of Real Analysis (2006)

Theodore G. Faticoni, Direct Sum Decompositions of Torsion-Free Finite Rank Groups (2007)

R. Sivaramakrishnan, Certain Number-Theoretic Episodes in Algebra (2006)

Aderemi Kuku, Representation Theory and Higher Algebraic K-Theory (2006)

Robert Piziak and P. L. Odell, Matrix Theory: From Generalized Inverses to Jordan Form (2007)

Norman L. Johnson, Vikram Jha, and Mauro Biliotti, Handbook of Finite Translation Planes (2007)

Direct Sum Decompositions of Torsion-Free Finite Rank Groups

Theodore G. Faticoni

Fordham University
Bronx, New York, U.S.A.

CRC Press

Taylor & Francis Group

Boca Raton   London   New York

CRC Press is an imprint of the
Taylor & Francis Group, an **informa** business

A CHAPMAN & HALL BOOK

CRC Press
Taylor & Francis Group
6000 Broken Sound Parkway NW, Suite 300
Boca Raton, FL 33487-2742

First issued in paperback 2019

ISBN-13: 978-1-58488-726-3 (hbk)
ISBN-13: 978-0-367-38932-1 (pbk)

Library of Congress Cataloging-in-Publication Data

Faticoni, Theodore G. (Theodore Gerard), 1954-
    Direct sum decompositions of torsion-free finite rank groups / Theodore G. Faticoni.
        p. cm. -- (Pure and applied mathematics)
    Includes bibliographical references and index.
    ISBN-13: 978-1-58488-726-3
    ISBN-10: 1-58488-726-5
    1. Torsion free Abelian groups. 2. Direct sum decompositions. I. Title.

QA180.F38 2006
512'.25--dc22   2006021040

Visit the Taylor & Francis Web site at
http://www.taylorandfrancis.com

and the CRC Press Web site at
http://www.crcpress.com

To my parents who gave me life.

To Adolph and Margaret Faticoni.

Contents

Preface

This book is directed at mathematicians who wish to study advanced topics in direct sum decompositions of abelian groups and their variants. I assume that the reader has had experience with the ring theory covered in [6], and with the abelian group theory covered in the text that inspired us to write this text, [10]. A read through the abelian group text [46] will also be helpful.

The use of *reduced torsion-free finite rank abelian groups* or of *rtffr abelian groups* can be traced back to solutions of linear equations with integer coefficients. We reinvent this point of view today when we study full free subgroups of a finite dimensional $\mathbb{Q}$-vector space.

Suppose that the rtffr group G has a direct sum decomposition

$$G_1^{(n_1)} \oplus \cdots \oplus G_t^{(n_t)} \tag{1}$$

for some integers $t, n_1, \ldots, n_t > 0$ and some indecomposable groups $G_1, \ldots, G_t$. To eliminate redundancies we require that $G_i \cong G_j \Rightarrow i = j$.

There are several theorems that give us conditions under which the integers $n_1, \ldots, n_t$ and the isomorphism classes of the groups $G_1, \ldots, G_t$ are unique for G. Two of the most important results of this kind are the Azumaya-Krull-Schmidt Theorem and the Baer-Kulikov-Kaplansky Theorem. Each states that if the groups G_i satisfy some condition or other then the integers $t, n_1, \ldots, n_t > 0$ and the isomorphism classes (G_i) of the groups G_i in (1) are unique to G. Thus one might expect this uniqueness to be common. This is not the case.

For example, G is called *completely decomposable* if G has a direct sum decomposition (1) in which G_i is a rank one rtffr group for each $i = 1, \cdots, t$. That is, $G_i \subset \mathbb{Q}$. The Baer-Kulikov-Kaplansky Theorem states that if (1) is a direct sum of rank one groups $G_1, \ldots, G_t$ then the integers $t, n_1, \ldots, n_t > 0$ and the isomorphism classes of the groups $G_1, \ldots, G_t$ are unique to G.

However, assume that $t, n_1, \ldots, n_t > 1$. Then several examples by B. Jónsson, A.L.S. Corner, and L. Fuchs and F. Loonstra (see [47] or the appendices) show that there are subgroups of finite index in G that possess direct sum decompositions whose indecomposable terms are not unique in any sensitive generalization of the term. Such a group H is called an *almost completely decomposable* group, or an *acd* group. Specifically, there are indecomposable *acd* groups G, H, A, B, C such that $G^{(2)} \cong H^{(2)}$ while G and H are not isomorphic, and $A \oplus B \cong A \oplus C$ while B and C are not isomorphic. Thus one can routinely construct subgroups H of finite index in G whose direct sum decompositions behave badly. Hence any theorems on direct sum decompositions of rtffr groups must respect this juxtaposition of uniqueness with nonuniqueness.

B. Jónsson proved a theorem, considered by some to be the *Fundamental Theorem of Torsion-Free Finite Rank Abelian Groups*, that allows us to salvage a course uniqueness of the indecomposable terms in the direct sum (1). The idea is to substitute the weaker *quasi-isomorphism* for isomorphism in the Azumaya-Krull-Schmidt Theorem. Groups X and Y are *quasi-isomorphic* if there are maps $f : X \to Y$ and $g : Y \to X$ and an integer $n > 0$ such that $fg = n1_Y$ and $gf = n1_X$. We write $X \cong Y$ when X is quasi-isomorphic to Y. The group X is *strongly indecomposable* if $X \cong Y$ implies that Y is indecomposable.

THEOREM: [B. Jónsson] Let G be an rtffr group.

1. There are integers $t, n_1, \ldots, n_t > 0$ and strongly indecomposable groups $G_1, \ldots, G_t$ such that $G_i \cong G_j$ implies that $i = j$ and such that $G \cong G_1^{m_1} \oplus \cdots \oplus G_t^{n_t}$.

2. Suppose that $G \cong H_1^{m_1} \oplus \cdots \oplus H_s^{m_s}$ for some integers s, m_1, ..., $m_s > 0$ and strongly indecomposable groups $H_1, \ldots, H_s$ such that $H_i \cong H_j$ implies that $i = j$. Then $s = t$, $n_i = m_i$ for each $i = 1, \ldots, t$, and $G_i \cong H_i$ after a permutation of the subscripts.

Thus $G^{(2)} \cong H^{(2)}$ implies that $G \cong H$, and $A \oplus B \cong A \oplus C$ implies that $B \cong C$. While Jónsson's Theorem accounts for some of the subgroup structure of G, it misses the finer properties that a direct sum decomposition can have. For example, there are indecomposable groups that are quasi-isomorphic to groups that possess a nontrivial direct sum decomposition. Thus abelian groupists have replaced indecomposable rtffr groups with *strongly indecomposable* rtffr groups, direct sum decompositions of rtffr groups with *quasi-direct sum decompositions* of rtffr groups, and uniqueness up to isomorphism is replaced with uniqueness up to *quasi-isomorphism*.

E. L. Lady [57] introduced another generalization of isomorphism called *near isomorphism* that is logically situated between isomorphism and quasi-isomorphism. Lady's first applications of near isomorphism include the fact that if $G^{(2)} \cong H^{(2)}$ then G and H are near isomorphic, and if $A \oplus B \cong A \oplus C$ then B and C are near isomorphic. Near isomorphism shows up in direct sums of many types of groups. For example, if G is a completely decomposable rtffr group, if $H \subset G$ is a subgroup, and if G/H is a finite p-group for some prime $p \in \mathbb{Z}$, then a direct sum $H \cong H_1 \oplus \cdots \oplus H_r$ of indecomposable groups H_i is unique up to near isomorphism. That is, if $H \cong K_1 \oplus \cdots \oplus K_s$ for some indecomposable groups K_j then $r = s$ and after a permutation of the subscripts H_i is near isomorphic to K_i for each $i = 1, \ldots, r$.

Near isomorphism is a group theoretic version of *the genus class* of lattices over orders. It is our point of view that near isomorphism and genus class are essentially the same measurement of algebraic properties, but of objects from different categories. Therefore, if two objects are near isomorphic or in the same genus class we will say that these objects are *locally isomorphic*.

Following the publication of the papers [14] and [15], it became popular to study direct sum decompositions of G by studying the right ideal structure of $\text{End}(G)$ with emphasis on the finitely generated projective right $\text{End}(G)$-modules. This book takes the point of view that a group G can be effectively studied by considering the finitely generated projective right $\text{End}(G)$-modules, the left $\text{End}(G)$-module G, and the ring

$$E(G) \quad = \quad \text{End}(G)/\mathcal{N}(\text{End}(G)).$$

D. M. Arnold [10] introduced a functor $A(\cdot)$ to abelian groups that transforms direct summands of G into finitely generated projective right $E(G)$-modules. Our fundamental approach uses $A(\cdot)$ as the primary tool for understanding direct sum decompositions of rtffr groups. Thus $A(\cdot)$ enables us to treat the direct sum decompositions of the rtffr group G as a direct sum of finitely generated projective modules over the Noetherian semi-prime rtffr ring $E(G)$.

One advantage of considering $E(G)$ over $\text{End}(G)$ is that the ring and group structure of $E(G)$ are simpler than that of $\text{End}(G)$. Specifically, if $S = \text{center}(E(G))$ then $E(G)$ is a finitely generated S-module. We will make extensive use of the localization theory of S to study $E(G)$.

There is a finite product $\overline{E}$ of classical maximal S-orders in which $E(G)$ has finite index. We call $\overline{E}$ an *integrally closed* ring. Several applications of some Algebraic Number Theory to the ideal structure of $\overline{E}$ will give us good results on the group structure of G. For example, if $h(G)$ is the number of isomorphism classes (H) of groups H that are locally isomorphic to G and if $h(\overline{E})$ is the number of isomorphism classes (J) of nonzero fractional ideals J of $\overline{E}$ then $h(\overline{E})$ divides $h(G)$.

While hereditary endomorphism rings have been scrutinized over the last 20 years, (see [54], and especially the papers of U. Albrecht), there seems to have been little attention paid to other pleasantly structured endomorphism rings. We will initiate an examination of the commutative property in $E(G)$. The reader may be surprised at the number of groups G for which $E(G)$ is commutative. Assuming that $E(G)$ is a commutative ring, we can prove several results on the uniqueness and existence of direct sum decompositions (1) of G.

Our investigations will show that G possesses some interesting properties when $\text{End}(G)$ is commutative and $\text{End}(G) \subset G \subset \mathbb{Q}\text{End}(G)$. Assuming that $E(G)$ is commutative our work shows that for each integer $n > 0$, G^n has a *locally unique decomposition*. That is,

1. $G^n = H_1 \oplus \cdots \oplus H_r$ where for each $i = 1, \ldots, r$, H_i is an indecomposable group such that G is locally isomorphic to $H_i \oplus H_i'$ for some group H_i', and

2. If $G^n = K_1 \oplus \cdots \oplus K_s$ for some indecomposable groups K_j then $r = s$ and after a permutation of the subscripts, H_i is locally isomorphic to K_i for $i = 1, \ldots, r$.

We show that G^n has a locally unique decomposition for each integer $n > 0$ if

1. G has a direct sum decomposition (1) in which the $G_1, \ldots, G_t$ are strongly indecomposable rtffr groups such that $G_i \cong G_j \Rightarrow i = j$, and

2. $E(G_i)$ is commutative for each $i = 1, \ldots, t$.

Condition 2, that $E(G_i)$ is commutative, is true whenever G is a strongly indecomposable rtffr group with square-free rank $\text{rank}(G_i)$. Thus there are plenty of rtffr groups that satisfy these two conditions. This result is an extension of the Baer-Kulikov-Kaplansky Theorem.

Let S and $\overline{E}$ be as above. There is a nonzero ideal $\tau \subset S$ called the *conductor of G* such that

$$\tau\overline{E} \subset E(G) \subset \overline{E}.$$

It is possible to define a localization functor $(\cdot)_\tau$ in a manner that is consistent with the traditional localization of a commutative ring at a prime ideal. Localization theory shows us that if $H \oplus H' \cong K \oplus K' \cong G^{(n)}$ for some integer $n > 0$ then H and K are locally isomorphic iff $A_\tau(H)$ and $A_\tau(K)$ are isomorphic right $E_\tau(G)$-modules, where $A_\tau(x) = (A(x))_\tau$ is the localization of $A(x)$ at τ. This gives us a different proof of Arnold's Theorem: if G is locally isomorphic to $H_1 \oplus H_2$ then $G = G_1 \oplus G_2$ where G_i is locally isomorphic to H_i for $i = 1, 2$.

An ideal P in a ring R is *primary* if R/P is a local ring with nilpotent Jacobson radical. That is, the Jacobson radical J of R/P is the unique maximal right ideal in R/P and $J^n = 0$ for some integer $n > 0$. The group G is a *semi-primary rtffr group* if there is a group $\overline{G}$ with a direct sum decomposition (1) such that

1. $E(G_i)$ is a Dedekind domain for each integer $i = 1, \ldots, t$ and such that

2. for each $i = 1, \ldots, t$ there is a primary ideal $\mathcal{N}(\text{End}(G_i)) \subset P_i \subset \text{End}(G_i)$ such that $P_1 G_1^{(n_1)} \oplus \cdots \oplus P_t G_t^{(n_t)} \subset G \subset \overline{G}$.

Semi-primary groups are generalizations of almost completely decomposable groups with primary regulating quotient. They include certain strongly indecomposable groups G such that $E(G)$ is commutative. Our results apply to subgroups of primary index in direct sums of strongly indecomposable bracket groups, strongly indecomposable strongly homogeneous groups, and strongly indecomposable Murley groups.

It is natural to ask if rtffr groups G satisfy a splitting property like that of free groups. We will characterize in Chapter 6 those rtffr groups G that satisfy the *Baer splitting property*. That is, we examine those G for which a surjection $\pi : G^{(c)} \longrightarrow G$ *splits* whenever c is a cardinal. The main results are Theorems 6.3.1 and 6.3.2 where we relate, *á la* U. Albrecht [3], the Baer splitting property for G to the vanishing of the tensor product $M \otimes_E G$ for nonzero right $\text{End}(G)$-modules M.

A discussion of $\mathcal{J}$-groups, $\mathcal{L}$-groups, and $\mathcal{S}$-groups follows. The group G is a $\mathcal{J}$-*group* if $G \cong H$ implies that $G \cong H$. The group G is an $\mathcal{L}$-*group* if $G \cong H$ implies that G is locally isomorphic to H. The group G is an $\mathcal{S}$-*group* if $G \cong H$ implies that G generates H. While it is clear that $\mathcal{J}$-group $\Rightarrow \mathcal{L}$-group $\Rightarrow \mathcal{S}$-group the converses require some power. We use several deep results from Analytic Number Theory to show that the most finitely faithful $\mathcal{S}$-groups are $\mathcal{J}$-groups, and that most rtffr $\mathcal{L}$-groups are $\mathcal{J}$-groups.

Over the past 30 years it has become fashionable to consider the structure of G as a left $\mathrm{End}(G)$-module. R. S. Pierce introduced the following clever way of labelling module properties on groups that we will adopt as our own. Given a module theoretic property *property* we say that G is an E-*property group* if G satisfies *property* as a left $\mathrm{End}(G)$-module. Thus E-projective groups are projective left $\mathrm{End}(G)$-modules, G is an E-finitely generated group if G is a finitely generated left $\mathrm{End}(G)$-module, an E-generator group is a generator as a left $\mathrm{End}(G)$-module, and an E-Noetherian group is a Noetherian left $\mathrm{End}(G)$-module. We will give a unified approach to rtffr groups G with these properties by showing that the group structure of G stems from a general structure theorem for finitely generated rtffr modules over an rtffr ring. Subsequently we show that several of these E-*properties* coincide. In the latter half of Chapter 9 we will examine the possible homological dimensions of the left $\mathrm{End}(G)$-module G. We show that for a given rtffr ring E and integer k less than the left global dimension of E there are rtffr groups P_k, F_k such that $E = \mathrm{End}(P_k) = \mathrm{End}(F_k)$ and whose projective and flat dimensions, respectively, are k. Our techniques must be significantly modified if they are to give us broad results on the injective dimension.

The text contains a number of exercises at the end of each chapter but the chapters themselves contain many statements like *the reader will prove that* .... Each of these is an intended unlisted exercise. The young abelian group theorist is advised to follow these directions. Also the author comes from the school of mathematics that emphasizes the use of examples. Examples guide our intuition and they steer us clear of falsehoods. Examples are used in this way throughout this text. Thus the reader should not be surprised at the number of examples used to motivate our discussions.

I would like to thank Fordham University for giving me the time and the resources needed to write this book. I would also like to thank my colleagues for carefully reading this book in manuscript form and for their subsequent comments. These people include the group of mathematicians that I was introduced to during my formative years. They are Ulrich Albrecht, Dave M. Arnold, R. A. Beaumont, Donna Beers, Anthony L. S. Corner, Manfred Dugas, Laszlo Fuchs, H. Pat Goeters, Anthony Giovanitti, Roger Hunter, E. Lee Lady, Ray Mines, DeAlberto Orsatti, Richard Pierce, K. M. Rangaswamy, James Reid, Fred Richman, Luigi Salce, Phil Schultz, Charles Vinsonhaler, Carol Walker, Elbert A. Walker, and William Wickless. Their encouragement over the years helped me forge this book.

Since my research style produces more TEX files than paper files, I am both author and technical typist on this project. Thus, I put quite a lot of wear on my keyboard during the writing process. Any mathematical errors of commission, of omission, or of a typographical nature are my responsibility. Certain results are attributed to your author but give no publication reference. This is because these results do not appear elsewhere in the literature.

Theodore G. Faticoni
Department of Mathematics
Fordham University
Bronx, New York 10458
faticoni@fordham.edu

Chapter 1

Notation and Preliminary Results

This text assumes that the reader is familiar with abelian groups and unital modules over associative rings with unity as contained in the texts by F.W. Anderson and K.R. Fuller [6], D.M. Arnold [10], and L. Fuchs [46]. We will reference but not prove those results that we feel fall outside of the line of thought of this book. I suggest that you use this chapter as a reference and nothing more. Skim through this chapter. Unless otherwise directed, do not attempt to plow through these results as though they were exercises.

1.1 Abelian Groups

Throughout, we let $\mathbb{Z}$ denote the ring of integers, $\mathbb{Q}$ denotes the field of rational numbers, and for a prime $p \in \mathbb{Z}$, $\mathbb{Z}_p$ is the localization of $\mathbb{Z}$ at p. We let E, R, S, T denote an associative ring with identity, hereafter referred to simply as a *ring*. Their modules are denoted by $G, H, K, L,$ M, N, U, V, W. (The side will be specified.) The term *group* means *abelian group*, (i.e., a $\mathbb{Z}$-module), and G, H, K denote groups. The group G is a *torsion-free group* if its torsion subgroup $\{x \in G \mid nx = 0$ for some integer $n \neq 0\}$ is zero. Thus G is torsion-free if $x \in G$ and $nx = 0$ for some integer $n \neq 0$ implies that $x = 0$. We view $\mathbb{Z}$ and $\mathbb{Q}$ as groups or as rings, as the setting requires.

If G and H are contained in a common $\mathbb{Q}$-vector space V, then we say that G and H are *quasi-equal* if there are nonzero integers n, m such that $nG \subset H$ and $mH \subset G$. In this case we write

$$G \doteq H.$$

Let G be a group. Let $\mathcal{I}$ be a set, let $c = \text{card}(\mathcal{I})$, and for each $i \in \mathcal{I}$ let $G_i \cong G$. The usual *direct sum* and *direct product* of c copies of G are denoted as

$$G^{(c)} = \oplus_{i \in \mathcal{I}} G_i$$
$$G^c = \prod_{i \in \mathcal{I}} G_i.$$

We say that the group G is *indecomposable* if $G = H \oplus K$ implies that $H = 0$ or $K = 0$. $\mathbb{Z}$, $\mathbb{Q}$, $\mathbb{Z}_p$, and $\mathbb{Z}/p^k\mathbb{Z}$ are indecomposable groups for primes $p \in \mathbb{Z}$ and integers $k > 0$. Given an integer $n > 0$ there is an indecomposable group $\mathbb{Z}^n \subset G \subset \mathbb{Q}^n$, so there are plenty of indecomposable torsion-free groups.

Let G be a torsion-free group. We let $\mathbb{Q}G$ denote *the divisible hull of G*. Since G is torsion-free $\mathbb{Q}G$ is a $\mathbb{Q}$-vector space spanned by G, or equivalently

$$\mathbb{Q}G = \mathbb{Q} \otimes_{\mathbb{Z}} G.$$

The *rank of G* is the cardinality of a maximal $\mathbb{Z}$-linearly independent subset of G, or equivalently

$$\text{rank}(G) = \mathbb{Q}\text{-dim}(\mathbb{Q}G).$$

With few exceptions, the torsion-free groups that we consider in this book have finite rank.

The group G is *reduced* if $\bigcap_{n>0} nG = 0$, or equivalently if $\text{Hom}(\mathbb{Q}, G) = 0$. An

rtffr group

is a *reduced torsion-free finite rank* group. As you read this book, pronounce *rtffr* as are-tee-ef-ef-are. A ring E is said to be an *rtffr ring* if its additive structure $(E, +)$ is an rtffr group. A right E-module M is said to be an *rtffr E-module* if its additive structure $(M, +)$ is an rtffr group.

The reader should prove the following fact as an exercise. If G is an rtffr group and if $H \subset G$ is a subgroup such that G/H is a *bounded group* (i.e., there is an integer $n \neq 0$ such that $nG \subset H$), then G/H is a finite group.

We say that G is *p-reduced* if $\bigcap_{k>0} p^k G = 0$, and we say that G is *p-local* if $pG \neq G$ and $qG = G$ for each prime $q \neq p \in \mathbb{Z}$. If $pG = G$ then we say that G is *p-divisible*. The group G is *semi-local* if there is a finite set of primes $\{p_1, \ldots, p_s\} \subset \mathbb{Z}$ such that $pG = G$ for each $p \notin \{p_1, \ldots, p_s\}$.

For example, $\mathbb{Z}$ and $\mathbb{Q}$ are rtffr groups, $\mathbb{Z}$ is reduced, and $\mathbb{Q}$ is divisible. The subgroup of $\mathbb{Q}$

$$\left\{ \frac{a}{n} \mid a \in \mathbb{Z} \text{ and } n \text{ is a square-free integer} \right\}$$

is reduced and torsion-free group of rank one. Moreover the subring of $\mathbb{Q}$

$$\mathbb{Z}_p = \left\{ \frac{a}{b} \mid a, b \in \mathbb{Z} \text{ and } \gcd(p, b) = 1 \right\}$$

is a reduced, p-reduced, torsion-free ring of rank one, but $q\mathbb{Z}_p = \mathbb{Z}_p$ for primes $q \neq p \in \mathbb{Z}$.

The ring of algebraic integers in an algebraic number field is an rtffr ring. The ring

$$\begin{pmatrix} \mathbb{Z} & 0 \\ \mathbb{Z} & \mathbb{Z} \end{pmatrix} = \left\{ \begin{pmatrix} a & 0 \\ b & c \end{pmatrix} \mid a, b, c \in \mathbb{Z} \right\}$$

is a reduced and torsion-free ring of rank three.

Let G and H be groups. As usual, we let

$\text{End}(G) = $ *the ring of group endomorphisms of* G.

Then $\text{End}(G)$ is the set of group homomorphisms $f : G \longrightarrow G$. We let $\text{Hom}(G, H)$ denote the group of group homomorphisms $f : G \longrightarrow H$. We consider G as a left $\text{End}(G)$-module by setting $f \cdot x = f(x)$ for $f \in \text{End}(G)$ and $x \in G$. Similarly $\text{Hom}(G, H)$ is a right $\text{End}(G)$-module if we set $gf = g \circ f$ for each $g \in \text{Hom}(G, H)$ and $f \in \text{End}(G)$. If G is an rtffr group, then $\text{End}(G)$ is an rtffr ring.

Given groups G and H let

$$S_G(H) = \sum \{ f(G) \mid \text{group maps } f : G \longrightarrow H \}.$$

We adopt the standard notation from [6]. Thus $N \otimes_R M$ denotes the tensor product of a right R-module N and a left R-module M. Since they are integral parts of our discussion we adopt a notation for a special tensor product and a special homset.

$$T_G(\cdot) = \cdot \otimes_{\text{End}(G)} G \quad \text{and} \quad H_G(\cdot) \;=\; \text{Hom}(G, \cdot).$$

There are natural transformations

$$\begin{aligned} \Theta &: T_G H_G(\cdot) \longrightarrow 1 \\ \Phi &: 1 \longrightarrow H_G T_G(\cdot) \end{aligned}$$

defined by

$$\begin{aligned} \Theta_H(f \otimes x) &= f(x) \\ \Phi_M(x)(\cdot) &= \cdot \otimes x. \end{aligned}$$

The reader should verify that these functors and associated natural transformations enjoy the following very useful identities.

$$\begin{aligned} 1_{H_G(H)} &= H_G(\Theta_H) \circ \Phi_{H_G(H)} \\ 1_{T_G(M)} &= \Theta_{T_G(M)} \circ T_G(\Phi_M) \end{aligned}$$

Let

$$\text{QHom}(G, H) = \mathbb{Q} \otimes \text{Hom}(G, H)$$

$$\text{QEnd}(G) = \mathbb{Q} \otimes \text{End}(G).$$

An element $f \in \text{QHom}(G, H)$ is called a *quasi-homomorphism*. The elements in $\text{QHom}(G, H)$ are linear transformations $f : \mathbb{Q}G \to \mathbb{Q}H$ such that $f(G) \doteq H$. Observe that f is a quasi-homomorphism iff there is an integer $n \neq 0$ such that $nf : G \to H$. We say that $f \in \text{QHom}(G, H)$ is a *quasi-isomorphism* if there is a map $g : \text{QHom}(H, G)$ such that $fg = 1_H$

and $gf = 1_G$. Equivalently $f : G \to H$ is a quasi-isomorphism iff there is an integer $n \neq 0$ and a group homomorphism $g : H \to G$ such that $fg = n1_H$ and $gf = n1_G$. We write

$$G \overset{.}{\cong} H$$

when G is quasi-isomorphic to H.

The following group is used in this text only as a source of illuminating examples. The ring of $\mathbb{Z}$-*adic integers* is denoted by

$$\widehat{\mathbb{Z}} = \lim_{\leftarrow n} \mathbb{Z}/n\mathbb{Z}.$$

Given a prime $p \in \mathbb{Z}$ the ring of *p-adic integers*

$$\widehat{\mathbb{Z}}_p = \lim_{\leftarrow k} \mathbb{Z}/p^k\mathbb{Z}$$

is a source for a number of important examples. To begin with, (1) $\widehat{\mathbb{Z}}$ and $\widehat{\mathbb{Z}}_p$ have uncountable rank, (2) for primes $p \in \mathbb{Z}$, $r_p(\widehat{\mathbb{Z}}) = r_p(\widehat{\mathbb{Z}}_p) = 1$, and (3) for primes $q \neq p \in \mathbb{Z}$, $\widehat{\mathbb{Z}}_p = q\widehat{\mathbb{Z}}_p$.

1.2 Associative Rings

Since some of our discussions deal with several rings at once we will need more than one symbol to denote a ring. For instance, E usually denotes an rtffr ring (a potential endomorphism ring of an rtffr group), and S denotes a commutative ring, often but not always $S = $ the center of $E = \text{center}(E) = \{x \in E \mid xy = yx \text{ for each } y \in E\}$.

Let E be a ring and let M be a right E-module. If $x \in M$ then the right ideal

$$\text{ann}_E(x) = \{r \in E \mid xr = 0\}$$

in E is called *the annihilator of x in E* and the ideal

$$\text{ann}_E(M) = \{r \in E \mid Mr = 0\}$$

is called *the annihilator of M in E*. For example given an integer $n > 0$, $\text{ann}_\mathbb{Z}(\mathbb{Z}/n\mathbb{Z}) = n\mathbb{Z}$. If $E = \begin{pmatrix} \mathbb{Z} & 0 \\ \mathbb{Z} & \mathbb{Z} \end{pmatrix}$ then let $e_{11} = \begin{pmatrix} 1 & 0 \\ 0 & 0 \end{pmatrix}$ and let $N = \begin{pmatrix} 0 & 0 \\ \mathbb{Z} & 0 \end{pmatrix}$. The reader can show that $\text{ann}_E(e_{11}) = \begin{pmatrix} 0 & 0 \\ \mathbb{Z} & \mathbb{Z} \end{pmatrix}$ and that $\text{ann}_E(N) = \begin{pmatrix} 0 & 0 \\ \mathbb{Z} & \mathbb{Z} \end{pmatrix}$.

In general, if $M = E/I$ for some right ideal $I \subset E$ then $1 + I \in E/I$ has annihilator $\text{ann}_E(1 + I) = I$ in E, and $\text{ann}_E(M)$ is the largest ideal of E that is contained in I. If $e^2 = e \in E$ then $\text{ann}_E(e) = (1 - e)E$. If M is a *simple* right E-module (that is, the only E-submodules of M are 0 and M), then $\text{ann}_E(x)$ is a maximal right ideal in E for each $x \neq 0 \in M$. If $E = \text{End}(G)$ then $\text{ann}_E(G) = 0$.

The Jacobson radical of E is the ideal $\mathcal{J}(E)$ defined as follows.

$$
\begin{aligned}
\mathcal{J}(E) &= \cap\{M \mid M \subset E \text{ is a maximal right ideal }\} \\
&= \cap\{M \mid M \subset E \text{ is a maximal left ideal }\} \\
&= \cap\{\text{ann}_E(K) \mid K \text{ is a simple left } E\text{-module }\} \\
&= \cap\{\text{ann}_E(K) \mid K \text{ is a simple right } E\text{-module }\} \\
&= \{r \in E \mid 1 + rx \text{ is a unit in } E\}
\end{aligned}
$$

For example, given a prime $p \in \mathbb{Z}$, $\mathcal{J}(\mathbb{Z}_p) = p\mathbb{Z}_p$. Given $n > 0 \in \mathbb{Z}$, let $n = p_1^{n_1} \cdots p_t^{n_t}$ be a product of powers $n_1, \ldots, n_t > 0$ of distinct primes $p_1, \ldots, p_t$. Let

$$
\mathbb{Z}_n = \left\{ \frac{a}{m} \,\middle|\, a, m \in \mathbb{Z}, m \text{ is relatively prime to } n \right\}.
$$

Then $\mathbb{Z}_n$ is a semi-local rtffr ring with Jacobson radical $\mathcal{J}(\mathbb{Z}_n) = (p_1 \cdots p_t)\mathbb{Z}_n$. Evidently $n \in \mathcal{J}(\mathbb{Z}_n)$. It is a healthy exercise for the reader to show that $\mathcal{J}(\mathbb{Z}) = 0$ and that $\mathcal{J} \begin{pmatrix} \mathbb{Z} & 0 \\ \mathbb{Z} & \mathbb{Z} \end{pmatrix} = \begin{pmatrix} 0 & 0 \\ \mathbb{Z} & 0 \end{pmatrix}.$

Given an rtffr ring E then the ring

$$E_n = \left\{ \frac{x}{m} \,\middle|\, x \in E \text{ and } m \in \mathbb{Z} \text{ is relatively prime to } n \right\}$$

is a torsion-free finite rank ring that may not be reduced. Note that $E_n \cong E \otimes_{\mathbb{Z}} \mathbb{Z}_n$.

LEMMA 1.2.1 *Let E be a semi-local rtffr ring and assume that $n = \prod\{$primes $p \in \mathbb{Z} \,|\, pE \neq E\}$. Then $n \in \mathcal{J}(E)$.*

Proof: Let M be a maximal right ideal in E. We will prove in Lemma 4.2.3 that since E is reduced there is a prime $p \in \mathbb{Z}$ such that $pE \subset M \neq E$, and thus $n \in M$. Consequently $n \in \cap\{M \,|\, $ M is a maximal ideal in $E\} = \mathcal{J}(E)$.

The ring R is *local* if any of the following equivalent properties hold.

1. R possesses a unique maximal right ideal M.

2. $\mathcal{J}(R)$ is the unique maximal right ideal of R.

3. $u \in R$ is a unit of R iff $u \notin \mathcal{J}(R)$.

For example, fields are local rings. Given a prime $p \in \mathbb{Z}$, $\mathbb{Z}_p$ and $\widehat{\mathbb{Z}}_p$ are local rings.

The right ideal $I \subset E$ is a *nil right ideal* if each $x \in I$ is nilpotent. That is, for each $x \in I$ there exists an integer n such that $x^n = 0$. The *nilradical of E* is the ideal $\mathcal{N}(E)$ that is defined as follows.

$$
\begin{aligned}
\mathcal{N}(E) &= \{x \in E \,|\, xE \text{ is a nil right ideal in } E\} \\
&= \{x \in E \,|\, Ex \text{ is a nil left ideal in } E\} \\
&= \sum\{I \,|\, I \text{ is a nil left ideal in } E\} \\
&= \text{ the largest nil ideal in } E
\end{aligned}
$$

Let $x \in E$. Since $x^n = 0$ implies that $1 - x$ is a unit in E (show that one, reader), $1 - xy$ is a unit of E for each $x \in \mathcal{N}(E)$ and $y \in E$. Thus

$$\mathcal{N}(E) \subset \mathcal{J}(E).$$

In the special case of an rtffr ring E we have the following relationships between $\mathcal{N}(E)$ and $\mathcal{J}(E)$.

LEMMA 1.2.2 *Let E be an rtffr ring.*

1. $\mathcal{N}(\mathbb{Q}E) = \mathcal{J}(\mathbb{Q}E)$ *is the largest nilpotent ideal in* $\mathbb{Q}E$.

2. $\mathcal{N}(E) = \mathcal{J}(\mathbb{Q}E) \cap E$ *and* $\mathbb{Q}\mathcal{N}(E) = \mathcal{J}(\mathbb{Q}E)$.

The reader may be surprised at the number of times we appeal to *Nakayama's Lemma* in this text.

LEMMA 1.2.3 [Nakayama] *(See [6, page 169].) Let E be a ring, let M be a right E-module and let $K \subset M$ be an E-submodule.*

1. *If M is finitely generated and if $K + M\mathcal{J}(E) = M$ then $K = M$.*

2. *If I is a nilpotent ideal in E such that $K + MI = M$ then $K = M$.*

In particular if $J \subset E$ is a right or a left ideal such that $J + \mathcal{J}(E) = E$ then $J = E$.

An *idempotent* is an element $e \in E$ such that $e^2 = e$. Often we will avoid the term idempotent and just write $e^2 = e$. Given a ring E and an $e^2 = e \in E$ then

$$eEe = \{exe \mid x \in E\}$$

is a ring with identity

$$1_{eEe} = e.$$

Although $eEe \subset E$, eEe is not a *unital* subring of E.

LEMMA 1.2.4 *Let E be a ring and let $e^2 = e \in E$. Then* $\mathrm{End}_E(eE) = eEe$.

Lifting theorems are a part of ideal theory. The following lifting results will be used throughout this text without fanfare.

LEMMA 1.2.5 *Let E be a ring.*

1. *If $u + \mathcal{J}(E)$ is a unit in $E/\mathcal{J}(E)$ then u is a unit in E.*

2. *If I is a nil ideal in E and if $(f + I)^2 = f + I \in E/I$ then there is an $e^2 = e \in E$ such that $e - f \in I$.*

1.3 Finite Dimensional $\mathbb{Q}$-Algebras

Our source for structure results on rtffr rings is [10, Chapters 9-14]. The ring E is *semi-prime* if I is an ideal of E and if $I^2 = 0$ then $I = 0$. Equivalently, E is semi-prime iff $\mathcal{N}(E) = 0$. Given a group G let

$$E(G) = \mathrm{End}(G)/\mathcal{N}(\mathrm{End}(G)).$$

Notice that $E(G)$ is in general a semi-prime ring. If G is an rtffr group then $E(G)$ is a Noetherian semi-prime rtffr ring that is finitely generated by its center, S. See [10, Theorems 9.4, 9.9]. It is hard to overstate the importance of the semi-prime ring $E(G)$ to our deliberations.

Let G and H be rtffr groups. We say that H *is a quasi-summand of G* if there is a group K such that $G \cong H \oplus K$. Equivalently, H is a quasi-summand of G iff there is an integer $n \neq 0$ and maps $f : G \to H$ and $g : H \to G$ such that $fg = n1_H$. The rtffr group G is *strongly indecomposable* if $G \cong H \oplus K$ implies that either $H = 0$ or $K = 0$. Equivalently G is a strongly indecomposable group iff each subgroup of finite index in G is indecomposable. Each subgroup of $\mathbb{Q}$ is strongly indecomposable as is $\mathbb{Z}$ and for primes $p \in \mathbb{Z}$, $\mathbb{Z}_p$ and $\widehat{\mathbb{Z}}_p$ are strongly indecomposable. Each pure subgroup of $\widehat{\mathbb{Z}}_p$ is strongly indecomposable so there are plenty of strongly indecomposable groups around.

A proof of the next lemma is found in [10, Theorem 9.10].

LEMMA 1.3.1 *Suppose that G is an rtffr group such that*

$$G = G_1^{(n_1)} \oplus \cdots \oplus G_t^{(n_t)}$$

for some integers $t, n_1, \ldots, n_t > 0$ and some strongly indecomposable groups $G_1, \ldots, G_t$ such that $G_i \cong G_j \Rightarrow i = j$. Then

$$E(G) \cong \mathrm{Mat}_{n_1}(E(G_1)) \times \cdots \times \mathrm{Mat}_{n_t}(E(G_t)). \qquad (1.1)$$

The next result is [46, page 149, Lemma 92.2].

LEMMA 1.3.2 *[J.D. Reid] Let G be an rtffr group. There is a bijection from the set of isomorphism classes of $\{(e\mathbb{Q}\mathrm{End}(G)) \,\big|\, e^2 = e \in \mathbb{Q}\mathrm{End}(G)\}$ onto the set of quasi-isomorphism classes $\{[H] \,\big|\, H$ is a quasi-summand of $G\}$. The bijection is given by $(e\mathbb{Q}\mathrm{End}(G)) \mapsto [eG]$.*

The following classification of strongly indecomposable groups is due to J. D. Reid [46, page 149, Proposition 92.3].

LEMMA 1.3.3 *[J.D. Reid] Let G be an rtffr group. The following are equivalent.*

1. *G is strongly indecomposable.*

2. *$\mathbb{Q}\mathrm{End}(G)$ is a local ring.*

3. *An endomorphism $f : G \to G$ is either nilpotent or an injection.*

Some of the ideal structure of $\mathbb{Q}E$ filters down to E.

LEMMA 1.3.4 *(See [10, Theorem 9.10].) Let E be an rtffr ring. The following are equivalent.*

1. *$\mathcal{N}(E) = 0$.*

2. *There are integers $t, n_1, \ldots, n_t > 0$ and division $\mathbb{Q}$-algebras $D_1, \ldots, D_t$ such that*

$$\mathbb{Q}E = \mathrm{Mat}_{n_1}(D_1) \times \cdots \times \mathrm{Mat}_{n_t}(D_t) \tag{1.2}$$

   as rings.

3. *Each right ideal of $\mathbb{Q}E$ is a direct summand of $\mathbb{Q}E$.*

LEMMA 1.3.5 *(See [10, Corollary 10.14].) Let E be a semi-prime rtffr ring. Then*

1. *If I is a right ideal of E, there is an $e^2 = e \in \mathbb{Q}E$ and an integer $n \neq 0$ such that neE has finite index in I and*

$$nE \subset neE \oplus n(1 - e)E \subset E \subset eE \oplus (1 - e)E.$$

2. *E is a Noetherian ring.*

3. *E is finitely generated as a module over its center center(E).*

1.4   Localization in Commutative Rings

Let us review some ideas about localization in commutative rings. We will refer to these facts without fanfare.

An element $c \in S$ is a *regular element* if c is not a zero divisor in S. If I is an ideal in the commutative ring S we let

$$\Gamma(I) = \{c \in S \,|\, c + I \text{ is a regular element in } S/I\}.$$

In general $\Gamma(I)$ is a multiplicatively closed subset of S that contains 1. For example, if S/I is a finite ring then $c \in \Gamma(I)$ iff $c + I$ is a unit in S/I. For $x \in S$ and $c, d \in \Gamma(I)$ we define an equivalence relation $\sim$ on $(x, c) \in S \times \Gamma(I)$ by $(x, c) \sim (y, d)$ iff $xd = yc$. Then the equivalence class of $(1, c)$ is denoted by c^{-1}, and xc^{-1} is the equivalence class of (x, c). By setting

$$K(I) = \{x \in S \mid xc = 0 \text{ for some } c \in \Gamma(I)\}$$

then we can form the *localization of S at I* denoted by

$$S_I \;=\; \{xc^{-1} \mid x \in S/K(I) \text{ and } c \in \Gamma(I)\}.$$

If S is an rtffr ring then $S_0 = \mathbb{Q}S$, However, not every localization of S is formed by inverting a set of integers.

Given an S-module X we write

$$X_I = X \otimes_S S_I.$$

Given any injection $f : X \to Y$ the canonical map

$$f_I : X_I \longrightarrow Y_I : xc^{-1} \longmapsto f(x)f(c)^{-1}$$

is an injection. Thus the inclusion map $J \subset S$ induces an injection $J_I \subset S_I$ onto an ideal in S_I. Indeed

$$I_I \subset \mathcal{J}(S_I)$$

since for each $x \in I_I$ and $s \in S_I$, $1 + xs \equiv 1 (\text{mod} I_I)$ and 1, being regular modulo I, maps onto a unit of S_I. In particular, if S is an rtffr ring and if $n \in \mathbb{Z}$ then

$$n \in \mathcal{J}(S_{nS}).$$

Note that S_{nS} is different from $S_n = S \otimes_{\mathbb{Z}} \mathbb{Z}_n$.

The following example will show us why S_n and S_{nS} may be different rings. Let $S = \mathbb{Z}_6 \times \mathbb{Z}_{15}$. With $n = 2$ we have

$$S_2 = S \otimes_{\mathbb{Z}} \mathbb{Z}_2 = (\mathbb{Z}_6 \times \mathbb{Z}_{15}) \otimes_{\mathbb{Z}} \mathbb{Z}_2 = \mathbb{Z}_2 \times \mathbb{Q}$$

while with $n = 3$ we have

$$S_3 = S \otimes_{\mathbb{Z}} \mathbb{Z}_3 = \mathbb{Z}_3 \times \mathbb{Z}_3.$$

But with $I = 2S$ we have the following. Since $(1,0)$ maps to a unit in $S/2S$ and since $(1,0)$ annihilates $0 \times \mathbb{Z}_{15}$, we have

$$K(2S) \;=\; 0 \times \mathbb{Z}_{15} = \{x \in S \mid xc = 0 \text{ for some } c \in \Gamma(2S)\}.$$

Thus

$$S_{2S} = (S/K(2S))_{2S} = \mathbb{Z}_2.$$

We observe that while $2 \in \mathcal{J}(S_{2S})$ we also have

$$(1,1)2 = (2,2) \notin \mathcal{J}(S_2) = 2\mathbb{Z}_2 \times 0.$$

If $I = 0$ then $\Gamma(0)$ is the set of regular elements in S and so $S_0 = \mathbb{Q}S$ is the classical ring of quotients of S.

If I is a prime ideal in S then S_I is a local ring with unique maximal ideal I_I. Furthermore, there is a natural isomorphism $S_I/I_I \cong (S/I)_I$, and S_I/I_I is the field of fractions of the integral domain S/I.

Let S be a commutative ring and let $I \subset S$ be an ideal. If S/I is *finite* then $\Gamma(I)$ maps to the set of units of S/I so that the canonical map

$$S \longrightarrow S_I : x \mapsto x + K(I)$$

induces an isomorphism

$$S/I \cong S_I/I_I \cong (S/I)_I.$$

Furthermore, if X is a finite S-module such that $XI = 0$ then the canonical map

$$X \longrightarrow X_I : x \mapsto x1^{-1}$$

is an isomorphism.

For example, if $p \in \mathbb{Z}$ is a prime and if G is an abelian group such that $p^k G = 0$ for some integer k then $G \cong G_p \cong G \otimes_{\mathbb{Z}} \mathbb{Z}_p$. Specifically if G is any group then $G/p^k G \cong G_p/p^k G_p$.

LEMMA 1.4.1 *Let I, J be ideals in a commutative integral domain S. Let X be an S-module.*

1. $(X_I)_J \cong (X_J)_I$.

2. *If $I \subset J$ and if S/I is finite then the natural surjection $S/I \to S/J$ takes units to units. Hence $(S_I)_J \cong (S_J)_I \cong S_J$.*

1.5   Local-Global Remainder

Any theorem that describes a property of M in terms of some property of M_I for each maximal ideal $I \subset S$ is called a *local-global property*. We use several local-global properties in this text.

THEOREM 1.5.1 [Local-Global Theorem] *(See [72, page 105].) Let S be a commutative ring and let M be an S-module. Then*

$$M = 0 \quad \Leftrightarrow \quad M_I = 0 \text{ for each maximal ideal } I \subset E.$$

Consequently $f : M \to N$ is a surjection iff the induced map $f_I : M_I \to N_I$ is a surjection for each maximal ideal $I \subset S$. Dually $f : M \to N$ is an injection iff the induced map $f_I : M_I \to N_I$ is an injection for each maximal ideal $I \subset S$. In particular, given $N, K \subset M$ then $K \subset N$ iff $K_I \subset N_I$ as S_I-submodules of M_I for each maximal ideal $I \subset S$. Another type of Local-Global Theorem relates maps $M \to N$ to maps $M_I \to N_I$ for maximal ideals I in S.

THEOREM 1.5.2 [Change of Rings] *(See [72, page 106].) Let E be an S-algebra for some commutative ring S, and let M be a finitely presented right E-module. If $\mathcal{C} \subset S$ is a multiplicatively closed subset of S containing 1 then for each S-module N*

1. $\mathrm{Hom}_E(M, N)[\mathcal{C}^{-1}] \cong \mathrm{Hom}_{E[\mathcal{C}^{-1}]}(M[\mathcal{C}^{-1}], N[\mathcal{C}^{-1}])$ *as groups.*

2. $\mathrm{Ext}^1_E(M, N)[\mathcal{C}^{-1}] \cong \mathrm{Ext}^1_{E[\mathcal{C}^{-1}]}(M[\mathcal{C}^{-1}], N[\mathcal{C}^{-1}])$ *as groups.*

In particular if M is a finitely presented right E-module then $\mathrm{End}_E(M)_I \cong \mathrm{End}_{E_I}(M_I)$. Furthermore, a finitely presented right E-module M is a projective E-module iff M_I is a projective right E_I-module for each maximal ideal I in S.

Beware: *The Change of Rings Theorem is false unless M is finitely presented.*

The following congruence result is used in a variety of settings in the sequel.

THEOREM 1.5.3 [Chinese Remainder Theorem] *(See [30, page 46].) Let S be a commutative ring, let $I_1, \ldots, I_t$ be ideals in S such that $I_i + I_j = S$ for $1 \leq i \neq j \leq t$, and let M be a right S-module. Let*

$$M^* = MI_1 \cap \cdots \cap MI_t.$$

1. Given $x_1, \ldots, x_t \in M$ there is an $x \in M$ such that

$$x - x_i \in MI_i \quad \text{for each } i = 1, \ldots, t.$$

2. The diagonal map

$$\delta : \frac{M}{M^*} \longrightarrow \bigoplus_{i=1}^{t} \frac{M}{MI_i}$$

such that $\delta(x + M^*) = (x + I_1 M, \ldots, x + MI_t)$ is an isomorphism.

3. $MI_1 \cap \cdots \cap MI_t = M[I_1 \cap \cdots \cap I_t]$.

Proof: See [30, Theorem 18.30-31] for a proof of parts 1 and 2.
3. Let $I = I_1 \cap \cdots \cap I_t$. By part 2 there is a canonical isomorphism

$$\prod_{i=1}^{t} \frac{S}{I_i} \cong \frac{S}{I}$$

so there are isomorphisms

$$\frac{M}{M^*} \cong \bigoplus_{i=1}^{t} \frac{M}{MI_i} \cong M \otimes_S \prod_{i=1}^{t} \frac{S}{I_i} \cong M \otimes_S \frac{S}{I} \cong \frac{M}{MI}.$$

Since these are canonical isomorphisms $M^* = MI$, as required by part 3.

In particular, if $t, n_1, \ldots, n_t > 0$ are integers, if $p_1, \ldots, p_t \in \mathbb{Z}$ are distinct primes, and if

$$n = p_1^{n_1} \cdots p_t^{n_t}$$

then given an abelian group G

$$nG = p_1^{n_1} G \cap \cdots \cap p_t^{n_t} G.$$

Consequently, there is a natural isomorphism

$$\delta : \frac{G}{nG} \longrightarrow \prod_{i=1}^{t} \frac{G}{p_i^{n_i} G}$$

such that

$$\delta(x + nG) = (x + p_1^{n_1} G, \ldots, x + p_t^{n_t} G)$$

of groups.

1.6 Integrally Closed Rings

See [10]. The commutative rtffr integral domain S is a *Dedekind domain* iff it satisfies one of the following equivalent conditions.

1. If $S \subset S' \subset \mathbb{Q}S$ is a ring and if S' is a finitely generated S-module then $S = S'$.

2. If $I \subset S$ is a nonzero ideal then I is invertible. That is, $II^* = S$ where $I^* = \{q \in \mathbb{Q}S \mid qI \subset S\}$.

3. Each nonzero ideal of S is a unique product of maximal ideals in S.

4. S is a hereditary Noetherian integral domain.

5. The localization S_M is a *discrete valuation domain* for each maximal ideal M in S. That is, there is an element $\pi \in S_M$ such that each ideal of S_M has the form $\pi^k S_M$ for some integer $k > 0$.

For example, $\mathbb{Z}$ is a Dedekind domain as is any *pid*. The ring of algebraic integers in an algebraic number field is a Dedekind domain.

The rtffr ring E is *integrally closed* if whenever $E \subset E' \subset \mathbb{Q}E$ is a ring such that E'/E is finite then $E = E'$. A *classical maximal order* is an integrally closed prime rtffr ring. A classical maximal order is finitely generated and torsion-free as a module over its center. The rtffr ring E is an integrally closed ring iff

$$E = E_1 \times \cdots \times E_t$$

where each E_i is a classical maximal order. An *E-lattice* is a finitely generated right E-submodule of the right E-module $\mathbb{Q}E^{(n)}$ for some $n \in \mathbb{Z}$.

Dedekind domains are classical maximal orders. If E is a classical maximal order and if U is an E-lattice then $\mathrm{Mat}_n(E)$ and $\mathrm{End}_E(U)$ are classical maximal orders.

The next results are [69, Theorem 21.1], [10, Corollary 11.5], [10, Theorem 11.8], and [69, Theorems 18.10, 27.1].

LEMMA 1.6.1 *Suppose that the rtffr ring E is a classical maximal order.*

1. *If I is a right ideal of finite index in E then $\mathcal{O}(I) = \{q \in \mathbb{Q}E \mid qI \subset I\}$ is a classical maximal order.*

2. E is an hereditary Noetherian ring. That is, each one-sided ideal in E is a finitely generated projective E-module.

3. For each integer $n \in \mathbb{Z}$, each right ideal in E_n is principal.

4. If U and V are E-lattices then U is locally isomorphic to V (see section 2.5) iff $\mathbb{Q}U \cong \mathbb{Q}V$ as right $\mathbb{Q}E$-modules. (Warning: This is true only of lattices over maximal orders.)

Semi-prime rtffr rings are closely connected to integrally closed rings as the following result shows. The next two results follow from [10, page 127].

LEMMA 1.6.2 Let E be a semi-prime rtffr ring. There is a finite set of primes $\{p_1, \ldots, p_s\} \subset \mathbb{Z}$ such that E_p is a classical maximal order for all $p \notin \{p_1, \ldots, p_s\}$.

LEMMA 1.6.3 Suppose that E is a semi-prime rtffr ring. There is an integrally closed ring $\overline{E} \subset \mathbb{Q}E$ and an integer $n \neq 0$ such that

$$n\overline{E} \subset E \subset \overline{E}$$

1.7   Semi-Perfect Rings

The ring E is *semi-perfect* if

1. $E/\mathcal{J}(E)$ is semi-simple Artinian and

2. Given an $\bar{e}^2 = \bar{e} \in E/\mathcal{J}(E)$ there is an $e^2 = e \in E$ such that $\bar{e} = e + \mathcal{J}(E)$. That is, *idempotents lift modulo $\mathcal{J}(E)$*.

See [6, Chapter 7 §27] for a complete discussion of semi-perfect rings and their modules. Fields, local rings, and Artinian rings are semi-perfect rings. $\mathbb{Z}$ and $\mathbb{Z}_6$ are not semi-perfect but $\mathbb{Z}_p$ and $\widehat{\mathbb{Z}}_p$ are semi-perfect rings for primes $p \in \mathbb{Z}$.

Suppose that E is semi-perfect. Each indecomposable projective right E-module P has the form eE for some $e^2 = e \in E$. Given $e^2 = e$, $f^2 = f \in E$ then

$$eE/e\mathcal{J}(E) \cong fE/f\mathcal{J}(E) \Leftrightarrow eE \cong fE.$$

LEMMA 1.7.1 Let E be a semi-perfect ring and let P be a projective right E-module. Then

1. P is indecomposable iff $\text{End}_E(P)$ is a local ring.

2. E is a direct sum of right E-modules with local endomorphism rings.

Thus direct sum decompositions of projective modules over semi-perfect rings are unique by the Azumaya-Krull-Schmidt Theorem 2.1.6.

We will also find it necessary to use the following results on lifting idempotents in semi-perfect rings.

LEMMA 1.7.2 Let E be a semi-perfect ring, and let $e^2 = e$, $f^2 = f \in E$ be such that $eE \cong fE$. There is a unit $u \in E$ such that $ueu^{-1} = f$.

Proof: Since $e^2 = e$ and $f^2 = f$ we can write

$$E = eE \oplus (1-e)E = fE \oplus (1-f)E$$

and since E is semi-perfect $(1-e)E \cong (1-f)E$ by The Azumaya-Krull-Schmidt Theorem 2.1.6. Thus there is an isomorphism $u : E \to E$ such that

$$
\begin{aligned}
ueu^{-1} &\in u(eE) = fE \quad \text{and} \\
u(1-e)u^{-1} &\in u((1-e)E) = (1-f)E.
\end{aligned}
$$

Evidently u is multiplication by a unit of E so that

$$
\begin{aligned}
1 &= u1u^{-1} \\
&= u(e + (1-e))u^{-1} \\
&= ueu^{-1} \oplus u(1-e)u^{-1} \\
&\in fE \oplus (1-f)E.
\end{aligned}
$$

Since 1 can be written in exactly one way as an element in a direct sum we must have $ueu^{-1} = f$. This completes the proof of the lemma.

LEMMA 1.7.3 Let E be semi-perfect and let $e^2 = e$, $f^2 = f \in E$. If

$$eE/e\mathcal{J}(E) \cong fE/f\mathcal{J}(E)$$

then $eE \cong fE$.

Proof: Since $e^2 = e$, eE is a projective right E-module so the isomorphism $eE/e\mathcal{J}(E) \cong fE/f\mathcal{J}(E)$ lifts to a map

$$\phi : eE \longrightarrow fE$$

such that

$$\ker \phi \subset e\mathcal{J}(E) \quad \text{and} \quad fE = \phi(eE) + f\mathcal{J}(E).$$

By Nakayama's Lemma 1.2.3, $fE = \phi(eE)$ and since $f^2 = f \in E$

$$eE = U \oplus \ker \phi$$

where $U \cong fE$. Inasmuch as $\ker \phi \subset e\mathcal{J}(E)$ another appeal to Nakayama's Lemma 1.2.3 shows us that $\ker \phi = 0$, whence $eE \cong fE$. This completes the proof.

LEMMA 1.7.4 *Let E be a semi-perfect ring, and let $J \subset \mathcal{J}(R)$ be an ideal. Then idempotents lift modulo J in E.*

Proof: Let $J \subset \mathcal{J}(E)$ and let $e \in E$ be such that $e^2 - e \in J \subset \mathcal{J}(E)$. Let $\overline{E} = E/J$ and given $x \in E$ let $\bar{x} = x + J \in \overline{E}$.

Because E is semi-perfect there is an $f^2 = f \in E$ such that $e - f \in \mathcal{J}(E)$. Then $\bar{e}, \bar{f} \in \overline{E}$ are idempotents such that $\bar{e} - \bar{f} \in \mathcal{J}(\overline{E}) = \mathcal{J}(E)/J$. We will show that there is a unit $\bar{u} \in \overline{E}$ such that $\bar{u}\bar{f}\bar{u}^{-1} = \bar{e}$.

We have $\bar{e} + \mathcal{J}(\overline{E}) = \bar{f} + \mathcal{J}(\overline{E})$ and $\bar{e}^2 = \bar{e}$, $\bar{f}^2 = \bar{f} \in \overline{E}$ so that

$$\frac{\bar{e}\overline{E}}{\bar{e}\mathcal{J}(\overline{E})} \cong (\bar{e} + \mathcal{J}(\overline{E}))\frac{\overline{E}}{\mathcal{J}(\overline{E})} = (\bar{f} + \mathcal{J}(\overline{E}))\frac{\overline{E}}{\mathcal{J}(\overline{E})} \cong \frac{\bar{f}\overline{E}}{\bar{f}\mathcal{J}(\overline{E})}.$$

By Lemma 1.7.2 there is a unit $\bar{u} \in \overline{E}$ such that $\bar{e} = \bar{u}\bar{f}\bar{u}^{-1}$.

Since units lift modulo $J \subset \mathcal{J}(E)$, $\bar{u} = u + J$ for some unit $u \in E$ and hence $ufu^{-1} \equiv e (\text{mod } J)$. Thus idempotents lift modulo J in E.

1.8 Exercise

1. Verify each result mentioned in this chapter by proving it yourself or by finding it in the references.

Chapter 2

Motivation by Example

The group G is an

rtffr group

if G is a *reduced torsion-free finite abelian group*. The topics that we will discuss in this text are properties of direct summands of rtffr groups G. These properties include the uniqueness of direct sum decompositions, as well as the properties of the left $\mathrm{End}(G)$-module G.

A direct sum decomposition

$$G = G_1 \oplus \cdots \oplus G_t \tag{2.1}$$

is said to be *indecomposable* if each G_i is indecomposable. If $p \in \mathbb{Z}$ is a prime and if $k_1, \ldots, k_t > 0$ are integers such that $G_i \cong \mathbb{Z}/p^{k_i}\mathbb{Z}$ for each $i = 1, \ldots, t$ then the decomposition (2.1) is indecomposable. If $G_i \subset \mathbb{Q}$ for each $i = 1, \ldots, t$ then the decomposition (2.1) is indecomposable.

Suppose that (2.1) is an indecomposable decomposition of G and suppose that we can also write

$$G \cong H \oplus K \cong H \oplus K'. \tag{2.2}$$

We will pursue several broad questions on the nature of direct sums. For instance, is the direct sum (2.1) unique in some sense? What can be said about K and K' in (2.2)? Is H a direct sum of copies of the G_i? Are the group structures of K and K' similar in some sense? Are there examples of badly behaved direct sum decompositions of G? Is there a simpler category in which the direct summands of G can be handled?

Suppose that we are given a surjection $\pi : G^{(n)} \to G$ for some integer $n > 0$. If G is a free group then π is *split*.   That is, is there a map $\phi : G \to G^{(n)}$ such that $\pi\phi = 1_G$, or equivalently, such that $\ker \pi$ a direct summand of $G^{(n)}$. Under what conditions on G will π be split?

2.1   Some Well-Behaved Direct Sums

Before embarking on our journey we need a series of examples. The experience that these examples bring us will help us to develop intuition about the above questions. The constructions can be found in D. Arnold's [10] or L. Fuchs' [46].

Let G be a group. The purpose of this section is to show that under some conditions direct sum decompositions of G are well behaved, in a sense that we will make precise below.

The group G is *indecomposable* if $G = H \oplus H'$ implies that $H = 0$ or $H' = 0$. For instance, $\mathbb{Z}$ is indecomposable, and given a prime $p \in \mathbb{Z}$, $\mathbb{Z}_p$, $\mathbb{Z}(p^\infty)$, and $\mathbb{Z}/p^k\mathbb{Z}$ for integers $k > 0$ are indecomposable groups. Any subgroup of $\mathbb{Q}$ is indecomposable. The following example shows us that indecomposable rtffr groups can have finite index in a decomposable group.

EXAMPLE 2.1.1 (See Lemma A.1.2.) Let $X, Y \subset \mathbb{Q}$ be groups such that $\operatorname{Hom}(X, Y) = 0 = \operatorname{Hom}(Y, X)$ and suppose that there is a prime $p \in \mathbb{Z}$ such that $pX \neq X$ and $pY \neq Y$. Choose elements $x \in X \setminus pX$ and $y \in Y \setminus pY$ and define

$$G = (X \oplus Y) + \frac{1}{p}(x + y)\mathbb{Z}.$$

Then G is an *indecomposable group of rank two* such that $G \doteq X \oplus Y$.

The group G is *strongly indecomposable* if each subgroup of finite index in G is indecomposable. See Section 1.3. Given a prime $p \in \mathbb{Z}$ let

$$\mathbb{Z}\left[\frac{1}{p}\right] \;=\; \left\{ \frac{a}{p^k} \,\middle|\, a, k \in \mathbb{Z} \right\}.$$

EXAMPLE 2.1.2 (See Example A.1.1.) Let $X, Y \subset \mathbb{Q}$ be groups such that $\operatorname{Hom}(X, Y) = 0 = \operatorname{Hom}(Y, X)$ and suppose that there is a prime $p \in \mathbb{Z}$ such that $pX \neq X$ and $pY \neq Y$. Choose elements $x \in X \setminus pX$ and $y \in Y \setminus pY$ and define

$$G = (X \oplus Y) + \mathbb{Z}[\tfrac{1}{p}](x + y).$$

Then G is a *strongly indecomposable group of rank two*.

EXAMPLE 2.1.3 Let $p \in \mathbb{Z}$ be a prime and let G be a pure subgroup of $\widehat{\mathbb{Z}}_p$. Then G is strongly indecomposable.

Proof: Say G is pure in $\widehat{\mathbb{Z}}_p$ and suppose that $G \doteq A \oplus B$. The reader will show that $G/pG \cong A/pA \oplus B/pB$. Then $pG = p\widehat{\mathbb{Z}}_p \cap G$ so that

$$G/pG = G/(p\widehat{\mathbb{Z}}_p \cap G) \cong (G + p\widehat{\mathbb{Z}}_p)/p\widehat{\mathbb{Z}}_p \subset \widehat{\mathbb{Z}}_p/p\widehat{\mathbb{Z}}_p \cong \mathbb{Z}/p\mathbb{Z}.$$

Thus G/pG is indecomposable, whence $A/pA = 0$ or $B/pB = 0$. But then A or B is a p-divisible subgroup of the p-reduced group $\widehat{\mathbb{Z}}_p$. Hence $A = 0$ or $B = 0$, and therefore G is indecomposable.

We say that G has a *unique decomposition* if

1. G has an indecomposable decomposition $G \cong G_1 \oplus \cdots \oplus G_t$ and

2. Given an indecomposable decomposition $G \cong G'_1 \oplus \cdots \oplus G'_s$ then $s = t$ and after a permutation of the subscripts, $G_i \cong G'_i$ for each $i = 1, \ldots, t$.

In this case we call $G_1 \oplus \cdots \oplus G_t$ *the unique decomposition of* G. The unique decomposition of an rtffr group G is necessarily indecomposable. Rtffr groups having unique decomposition are considered to be rare.

EXAMPLE 2.1.4 1. The *Fundamental Theorem of Abelian Groups* states that if $p \in \mathbb{Z}$ is a prime and if G is a finite p-group then G has a unique decomposition

$$G = \mathbb{Z}/p^{n_1}\mathbb{Z} \oplus \cdots \oplus \mathbb{Z}/p^{n_t}\mathbb{Z}$$

for some integers $0 < n_1 \leq \cdots \leq n_t$.

2. Finitely generated free groups have unique decomposition as do the (possibly mixed) divisible groups.

3. A vector space over a field has a unique decomposition as a group.

EXAMPLE 2.1.5 Suppose $\text{End}(G)$ is right Artinian. Because $\text{End}(G)$ is then semi-perfect the indecomposable decomposition $G = e_1 G \oplus \cdots \oplus e_t G$ is unique.

An rtffr endomorphism ring $\text{End}(G)$ contains a complete set of primitive orthogonal indecomposable idempotents $\{e_1, \ldots, e_t\}$ because $\mathbb{Q}\text{End}(G)$ is Artinian. Then G has an indecomposable decomposition $G = e_1 G \oplus \cdots \oplus e_t G$. However this decomposition need not be unique.

One of the more interesting properties associated with direct sums is the *refinement property*. The group G has the refinement property if

1. G possesses an indecomposable decomposition

$$G = G_1 \oplus \cdots \oplus G_t$$

   and

2. If $G^{(n)} \cong H \oplus K$ for some integer $n > 0$ then there are integers $h_1, \ldots, h_t \geq 0$ such that

$$H = G_1^{(h_1)} \oplus \cdots \oplus G_t^{(h_t)}.$$

It is generally believed that rtffr groups with the refinement property are rare.

A theorem on the uniqueness of indecomposable decompositions that will be essential to our investigations is the *Azumaya-Krull-Schmidt Theorem,* hereafter referred to as the AKS-Theorem. A proof can be found in [6]. The AKS Theorem represents the best conditions that one could hope for concerning the uniqueness of direct sum decompositions.

THEOREM 2.1.6 [Azumaya-Krull-Schmidt]   *Let $G_1, \cdots, G_t$ be left E-modules such that $\text{End}_E(G_i)$ is a local ring for each $i = 1, \ldots, t$ and let $G = G_1 \oplus \cdots \oplus G_t$.*

1. *$G_1 \oplus \cdots \oplus G_t$ is the unique decomposition for G.*

2. *G has the refinement property.*

A consequence of the AKS Theorem 2.1.6 is that if G is a direct sum of modules with local endomorphism rings then $G \oplus K \cong G \oplus K'$ for some modules K and K' implies that $K \cong K'$. See [30, Cancellation Theorem 21.2]. Some examples will help us see how to use the AKS Theorem 2.1.6.

EXAMPLE 2.1.7 Let $p_1, \ldots, p_t \in \mathbb{Z}$ be a finite list of *distinct* primes and for each $i = 1, \ldots, t$ let G_i be the localization $G_i = \mathbb{Z}_{p_i}$. Then $\text{End}(G_i)$ is a local commutative ring for each i and hence the AKS Theorem 2.1.6 states that $G = G_1 \oplus \cdots \oplus G_t$ has a unique decomposition. The reader can show that $\text{End}(G)$ is commutative in this case.

EXAMPLE 2.1.8 If $G_1, \ldots, G_t$ are simple right modules over some ring R then $\text{End}_R(G_i)$ is a division ring for each $i = 1, \ldots, t$. The AKS Theorem 2.1.6 states that $G = G_1 \oplus \cdots \oplus G_t$ is the unique decomposition of G, and that G has the refinement property.

EXAMPLE 2.1.9 An indecomposable projective module over a semi-perfect ring has a local endomorphism ring. Thus the AKS Theorem 2.1.6 applies to direct sums of indecomposable projective modules over a semi-perfect ring.

J.D. Reid's Lemma 1.3.3 states that the rtffr group G is strongly indecomposable iff $\mathbb{Q}\text{End}(G)$ is a local ring, so it is reasonable to expect that some form of the AKS Theorem 2.1.6 applies to direct sums of strongly indecomposable rtffr groups. The next theorem is referred to in the literature as Jónsson's Theorem. It is generally accepted that this is the best possible existence and uniqueness result that we can expect for the general rtffr group G. See [46, page 150, Theorem 92.5].

THEOREM 2.1.10 [B. Jónsson] *Let G be an rtffr group. There is a finite list of $G_1, \ldots, G_t$ strongly indecomposable groups such that*

1. $G \overset{.}{\cong} G_1 \oplus \cdots \oplus G_t$.

2. *If $G \overset{.}{\cong} G_1' \oplus \cdots \oplus G_s'$ and if each G_i' is strongly indecomposable then $s = t$ and after a permutation of the indices $G_i \cong G_i'$ for each $i = 1, \ldots, t$.*

3. *Given an integer $n > 0$ and a direct sum $G^{(n)} \overset{.}{\cong} H \oplus H'$ then there is a subset $\mathcal{I} \subset \{1, \ldots, t\}$ and integers $h_i > 0$, $i \in \mathcal{I}$, such that $H \overset{.}{\cong} \oplus_{i \in \mathcal{I}} G_i^{(h_i)}$.*

4. *If $G \oplus H \overset{.}{\cong} G \oplus H'$ for some rtffr groups H and H' then $H \overset{.}{\cong} H'$.*

Proof: We sketch a proof due to E.A. Walker [86]. The category of quasi-homomorphisms $\mathbb{Q}\mathbf{Ab}$ is the category

1. whose objects are the torsion-free groups of finite rank, and

2. whose homsets are given by $\mathbb{Q}\mathrm{Hom}(\star, \star)$.

The rtffr group G is strongly indecomposable iff G is indecomposable in $\mathbb{Q}\mathbf{Ab}$, and G is strongly indecomposable iff $\mathbb{Q}\mathrm{End}(G)$ is a local Artinian ring, (Lemma 1.3.3). Thus each object in $\mathbb{Q}\mathbf{Ab}$ is a direct sum of objects with local endomorphism rings. Then by a version of the AKS Theorem 2.1.6 for objects in additive categories in which idempotents split, each object in $\mathbb{Q}\mathbf{Ab}$ has a unique decomposition $G \cong G_1 \oplus \cdots \oplus G_t$ in $\mathbb{Q}\mathbf{Ab}$ where each G_i is strongly indecomposable. Since two rtffr groups are isomorphic in $\mathbb{Q}\mathbf{Ab}$ iff they are quasi-isomorphic groups, $G \doteq G_1 \oplus \cdots \oplus G_t$ as groups. This completes the proof.

We observe that the AKS Theorem 2.1.6 has been our only tool to this point for demonstrating the existence of a unique decomposition. The next result is a theorem on uniqueness and the refinement property of indecomposable decompositions of a certain type of group. Its proof is interesting in that it does not refer to the AKS Theorem 2.1.6. See [46, page 112, Proposition 86.1].

The group G is called *completely decomposable* if there are *rank one* groups $G_1, \ldots, G_t$ such that $G = G_1 \oplus \cdots \oplus G_t$. The group H has rank one if $H \subset \mathbb{Q}$.

THEOREM 2.1.11 [Baer-Kulikov-Kaplansky] *Assume that* $G \cong G_1 \oplus \cdots \oplus G_t$ *where each* G_i *is a rank one group.*

1. $G_1 \oplus \cdots \oplus G_t$ *is the unique decomposition of* G.

2. G *has the refinement property.*

The group G is an *acd group* (almost completely decomposable group) if there is a completely decomposable group C and an integer $n > 0$ such that $nC \subset G \subset C$. See A. Mader's text [60] for everything concerning *acd* groups. There is at least one type of *acd* group whose direct sum decompositions are well behaved. See [10, Theorem 2.3] for a proof.

THEOREM 2.1.12 [R.A. Beaumont and R.S. Pierce] *Suppose that* $G \doteq G_1^{n_1} \oplus \cdots \oplus G_t^{n_t}$ *for some integers* $t, n_1, \ldots, n_t > 0$ *and a chain* $G_1 \subset \ldots \subset G_t$ *of rank one groups such that* $G_i \cong G_j$ *iff* $i = j$. *Then* $G \cong G_1^{n_1} \oplus \cdots \oplus G_t^{n_t}$.

Theorem 2.2.1, however, will show us that *acd* groups can have badly behaved direct sum decompositions.

2.2 Some Badly-Behaved Direct Sums

Now that we have established some useful conditions under which direct sums are well behaved let us see what happens in more general situations.

Every torsion-free group of finite rank has an *indecomposable* decomposition (i.e., the given group can be written as a direct sum of indecomposable groups) but this decomposition need not be unique. (See Theorem 2.2.1 below.) The purpose of this section is to give some examples that illustrate the unruly behavior of indecomposable decompositions of some rtffr groups.

In the 1960s several examples came to light illustrating that within the class of *acd* groups there are examples of direct sum decompositions that are not unique in any sense of the word. Even though the completely decomposable groups possess a unique decomposition (Theorem 2.1.11), an indecomposable decomposition of an *acd* group may not be unique.

THEOREM 2.2.1 [A.L.S. Corner] *See Example B.1.1. Let $n \geq k \geq 1$ be integers. There is a group $G = G(n)$ of rank n such that for each partition $r_1 + \cdots + r_k$ of n into k positive summands r_j there is an indecomposable direct sum decomposition*

$$G = G_1 \oplus \cdots \oplus G_k$$

such that $r_j = \operatorname{rank}(G_j)$ for each $j = 1, \ldots, k$.

For instance, suppose that p, p_1, p_2, p_3 are distinct primes, let A_1, A_2, A_3 be rank one groups such that $p_i A_i \neq A_i \neq pA_i$ for each $i = 1, 2, 3$, and let $m = p_1 p_2 p_3$. With $n = 4$ and $k = 2$ the above construction shows us that there is a group

$$m(A_1^2 \oplus A_2 \oplus A_3) \subset G \subset A_1^2 \oplus A_2 \oplus A_3$$

that possesses *indecomposable* decompositions

$$G = G_1 \oplus G_2 = H_1 \oplus H_2$$

such that $\operatorname{rank}(G_1) = 3$, $\operatorname{rank}(H_1) = \operatorname{rank}(H_2) = 2$, and $\operatorname{rank}(G_2) = 1$. Thus G does not possess a unique decomposition.

The nonuniqueness of indecomposable direct sum decompositions of rtffr groups is also seen in the following examples.

THEOREM 2.2.2 [L. Fuchs and F. Loonstra] *Examples C.1.2 and D.1.1.*

1. There are acd groups G_1, G_2, G_3 of ranks 1, 2, 2, respectively, such that

$$G_1 \oplus G_2 \cong G_1 \oplus G_3$$

   while $G_2 \not\cong G_3$.

2. There are indecomposable acd groups G_4, G_5 of rank 2 such that $G_4^{(2)} \cong G_5^{(2)}$ while $G_4 \not\cong G_5$. Thus $G = G_4^{(2)}$ does not possess a unique decomposition and does not satisfy the refinement property.

The next example illustrates that even if $\mathrm{End}(G_1) = \mathrm{End}(G_2) = \mathbb{Z}$ then indecomposable decompositions of $G = G_1 \oplus G_2$ can fail to be unique.

THEOREM 2.2.3 *[10, page 63, Example 6.9]. There are quasi-isomorphic indecomposable groups G_1, G_2, H_1, H_2 such that*

$$G_1 \oplus G_2 \cong H_1 \oplus H_2,$$

and $\mathrm{End}(X) \cong \mathbb{Z}$ for $X = G_1, G_2, H_1, H_2$, while H_2 is not isomorphic to $G_1, G_2,$ or H_1.

Thus in the AKS Theorem 2.1.6 the hypothesis that $\mathrm{End}(G_i)$ is a local ring for each G_i cannot be weakened to the condition that $\mathrm{End}(G_i)$ is a *pid*.

It is best if we use the above examples as guideposts to help us anticipate limitations in our results. For instance, while any theorem on the direct sum decompositions of *acd* groups might aspire to the conclusions of the AKS Theorem 2.1.6 or of the Baer-Kulikov-Kaplansky Theorem 2.1.11, it must also respect the badly behaved direct sum decompositions in the above section.

2.3 Corner's Theorem

The fact that *any* rtffr ring E is the endomorphism ring of an rtffr group G will allow us to construct rtffr groups possessing a variety of left E-module structures. These constructions are found in the Appendices.

The group G is *locally free* if G_p is a free $\mathbb{Z}_p$-module for each prime $p \in \mathbb{Z}$. M.C.R. Butler [21] constructs groups that have the same torsion-free rank as their endomorphism ring. See [10, Theorem 2.14].

THEOREM 2.3.1 [M.C.R. Butler] *See Example I.1.1. Let E be an rtffr ring whose additive structure $(E, +)$ is a locally free group. There is an rtffr group G such that $E \subset G \subset \mathbb{Q}E$ as left E-modules and $E \cong \text{End}(G)$ as rings.*

Let E be an rtffr ring. The left E-module M is called a *Corner E-module*, or when E is understood simply a *Corner module*, if its additive structure $(M, +)$ is a countable reduced torsion-free group. Similarly a *Corner ring* is an associative ring E with identity such that E is a Corner module. There is no confusion as to the orientation of M as all Corner modules are left modules. Call the group G a *Corner group* if G is a countable reduced torsion-free group.

One of the questions that we will address is *Is there a functorial means of characterizing the properties of G in terms of* $\text{End}(G)$? Dually, we can ask *Is there a functorial means that will allow us to characterize the properties of* $\text{End}(G)$ *in terms of properties of G.* Specifically, since G is a left $\text{End}(G)$-module we can ask about the module theoretic properties of G as a left $\text{End}(G)$-module. Corner's Theorem provides a warehouse of flexible examples that will allow us to gauge how frequently certain properties occur. See Theorem F.1.1 where Corner's Theorem follows from a more general result. See also [46, Theorem 110.1].

THEOREM 2.3.2 [A.L.S. Corner] *Let E be a Corner ring. There is a group G such that $E \cong \text{End}(G)$ as rings. If $\text{rank}(E) = n$ then we can choose G so that $\text{rank}(G) = 2n$.*

If E is an rtffr ring then the group G constructed in Corner's Theorem 2.3.2 is a pure and dense subgroup

$$E \subset G = \langle E, E\pi \rangle \subset \widehat{E}$$

for some transcendental element $\pi \in \widehat{\mathbb{Z}}$. Then G fits into a short exact sequence

$$0 \to E \longrightarrow G \longrightarrow \mathbb{Q}E \to 0$$

of left E-modules. The next two results will extend Corner's Theorem by replacing E and $\mathbb{Q}E$ with more general left E-modules while still retaining $E = \text{End}(G)$.

Given a Corner ring E and a Corner E-module M let

$$\mathcal{O}(M) = \{ q \in \mathbb{Q}E \mid qM \subset M \}.$$

The reader can show that under these hypotheses if $\text{ann}_E(M) = 0$ then $\mathcal{O}(M)$ is a Corner ring. See Theorem E.1.1 for a detailed proof.

THEOREM 2.3.3 [36, T.G. Faticoni] *Let E be a Corner ring and let M be a Corner module. There is a short exact sequence*

$$0 \to M \longrightarrow G \longrightarrow \mathbb{Q}C \to 0 \qquad (2.3)$$

such that $\mathcal{O}(M) \cong \text{End}(G)$ and where $C \cong \oplus_K K$ where K ranges over the cyclic E-submodules of $M^{(\aleph_0)}$.

For instance our choices for M in the above Theorem include any of the following left E-modules.

1. $M =$ an E-submodule of a countably generated free E-module such that $\text{ann}_E(M) = 0$.

2. $M = K \oplus E$ where K is a Corner module. The summand E ensures that $E = \mathcal{O}(M)$.

We will find the following result useful in constructing rtffr groups G whose left $\text{End}(G)$-module structure is specified. The left E-module M is an *rtffr E-module* if its additive structure $(M, +)$ is an rtffr group. See Theorem F.1.2 for a proof of the following result.

THEOREM 2.3.4 [43, T.G. Faticoni and H.P. Goeters] *Let E be an rtffr ring and let M be an rtffr E-module. There is an rtffr group G and a short exact sequence*

$$0 \to M \longrightarrow G \longrightarrow \mathbb{Q}E \oplus \mathbb{Q}E \to 0 \qquad (2.4)$$

such that $\mathcal{O}(M) \cong \text{End}(G)$.

It is clear from (2.4) that $\text{rank}(G)$ is finite.

EXAMPLE 2.3.5 Let E be a Corner ring.

1. Given a Corner module M then $E = \mathcal{O}(M \oplus E)$. By Theorem 2.3.4 there is a short exact sequence

$$0 \to M \oplus E \longrightarrow G \longrightarrow \mathbb{Q}E \oplus \mathbb{Q}E \to 0$$

   of left E-modules in which $E \cong \text{End}(G)$. One might choose $M \cong E^{(r)}$ for some integer $r > 0$.

2. If we let $M = E$ then $E = \mathcal{O}(M)$ so there is a short exact sequence

$$0 \to E \longrightarrow G \longrightarrow \mathbb{Q}E \oplus \mathbb{Q}E \to 0$$

   of left E-modules such that $E = \text{End}(G)$. For all practical modern purposes this is Corner's Theorem 2.3.3.

EXAMPLE 2.3.6 If we are willing to vary E but retain $\mathbb{Q}E$ then we can use Theorem 2.3.4 as follows.

   Let A be a finite dimensional $\mathbb{Q}$-algebra, let V be any finitely generated left A-module such that $\text{ann}_A(V) = 0$, and let M be a free *subgroup* of V such that $\mathbb{Q}M = V$. Then $E = \mathcal{O}(M) = \{q \in A \mid qM \subset M\}$ is a subring of A such that $(E, +)$ is a finitely generated free group and $\mathbb{Q}E = A$. An appeal to Theorem 2.3.4 produces a short exact sequence (2.4) in which $E \cong \text{End}(G)$.

   Thus we should avoid conjectures like "every group has a proper nonzero direct summand," "rank of G and rank of $\text{End}(G)$ are related by some simple function," and "indecomposable groups have local endomorphism rings." Such buffoonery is exposed by Butler's construction and Corner's Theorem.

2.4 Arnold-Lady-Murley Theorem

While the examples in the previous sections give us an idea that characterizing module theoretic properties of G in terms of $\text{End}(G)$ will not be easy, the Theorem of D.M. Arnold and E. L. Lady provides us with a friendly setting for studying direct sum decompositions of G. Finitely generated projective modules, at least on the surface, seem to be easier to work with than do torsion-free groups of finite rank.

2.4.1   Category Equivalence

Let G be an rtffr group. Recall the additive functors $H_G(\cdot)$ and $T_G(\cdot)$ and the associated natural transformations Θ and Φ from Chapter 1. Observe that $H_G(\cdot)$ takes groups to right $\text{End}(G)$-modules, while $T_G(\cdot)$ takes right $\text{End}(G)$-modules to groups. The first result classifies the direct summands of $G^{(n)}$ for integers $n > 0$ in terms of a finitely generated projective right $\text{End}(G)$-module.

We will let

$$\mathbf{P}_o(G) = \{H \mid G^{(n)} \cong H \oplus K \text{ for some integer}$$
$$n > 0 \text{ and some group } K\}.$$

$$\mathbf{P}_o(\text{End}(G)) = \text{ the set of finitely generated projective}$$
$$\text{right } \text{End}(G)\text{-modules}.$$

We consider $\mathbf{P}_o(G)$ as a category of groups and we consider $\mathbf{P}_o(\text{End}(G))$ as a category of right $\text{End}(G)$-modules.

THEOREM 2.4.1 [D. Arnold and L. Lady] *See [10, page 47, Theorem 5.1]. If G is an rtffr group then the functors*

$$H_G(\cdot) \quad : \quad \mathbf{P}_o(G) \longrightarrow \mathbf{P}_o(\text{End}(G))$$
$$T_G(\cdot) \quad : \quad \mathbf{P}_o(\text{End}(G)) \longrightarrow \mathbf{P}_o(G)$$

are inverse category equivalences.

Proof: For the sake of the argument let $E = \text{End}(G)$. Since Θ and Φ are natural transformations $\Theta_{H \oplus K} = \Theta_H \oplus \Theta_K$ for groups H and K, and $\Phi_{M \oplus N} = \Phi_M \oplus \Phi_N$ for right $\text{End}(G)$-modules M and N. Moreover, since G is a group, for each integer $n > 0$ there is a group homomorphism

$$\Theta_{G^{(n)}} = \oplus_n \Theta_G : T_G H_G(G^{(n)}) \longrightarrow G^{(n)}.$$

Thus given $H \oplus K \cong G^{(\mathcal{I})}$ we can prove that Θ_H is an isomorphism if we can prove that Θ_G is an isomorphism. Similarly to show that Φ_M is an isomorphism it suffices to show that Φ_E is an isomorphism.

Consider the map

$$\Phi_E : E \to H_G T_G(E).$$

Notice that $H_G T_G(E) \cong E$ with generator the map $f : G \to T_G(E)$ such that $f(x) = 1_G \otimes x$ for each $x \in G$. Then $\Phi_E(1) = f$ which implies that Φ_E is an isomorphism.

The reader proved in Chapter 1 that

$$\Theta_{T_G(E)} \circ T_G(\Phi_E) = 1_{T_G(E)}.$$

Then $\Theta_{T_G(E)}$ is an isomorphism since Φ_E and $T_G(\Phi_E)$ are isomorphisms. Inasmuch as $G \cong T_G(E)$, Θ_G is an isomorphism. Given our reductions, the proof is complete.

Let $\mathbf{P}(G) = \{$groups $H \mid H \oplus H' \cong G^{(c)}$ for some group H' and some cardinal $c\}$. Let $\mathbf{P}(\text{End}(G))$ be the category of projective right End(G)-modules.

EXAMPLE 2.4.2 Let $G = \oplus_p \mathbb{Z}_p$ where p ranges over the primes in $\mathbb{Z}$. Notice that G is not an rtffr group, that $\text{End}(G) = \prod_p \mathbb{Z}_p$ is a semi-hereditary ring, and that G is a projective (=flat) left End(G)-module. Let $I = G$. Then I is a projective ideal in End(G) that is not finitely generated and such that $T_G(I) = IG = G$. Inasmuch $I \not\cong \text{End}(G)$ we have shown that $T_G(\cdot) : \mathbf{P}(\text{End}(G)) \to \mathbf{P}(G)$ is not a category equivalence.

EXAMPLE 2.4.3 Let $G = \mathbb{Z}^{(\aleph_0)}$ and let $E = \text{End}(G)$. Then G is not an rtffr group but G is a cyclic projective left E-module, so $T_G(I) \cong IG$ for each right ideal $I \subset \text{End}(G)$. If we let $I = \{f \in A \mid f(G)$ has finite rank$\}$ then $T_G(I) = IG = G = T_G(\text{End}(G))$ but $I \not\cong \text{End}(G)$. Once again $T_G(\cdot) : \mathbf{P}(\text{End}(G)) \to \mathbf{P}(G)$ is not a category equivalence.

See [32] for generalizations of Theorem 2.4.1 to *self-small right R-modules* G over some ring R.

2.4.2 Functor $A(\cdot)$

There is another functor and associated ring that we will single out. Let G be an rtffr group and let

$$E(G) = \text{End}(G)/\mathcal{N}(\text{End}(G)).$$

While $H_G(\cdot)$ takes direct summands of G to cyclic projective right End(G)-modules the ring End(G) can itself have complicated structure. One of the strengths of the functor $A(\cdot)$ below is that we can translate between direct summands of G and cyclic projective right $E(G)$-modules over the Noetherian semi-prime rtffr ring $E(G)$. If E is an rtffr ring then two E-modules M and N are *quasi-isomorphic* if there is an integer $m \neq 0$ and E-module maps $f : M \to N$, $g : N \to M$ such that $fg = m1_N$ and $gf = m1_M$.

THEOREM 2.4.4 *Let G be an rtffr group. Let*

$$A(\cdot) : \mathbf{P}_o(G) \longrightarrow \mathbf{P}_o(E(G))$$

be defined by

$$A(\cdot) = \mathrm{Hom}(G, \cdot) \otimes_{\mathrm{End}(G)} \mathrm{End}(G)/\mathcal{N}(\mathrm{End}(G)).$$

1. *$A(\cdot)$ is an additive full functor that preserves direct sums. That is, given an $E(G)$-module map $f : A(H) \to A(H')$ there is a group map $\phi : H \to H'$ such that $A(\phi) = f$.*

2. *$A(\cdot)$ induces a bijection*

$$\alpha : \{(H) \mid H \in \mathbf{P}_o(G)\} \longrightarrow \{(P) \mid P \in \mathbf{P}_o(E(G))\}$$

   between the set of isomorphism classes (H) of $H \in \mathbf{P}_o(G)$ and the set of isomorphism classes (P) of $P \in \mathbf{P}_o(A(G))$.

3. *$A(\cdot)$ induces a bijection between the set of quasi-isomorphism classes $[H]$ of $H \in \mathbf{P}_o(G)$ and the set of quasi-isomorphism classes $[P]$ of finitely generated projective right $E(G)$-modules.*

Proof: 1. Say $f : A(H) \to A(H')$ is an $E(G)$-module map. By the Arnold-Lady Theorem 2.4.1, $\mathrm{H}_G(H)$ is a finitely generated projective $\mathrm{End}(G)$-module so f lifts to a map $g : \mathrm{H}_G(H) \to \mathrm{H}_G(H')$. Because $\mathrm{H}_G(\cdot)$ is a category equivalence on $\mathbf{P}_o(G)$ (Arnold-Lady Theorem 2.4.1), there is a group map $\phi : H \to H'$ such that $\mathrm{H}_G(\phi) = g$. The reader can verify that $A(\phi) = f$.

Parts 2 and 3 follow from [10, page 112, Corollary 9.6].

2.4.3   Elementary Uses of the Functors

The results in this subsection illustrate how one uses the above functors to translate properties of direct summands of G into properties of finitely generated projective modules over a ring. We will include the proofs since they are the first of their kind in this text.

PROPOSITION 2.4.5 *Let G be an rtffr group. The following are equivalent.*

1. *If $G^{(n)} = H \oplus H'$ for some integer $n > 0$ then there is an integer m such that $H \cong G^{(m)}$.*

2. *Each finitely generated projective right $E(G)$-module is free.*

Proof: Suppose that part 2 is true, let $n > 0$ be an integer, and let $G^{(n)} = H \oplus K$ as groups. By Theorem 2.4.4

$$E(G)^{(n)} \cong A(G^{(n)}) \cong A(H) \oplus A(K)$$

as projective right $E(G)$-modules. By part 2

$$A(H) \cong E(G)^{(m)}$$

for some integer m so that $H \cong G^{(m)}$ by Theorem 2.4.4. The converse is handled in an analogous manner so the proof is complete.

The above work shows us that properties of direct sum decompositions of the rtffr group G are equivalent to properties of direct sum decompositions of free right $\text{End}(G)$-modules. This is one example where good results come from the Arnold-Lady Theorem 2.4.1. The general idea is that projective $E(G)$-modules are better behaved than direct summands of $G^{(n)}$ for integers $n > 0$.

Since projective $E(G)$-modules are better behaved than those over $\text{End}(G)$ it is natural to ask for an AKS-like Theorem for direct sums of groups whose endomorphism rings are *pid*'s. Recall the *refinement property* and *unique decomposition* from section 2.1.

THEOREM 2.4.6 *Suppose that $G_1, \ldots, G_t$ are indecomposable rtffr groups such that for each $i \neq j$, the composite of each pair of maps $G_i \to G_j \to G_i$ is in $\mathcal{N}(\text{End}_R(G_i))$. Assume that $E(G_i)$ is a pid for each $i = 1, \ldots, t$. If we let $G \cong G_1 \oplus \cdots \oplus G_t$ then*

1. *G has the refinement property.*

2. *G has a unique decomposition.*

3. *If $G^{(n)} \cong H \oplus K \cong H \oplus K'$ for some integer $n > 0$ and some rtffr groups K and K' then $K \cong K'$.*

Proof: We begin the proof with some general comments about finitely generated projective right $E(G)$-modules.
By Lemma 1.3.1 and Theorem 2.4.4

$$E(G) = E(G_1) \times \cdots \times E(G_t)$$

where $E(G_i) = A(G_i)$ is indecomposable for each $i = 1, \ldots, t$.

Furthermore since $E(G_i)$ is a *pid*, given a finitely generated projective right $E(G)$-module P there are integers $p_1, \ldots, p_t \geq 0$ such that

$$P \cong E(G_1)^{(p_1)} \oplus \cdots \oplus E(G_t)^{(p_t)}.$$

The uniqueness of the rank of free modules over a *pid* implies that the integers $p_1, \ldots, p_t$ are unique to P.

1. Let $H \in \mathbf{P}_o(G)$. Since $A(H) \in \mathbf{P}_o(E(G))$ we have

$$
\begin{aligned}
A(H) &\cong E(G_1)^{(h_1)} \oplus \cdots \oplus E(G_t)^{(h_t)} \\
&\cong A(G_1)^{(h_1)} \oplus \cdots \oplus A(G_t)^{(h_t)} \\
&\cong A(G_1^{(h_1)} \oplus \cdots \oplus G_t^{(h_t)})
\end{aligned}
$$

for some integers $h_1, \ldots, h_t \geq 0$. Thus

$$H \cong G_1^{(h_1)} \oplus \cdots \oplus G_t^{(h_t)}$$

by Theorem 2.4.4, and hence G has the refinement property.

2. Suppose that

$$G = G_1' \oplus \cdots \oplus G_s'$$

for some indecomposable right E-modules $G_1', \ldots, G_s'$. Then

$$E(G) = E(G_1) \times \cdots \times E(G_t) = A(G_1') \oplus \cdots \oplus A(G_s').$$

Consider the indecomposable central idempotent $e_i \in E(G)$ that corresponds to $E(G_i)$. Then $e_i A(G_j')$ is a direct summand of $A(G_j')$, and since $A(G_j')$ is indecomposable, there is a permutation of the subscripts of $A(G_1'), \ldots, A(G_s')$ such that

$$e_i A(G_i') = A(G_i') \subset E(G_i)$$

for $i = 1, \ldots, t$. Since $E(G_i)$ is an indecomposable ideal in $E(G)$ and since $e_i A(G_i')$ is a direct summand of $E(G_i)$ we have $e_i A(G_i') = E(G_i) = A(G_i)$. Then by Theorem 2.4.4,

$$G_i' \cong G_i.$$

Since t is finite, $s = t$.

3. Cancellation follows from the uniqueness result in part 2 by counting indecomposable summands of K and K'. The reader should prove this if he/she has not already done so. This completes the proof.

Given a subgroup $H \subset \mathbb{Q}$, $\text{End}(H)$ is a *pid*. Thus the Baer-Kulikov-Kaplansky Theorem 2.1.11 is an immediate consequence of Theorem 2.4.6.

We close this section with elementary translations of the refinement property and uniqueness of decomposition for G.

PROPOSITION 2.4.7 *Let* G *be an rtffr group. The following are equivalent.*

1. G *has the refinement property as a group.*

2. $\text{End}(G)$ *has the refinement property as a right* $\text{End}(G)$*-module.*

3. $E(G)$ *has the refinement property as a right* $E(G)$*-module.*

PROPOSITION 2.4.8 *Let* G *be an rtffr group. The following are equivalent.*

1. G *has a unique decomposition.*

2. $\text{End}(G)$ *has a unique decomposition as a right* $\text{End}(G)$*-module.*

3. $E(G)$ *has a unique decomposition as a right* $E(G)$*-module.*

2.5 Local Isomorphism

Let E be an *rtffr ring* and let M and N be *rtffr E-modules*. We will say that M and N are *locally isomorphic* if for each integer $n > 0$ there is an integer $m \neq 0$ and E-module maps $f_n : M \to N$, $g_n : N \to M$ such that $\gcd(m, n) = 1$, $f_n g_n = m 1_N$, and $g_n f_n = m 1_M$. Since we have not ruled out $E = \mathbb{Z}$ we have defined *local isomorphism* for rtffr groups G and H.

Locally isomorphic rtffr groups are called *nearly isomorphic* in [10] and locally isomorphic *E-lattices* are said to be *in the same genus class* in [10, 69]. Our introduction of the new terminology is justified by the fact that we are using one term to describe two essentially identical concepts on the category of *rtffr E-modules*.

The next lemma gives us the relationship between local isomorphism of rtffr groups and the local isomorphism of finitely generated projective modules. Its proof rests on the fact that an additive functor takes multiplication by an integer to multiplication by an integer. It is left as an exercise.

LEMMA 2.5.1 *Let* G *be an rtffr group, let* H, $K \in \mathbf{P}_o(G)$ *be rtffr groups, and let* $E = \text{End}(G)$*. The following are equivalent.*

1. H *and* K *are locally isomorphic as groups.*

2. $H_G(H)$ and $H_G(K)$ are *locally isomorphic as right E-modules.*

Local isomorphism appears in several useful equivalent forms. We leave it to the reader to prove a similar result for finitely generated rtffr modules over an rtffr ring. Recall that $X_n = X \otimes_{\mathbb{Z}} \mathbb{Z}_n$ for integers $n \neq 0$.

LEMMA 2.5.2 *Let G and H be rtffr groups. The following are equivalent.*

1. *G and H are locally isomorphic.*

2. *For each integer $n \neq 0$ there are $f_n \in \operatorname{Hom}_E(G, H)_n$ and $g_n \in \operatorname{Hom}_E(H, G)_n$ such that $f_n g_n = 1_H$ and $g_n f_n = 1_G$.*

3. *For each prime $p \in \mathbb{Z}$ there are group maps $f_p \in \operatorname{Hom}_E(G, H)_p$ and $g_p \in \operatorname{Hom}_E(H, G)_p$ such that $f_p g_p = 1_H$, and $g_p f_p = 1_G$.*

4. *For each integer $n \neq 0$ there are there are $f_n \in \operatorname{Hom}_E(G, H)$ and $g_n \in \operatorname{Hom}_E(H, G)$ such that $f_n g_n - 1_H \in n\operatorname{End}(H)$ and $g_n f_n - 1_G \in n\operatorname{End}(G)$.*

Proof: $1 \Rightarrow 2$ Given part 1 and an integer $n \neq 0$ there is an integer $m \neq 0$, and maps $f_n : G \to H$, and $g_n : H \to G$ such that $\gcd(m, n) = 1$, $f_n g_n = m 1_H$, $g_n f_n = m 1_G$. Since $\gcd(m, n) = 1$, m is a unit in $\mathbb{Z}_n$, and so $\frac{1}{m} f_n \in \operatorname{Hom}_E(G, H)_n$. Then $(\frac{1}{m} f_n) g_n = 1_H$ and $g_n(\frac{1}{m} f_n) = 1_G$, which proves part 2.

$2 \Rightarrow 3$ is clear.

$3 \Rightarrow 4$ Let $n = p_1^{n_1} \cdots p_t^{n_t} \neq 0$ be an integer where $p_1, \ldots, p_t$ are the distinct prime divisors of n. Fix an $i \in \{1, \ldots, t\}$. By part 3, there are maps $f_i \in \operatorname{Hom}_E(G, H)_{p_i}$ and $g_i \in \operatorname{Hom}_E(H, G)_{p_i}$ such that $f_i g_i = 1_H$, and $g_i f_i = 1_G$. Since

$$\frac{X_p}{p^k X_p} \cong \frac{X + p^k X_p}{p^k X_p}$$

for p-reduced groups X and prime powers $p^k \in \mathbb{Z}$ there are maps $f_i' \in \operatorname{Hom}(G, H)$ and $g_i' \in \operatorname{Hom}(H, G)$ such that

$$f_i' - f_i \ \in \ p_i^{n_i} \operatorname{Hom}_E(G, H)_{p_i}$$
$$g_i' - g_i \ \in \ p_i^{n_i} \operatorname{Hom}_E(H, G)_{p_i}.$$

($X = \operatorname{Hom}(G, H)$ in this case.) By the Chinese Remainder Theorem 1.5.3 there are $f \in \operatorname{Hom}(G, H)$ and $g \in \operatorname{Hom}(H, G)$ such that

$$f - f_i' \ \in \ p_i^{n_i} \operatorname{Hom}_E(G, H)$$
$$g - g_i' \ \in \ p_i^{n_i} \operatorname{Hom}_E(H, G)$$

for each $i = 1, \cdots, t$. Then

$$gf - gf_i' \in p_i^{n_i} \mathrm{End}_E(G)$$
$$gf_i' - g_i f_i \in p_i^{n_i} \mathrm{End}_E(G)_{p_i}$$

so that

$$
\begin{aligned}
gf - 1_G &= gf - g_i f_i \\
&= (gf - gf_i') + (gf_i' - g_i' f_i') + (g_i' f_i' - g_i f_i) \\
&\in p_i^{n_i} \mathrm{End}_E(G)_{p_i}.
\end{aligned}
$$

Inasmuch as gf and $1_G \in \mathrm{End}(G)$

$$gf - 1_G \in p_i^{n_i} \mathrm{End}_E(G)_{p_i} \cap \mathrm{End}(G) = p_i^{n_i} \mathrm{End}(G).$$

Since $i \in \{1, \ldots, t\}$ was arbitrarily selected, part 3 of the Chinese Remainder Theorem 1.5.3 shows us that

$$gf - 1_G \in \bigcap_{i=1}^{t} p_i^{n_i} \mathrm{End}_E(G) = n \mathrm{End}_E(G).$$

This proves part 4.

4 $\Rightarrow$ 1 Choose an integer $n \neq 0$ such that $gf - 1_G \in n\mathrm{End}(G)$ and $fg - 1_H \in n\mathrm{End}(H)$. Because G is an rtffr group we may choose n so large that $\mathrm{End}(G)_n$ is a reduced ring. Then $n \in \mathcal{J}(\mathrm{End}(G)_n)$ (try proving that fact, reader), and hence fg is a unit of $\mathrm{End}(G)_n$. Let $u = (fg)^{-1}$ and choose an integer m such that $\gcd(m, n) = 1$ and $mu \in \mathrm{End}(G)$. Then $muf \in \mathrm{Hom}(G, H)$ and $(muf)g = m1_G$. Since $\mathbb{Q}G$ and $\mathbb{Q}H$ are finite dimensional vector spaces and since $[g(uf)]^2 = g(uf) \in \mathrm{End}(H)$ is an injection, $g(uf) = 1_{\mathbb{Q}H}$, and hence $g(muf) = m1_H$. This proves part 1 and completes the proof.

The proof of the next result is an exercise.

LEMMA 2.5.3 Let E be a semi-prime rtffr ring and let U and V be E-lattices. If E is a semi-local ring then U and V are locally isomorphic iff $U \cong V$.

Proof: Suppose that U and V are locally isomorphic and say that there is an integer $n \neq 0$ such that $pE = E$ for each prime $p \in \mathbb{Z}$ not dividing n. There is by definition an integer m and maps $f_n : U \to V$ and $g_n : V \to U$ such that $\gcd(m, n) = 1$, $f_n g_n = m1_V$, and $g_n f_n = m1_U$. Then m is a unit in $\mathbb{Z}_n$ so that f_n and g_n are isomorphisms. The converse is clear so the proof is complete.

LEMMA 2.5.4 *Let E be an rtffr ring. The following are equivalent for finitely generated projective rtffr E-modules M and N.*

1. M and N are locally isomorphic.

2. $M_p \cong N_p$ as right E_p-modules for each prime $p \in \mathbb{Z}$.

3. $M_n \cong N_n$ as right E_n-modules for each integer $n \neq 0 \in \mathbb{Z}$.

4. For each integer $n \neq 0$ there are injections $f_n : M \to N$, and $g_n : N \to M$ such that $f_n(M) + nN = N$ and $g_n(N) + nM = M$.

Proof: $1 \Rightarrow 2 \Rightarrow 3$ follow as they did in Lemma 2.5.2.

$3 \Rightarrow 4$ Assume part 3 and let $n \neq 0$ be an integer. There are $f_n \in \operatorname{Hom}_E(M_n, N_n)$ and $g_n \in \operatorname{Hom}_E(N_n, M_n)$ such that $f_n g_n = 1_{M_n}$. Since M and N are finitely generated, we may choose an integer m such that $\gcd(m, n) = 1$, $mf_n \in \operatorname{Hom}_E(M, N)$, $mg_n \in \operatorname{Hom}_E(N, M)$, and $(mf_n)(mg_n) = m^2 1_M$. Inasmuch as m is a unit in $\mathbb{Z}_n$ the isomorphism

$$\frac{X}{nX} \cong \frac{X_n}{nX_n} \cong \frac{X + nX_n}{nX_n}$$

and our choice of f_n and g_n show us that

$$(mf_n)(M) + nN = N \text{ and } (mg_n)(N) + nM = M.$$

This proves part 4.

$4 \Rightarrow 1$ Assume part 4 and let $n \neq 0$ be an integer. Choose E-module maps $f : M \to N$ and $g : N \to M$ such that $f(M) + nN = N$ and $g(N) + nM = M$. Since E is an rtffr ring we can assume without loss of generality that n is so large that the localization E_n is a reduced group. Then $n \in \mathcal{J}(E_n)$ and since M_n and N_n are finitely generated right E-modules

$$N_n = f(M_n) + nN_n = f(M_n)$$

by Nakayama's Lemma 1.2.3. Similarly $g(N_n) = M_n$, hence $\operatorname{rank}(M_n) = \operatorname{rank}(N_n)$, whence $f : M_n \to N_n$ is an isomorphism. Let $h : N_n \to M_n$ be the inverse of f. Inasmuch as M and N are finitely generated there is an integer $m \neq 0$ such that $\gcd(m, n) = 1$ and $mh(N) \subset M$. Then f and mh are maps between M and N such that $f(mh) = m1_N$ and $(mh)f = m1_M$. This completes the proof.

An extreme case of the above lemma is found if E is a *classical maximal order*. Recall that an E-lattice is a finitely generated left E-module M of $\mathbb{Q}E^{(n)}$ for some integer n.

LEMMA 2.5.5 *Let E be an integrally closed semi-prime rtffr ring and let M, N be E-lattices. Then M and N are locally isomorphic iff $\mathbb{Q}M \cong \mathbb{Q}N$ as right $\mathbb{Q}E$-modules.*

Proof: Since the integrally closed ring E equals $E_1 \times \cdots \times E_t$ for some classical maximal orders and since $\mathbb{Q}M \cong \mathbb{Q}N$ iff $\mathbb{Q}ME_i \cong \mathbb{Q}NE_i$ as $\mathbb{Q}E_i$-modules for each $i = 1, \ldots, t$, we have reduced the problem to the case where E is a classical maximal order. Given our reduction, suppose that $\mathbb{Q}E$ is a prime Artinian ring. Then $\mathbb{Q}M \cong \mathbb{Q}N$ iff M is locally isomorphic to N by [10, Corollary 12.4]. This completes the proof.

The next two results relate local isomorphism to properties of direct sums. The argument is originally due to E. L. Lady. See [16].

LEMMA 2.5.6 *Let E be an rtffr ring, and let G and H be rtffr E-modules. If G is locally isomorphic to H then $G \oplus G \cong H \oplus K$ for some group K.*

Proof: Let $n \neq 0$ be any integer. Since H is locally isomorphic to G there is an integer $m \neq 0$ and group maps $f_n : G \to H$ and $g_n : H \to G$ such that $\gcd(m, n) = 1$ and $g_n f_n = m 1_G$. Again there is an integer $k \neq 0$ and maps $f_m : G \to H$ and $g_m : H \to G$ such that $\gcd(k, m) = 1$ and $g_m f_m = k 1_G$. Since $\gcd(k, m) = 1$ there are integers a and b such that $am + bk = 1$. Consider the maps

$$\sigma : G \oplus G \longrightarrow H : x \oplus y \longmapsto a f_n(x) + b f_m(y)$$

$$\jmath : H \longrightarrow G \oplus G : z \longmapsto g_n(z) \oplus g_m(z).$$

Then

$$\sigma\jmath(z) = a f_n g_n(z) + b f_m g_m(z) = (am + bk)(z) = z$$

and so $G \oplus G \cong H \oplus \ker \sigma$. This completes the proof.

The next result shows us that local isomorphism is tied to the cancellation property.

THEOREM 2.5.7 *Let H, K, and K' be rtffr groups.*

1. *If $H \oplus K$ is locally isomorphic to $H \oplus K'$ then K is locally isomorphic to K'.*

2. *If $n > 0$ is an integer and if $H^{(n)} \cong K^{(n)}$ then H is locally isomorphic to K.*

Proof: 1. Suppose that $H \oplus K$ is locally isomorphic to $H \oplus K'$. By Lemma 2.5.1 we may assume without loss of generality that H, K, K' are projective right modules over an rtffr ring E such that $E = H \oplus K$.

Let $p \in \mathbb{Z}$ be a prime. Since E is an rtffr ring we can choose an integer multiple of p, say $n \neq 0$, such that $\cap_{k=1}^{\infty} n^k E = 0$. Then

$$E_n \cong H_n \oplus K_n \cong H_n \oplus K'_n$$

as right E_n-modules and consequently

$$\frac{E_n}{nE_n} \cong \frac{H_n}{nH_n} \oplus \frac{K_n}{nK_n} \cong \frac{H_n}{nH_n} \oplus \frac{K'_n}{nK'_n}$$

as projective right E_n/nE_n-modules. Since E is torsion-free of finite rank, E_n/nE_n is finite, hence Artinian. Then an application of the AKS Theorem 2.1.6 implies that there is an isomorphism $\bar{f} : K_n/nK_n \to K'_n/nK'_n$. There are natural isomorphisms

$$\frac{K}{nK} \cong \frac{K_n}{nK_n} \xrightarrow{\bar{f}} \frac{K'}{nK'} \cong \frac{K'_n}{nK'_n}.$$

Because K and K' are projective $\bar{f}$ lifts to a map $f : K \to K'$ such that $f(K) + nK' = K'$. Similarly there is a map $g : K' \to K$ such that $g(K') + nK = K$. Then by Lemma 2.5.4(4), K is locally isomorphic to K'. Given our reductions we have completed the proof of part 1.

Part 2 follows in a manner similar to that of part 1. This completes the proof of the lemma.

There are at least two general conditions on G that imply a cancellation property for $H \oplus K$. The group G is *semi-local* if there are at most finitely many primes $p \in \mathbb{Z}$ such that $pG \neq G$.

COROLLARY 2.5.8 *Let H and K be rtffr groups.*

1. *If K is semi-local then $H \oplus K \cong H \oplus K' \Rightarrow K \cong K'$.*

2. *If each right ideal of $\mathrm{End}(K)$ is principal then $H \oplus K \cong H \oplus K' \Rightarrow K \cong K'$.*

Proof: 1. Suppose that $H \oplus K \cong H \oplus K'$. Since K is semi-local there is an integer $n \neq 0$ such that $nK \neq K$ and $pK = K$ for each prime $p \in \mathbb{Z}$ such that $\gcd(p, n) = 1$. By Theorem 2.5.7, K is locally isomorphic to K' so there is an integer $m \neq 0$ and injections $f_n : K \to K'$ and $g_n : K' \to K$ such that $\gcd(m, n) = 1$, $f_n g_n = m 1_{K'}$, and such that

$g_n f_n = m1_K$. Inasmuch as $mK = K$ we see that $\frac{1}{m}g_n : K' \to K$ is a map such that $f_n(\frac{1}{m}g_n) = 1_{K'}$ and $(\frac{1}{m}g_n)f_n = 1_K$. That is, $K \cong K'$ as required.

2. Suppose that $H \oplus K \cong H \oplus K'$ and that each right ideal of $\mathrm{End}(K)$ is principal. By Theorem 2.5.7, K is locally isomorphic to K' so there is an injection $g : K' \to K$. The reader can show that $\mathrm{Hom}(K, K')K = K'$. Identify $\mathrm{Hom}(K, K') \cong g\mathrm{Hom}(K, K')$ with a right ideal of $\mathrm{End}(K)$. By the hypothesis on $\mathrm{End}(K)$ there is an $f \in \mathrm{Hom}(K, K')$ such that $g\mathrm{Hom}(K, K') = f\mathrm{End}(K)$. Thus when considered as subgroups of $\mathbb{Q}K$

$$g^{-1}f(K) = g^{-1}[f\mathrm{End}(K)]K = g^{-1}[g\mathrm{Hom}(K, K')]K = K'.$$

Since H, K, K' have finite rank, $\mathrm{rank}(K) = \mathrm{rank}(K')$, so that the surjection $g^{-1}f : K \to K'$ is an isomorphism. This completes the proof.

We can use local isomorphism to establish some kind of uniqueness for the direct sum decompositions of an rtffr group.

Let E be an rtffr ring and let G be an rtffr left E-module. We say that G has a *locally unique decomposition* if

1. There is a direct sum decomposition

$$G = G_1 \oplus \cdots \oplus G_t \qquad (2.5)$$

   of indecomposable E-modules $G_1, \ldots, G_t$, and

2. Given a direct sum decomposition

$$G = G'_1 \oplus \cdots \oplus G'_s$$

   of indecomposable E-modules $G'_1, \ldots, G'_s$ then $s = t$ and after a possible permutation of the indices G'_i is locally isomorphic to G_i.

In case G satisfies the above properties then $G_1 \oplus \cdots \oplus G_t$ is called *the locally unique decomposition of G*.

Observe that we do not require that the $G_1, \ldots, G_t$ are distinct groups in the above definition. Also, if $E = \mathbb{Z}$ then we have defined the *locally unique decomposition* of an rtffr group G.

EXAMPLE 2.5.9 A locally unique decomposition that is not a unique decomposition is constructed as follows. Let E be a Dedekind domain with class number $h > 1$. There is an ideal $I \not\cong E$ such that $I \cdots I \cong E$ with h factors. Then

$$M = I^{(h)} \cong E^{(h-1)} \oplus (I \cdots I) \cong E^{(h)}.$$

Thus M does not have a unique decomposition.

However, the decomposition is locally unique since at each prime $p \in \mathbb{Z}$, E_p is a *pid* (see [10] or [69]). Each direct summand of the projective E_p-module M_p is free. In particular any indecomposable direct summand U of M_p is isomorphic to E_p, whence M_p has a unique decomposition.

Let E be an rtffr ring and let G be an rtffr left E-module. We say that G has the *local refinement property* if

1. There is a direct sum decomposition (2.5) for some indecomposable left E-modules $G_1, \ldots, G_t$, and

2. Given $H \in \mathbf{P}_o(G)$ then

$$H = H_1 \oplus \cdots \oplus H_s$$

for some indecomposable left E-modules $H_1, \ldots, H_s$ such that each H_i is locally isomorphic to some element of $\{G_1, \ldots, G_t\}$.

Observe that the G_i in the direct sum (2.5) need not be distinct. If $E = \mathbb{Z}$ then we have defined what we mean by the *local refinement property* for the rtffr group G.

A large portion of the energy in this book is expended on the study of (locally) unique decompositions and the (local) refinement properties. Locally unique decomposition follows from the local refinement property.

LEMMA 2.5.10 *Let E be an rtffr ring and let G be an rtffr left E-module. If G has the local refinement property then G has a locally unique decomposition.*

Proof: Suppose that G has the local refinement property. Then G has a direct sum decomposition (2.5) for some indecomposable left E-modules $G_1, \ldots, G_t$. Suppose that

$$G = G'_1 \oplus \cdots \oplus G'_s \qquad (2.6)$$

for some indecomposable left E-modules $G'_1, \ldots, G'_s$. Since G has the local refinement property, G'_1 is locally isomorphic to some element of $\{G_1, \ldots, G_t\}$.

We will show that $s = t$ and that $G_i \cong G'_i$ for each $i = 1, \ldots, t$ after a permutation of the subscripts. We can assume without loss of generality that G'_1 is locally isomorphic to G_1 and we know that

$$G = G_1 \oplus (G_2 \oplus \cdots \oplus G_t) \quad \text{and} \quad G_1 \oplus (G'_2 \oplus \cdots \oplus G'_s)$$

are (locally) isomorphic. Then by Theorem 2.5.7

$$G_2 \oplus \cdots \oplus G_t \quad \text{and} \quad G'_2 \oplus \cdots \oplus G'_s$$

are locally isomorphic. By a simple induction on rank(G), $s - 1 = t - 1$, $s = t$, and after a possible reindexing, G_i is locally isomorphic to G'_i for each $i = 2, \ldots, t$. This concludes the proof.

We end this chapter by showing that at least one type of group has a unique decomposition. An idempotent $e \in E$ is *central* if $xe = ex$ for each $x \in E$.

LEMMA 2.5.11 *Let $G = G_1 \oplus \cdots \oplus G_t$ be an rtffr group, for each integer $i = 1, \ldots, t$ let $e_i^2 = e_i \in \mathrm{End}(G)$ be central indecomposable idempotents such that $e_i(G) = G_i$ and such that $1 = e_1 + \oplus \cdots \oplus e_t$. Then $G_1 \oplus \cdots \oplus G_t$ is the unique decomposition of G.*

Proof: Let $G = G_1 \oplus \cdots \oplus G_t$ be the given indecomposable decomposition and let $\{e_1, \ldots, e_t\} \subset \mathrm{End}(G)$ be the associated complete set of primitive orthogonal indecomposable central idempotents. Suppose that $G = G'_1 \oplus \cdots \oplus G'_s$ for some indecomposable groups G'_i. Let $\{f_1, \ldots, f_s\}$ be the complete set of orthogonal indecomposable idempotents corresponding to $G'_1 \oplus \cdots \oplus G'_s$. Because e_i is central $e_i f_j = f_j e_i$ for each i, j. It follows that

$$e_1 = e_1 1_G = e_1 f_1 \oplus \cdots \oplus e_1 f_s$$

is a decomposition of the indecomposable e_1 into orthogonal idempotents, so that $e_1 = e_1 f_i$ for some i. In a similar manner, since G'_i is indecomposable $e_1 f_i = f_i$. That is $e_1 = f_i$. Reindexing if necessary we can assume that we have chosen $f_1, \ldots, f_t$ such that $e_i = f_i$ for each $i = 1, \ldots, t$. Since

$$e_1 \oplus \cdots \oplus e_t = 1_G = f_1 \oplus \cdots \oplus f_t$$

$f_{t+1} = \ldots = f_s = 0$. Therefore $G_i = e_i G = f_i G = G'_i$ and the proof is complete.

We will find that commuting endomorphisms of rtffr groups G occur with some regularity.

2.6   Exercises

$p \in \mathbb{Z}$ is a prime, $n, m > 0$ are integers, c and d are cardinals. G and H are groups, E is a ring, M is a right E-module, and $f : G \to H$ is a map.

1. If G is an rtffr group then for each index set $\mathcal{I}$ and each $f : G \to G^{(\mathcal{I})}$ there is a finite subset $\mathcal{J} \subset \mathcal{I}$ such that $g(G) \subset G^{(\mathcal{J})}$.

2. The linear transformation $f : \mathbb{Q}G \to \mathbb{Q}H$ is in $\mathbb{Q}\mathrm{Hom}(G, H)$ iff $f(G) \doteq H$.

3. $\mathrm{Hom}(G, H)$ is a right $\mathrm{End}(G)$-module.

4. Let $p \in \mathbb{Z}$ be prime.

    (a) $\mathbb{Z}/p^n\mathbb{Z}$, $\mathbb{Z}(p^\infty)$, and $\widehat{\mathbb{Z}}_p$ are indecomposable groups.

    (b) Any subgroup of $\mathbb{Q}$ is indecomposable.

5. Let $p \in \mathbb{Z}$ be prime. Show that for each integer $n > 0$, $\mathrm{End}(\mathbb{Z}_p) \cong \mathbb{Z}_p$ and $\mathrm{End}(\mathbb{Z}/p^n\mathbb{Z}) \cong \mathbb{Z}/p^n\mathbb{Z}$ are local rings.

6. Show that if G is an rtffr group then $G/f(G)$ is finite for each injection $f : G \to G$.

7. Show that if G is an rtffr group then a bounded quotient G/H is finite.

8. The rtffr group G is strongly indecomposable iff G is an indecomposable object in $\mathbb{Q}\mathbf{Ab}$.

9. A pure subgroup of $\widehat{\mathbb{Z}}_p$ is strongly indecomposable.

10. Let $p \in \mathbb{Z}$ be prime. Show that the subgroups of $\mathbb{Z}(p^\infty)$ form a well ordered chain and show that $\mathrm{End}(\mathbb{Z}(p^\infty)) \cong \widehat{\mathbb{Z}}_p$.

11. Let Q be the field of fraction of the ring $\mathbb{Z}[[x]]$ of power series. Determine the $\mathbb{Z}[[x]]$-module $Q/\mathbb{Z}[[x]]$ and its $\mathbb{Z}[[x]]$-endomorphism ring.

12. Let $S \subset \mathbb{Q}$ be a ring. Then each element of S is an integral multiple of a unit in S.

13. Let E be a Corner ring, let M be a Corner E-module, and let K be a right E-module. Let (2.4) be the short exact sequence constructed in Theorem 2.3.3. Show that $\mathrm{Tor}_E^k(K, M) \cong \mathrm{Tor}_E^k(K, G)$ for each integer $k \geq 1$. Conclude that $\mathrm{fd}_E(M) = \mathrm{fd}_E(G)$.

14. Let E be a Corner ring, let M be a Corner E-module, and let N be a left E-module. Let (2.4) be the short exact sequence constructed in Theorem 2.3.3. Show that $\operatorname{Ext}_E^k(N, M) \cong \operatorname{Ext}_E^k(N, G)$ for each integer $k \geq 2$. Conclude that $\operatorname{pd}_E(M) = \operatorname{pd}_E(G)$.

15. Show that $\mathbb{Z}(p^\infty)$ is an E-torsion group.

16. This more of a project than an exercise. Prove that there is a Corner group G such that $IG \neq G$ for each finitely generated right ideal $I \subset \operatorname{End}(G)$ but such that $K \otimes_{\operatorname{End}(G)} G = 0$ for some nonzero finitely generated K.

17. Let $G = G_1 \oplus \cdots \oplus G_t$ be an abelian group where

$$E(G_i) = \operatorname{End}(G_i)/\mathcal{N}(\operatorname{End}(G_i))$$

is a *pid* for each $i = 1, \ldots t$. Show that if $H \oplus H' \cong G^{(c)}$ for some cardinal c then $H \cong G_1^{(h_1)} \oplus \cdots \oplus G_t^{(h_t)}$ for some cardinals $h_1, \ldots, h_t$.

18. If G and H are locally isomorphic then $\operatorname{Hom}(G, H)G = H$.

19. State and prove a result like Lemma 2.5.6 for locally isomorphic E-lattices G and H.

20. Prove part 2 of Lemma 2.5.7. If $n > 0$ is an integer and if $G^{(n)} \cong H^{(n)}$ then G and H are locally isomorphic.

2.7 Questions for Future Research

In this section I will give some questions that came up during the writing process. These questions can be used to give the reader an eye toward future research in the area of reduced torsion-free finite rank abelian groups, better known as *rtffr groups*.

Let G be an rtffr group, let E be an rtffr ring, and let $S = \operatorname{center}(E)$ be the center of E. Let τ be the conductor of G. There is no loss of generality in assuming that G is strongly indecomposable.

1. Extend the ideas of this book to self-small mixed groups.

2. Lift ideal and module theoretic properties from $E/\mathcal{N}(E)$ to E and conversely.

3. Lift group or ring theoretic properties from the $\mathbb{Z}$-adic completion $\widehat{G}$ of G to G. See [40].

4. Extend the Arnold-Lady-Murley Theorem to a larger class of groups (i.e., a class that contains $\mathbf{P}_o(G)$). See [32].

5. Study the group theoretic of those G such that $\text{End}(G)$ is an integrally closed ring. That is, $\text{End}(G)$ is a product of classical maximal orders.

6. Classify the groups G such that $\text{End}(G)$ is a semi-perfect ring.

7. In group theoretic terms, describe those groups G that have the Azumaya-Krull-Schmidt Property. For example, The Baer-Kulikov-Kaplansky Theorem shows that the completely decomposable groups satisfy a strong version of this property. If $\text{End}(G)$ is semi-perfect then G has the Azumaya-Krull-Schmidt Property.

8. Characterize the radical ideals in $\text{End}(G)$. For example, the Jacobson radical and the nil radical.

9. Give a correspondence between ideals of $\text{End}(G)$ and the subgroups of G. Ideally, the subgroups of G should be the fully invariant G-generated subgroups of G.

10. Let E be an rtffr ring. Characterize those (maximal) right ideals $I \subset E$ such that $IG = G$ for some group G such that $E = \text{End}(G)$. See [40].

Chapter 3

Local Isomorphism Is Isomorphism

---

rtffr means *reduced torsion-free finite rank.*

The functor $A(\cdot)$ strikes us as interesting. Thus we will make extensive use of the ideal structure of the semi-prime rtffr ring

$$A(G) = E(G) = \operatorname{End}(G)/\mathcal{N}(\operatorname{End}(G)).$$

In fact $A(\cdot)$ becomes more important in this book than $\mathrm{H}_G(\cdot)$. This is different from the tone of the literature at the time of this writing. We will show that under a certain commutative localization functor $L_\tau(\cdot)$ the local isomorphism of groups G and H translates into the isomorphism of right $E(G)_\tau$-modules.

3.1   Integrally Closed Rings

Recall that the ring E is *integrally closed* if given a ring $E \subset E' \subset \mathbb{Q}E$ such that E'/E is finite then $E = E'$. A prime integrally closed rtffr ring is called a *classical maximal order*.

By Lemma 1.6.3 if E is a semi-prime rtffr ring then $\mathbb{Q}E$ is semi-simple Artinian. In this case there is an integrally closed ring $\overline{E} \subset \mathbb{Q}E$ and an integer $n > 0$ such that

$$n\overline{E} \subset E \subset \overline{E}.$$

The integrally closed ring $\overline{E}$ is a finite product

$$\overline{E} = \overline{E}_1 \times \cdots \times \overline{E}_t \tag{3.1}$$

of classical maximal orders $\overline{E}_i$ in the simple Artinian rings $\mathbb{Q}\overline{E}_i$. Let $S = \text{center}(E)$, let $\overline{S}_i = \text{center}(\overline{E}_i)$ for each $i = 1, \ldots, t$, and let

$$\overline{S} = \overline{S}_1 \times \cdots \times \overline{S}_t. \tag{3.2}$$

Then $\overline{S} = \text{center}(\overline{E})$, $\overline{S}_i$ is a Dedekind domain for each $i = 1, \ldots, t$, and

$$n\overline{S} \subset S \subset \overline{S}.$$

It follows that S and $\overline{S}$ are *commutative Noetherian semi-prime rtffr rings* such that $\overline{S}/S$ is finite and such that $\mathbb{Q}S = \mathbb{Q}\overline{S}$.

LEMMA 3.1.1 Let $I \subset E$ be a right ideal that contains a regular element of E. Then E/I is a finite group. In particular at most finitely many (maximal) ideals of E contain I.

Proof: Say that $c \in I \subset E$ is regular. Then c is a unit in $\mathbb{Q}E$ so it has an inverse $d \in \mathbb{Q}E$. There is an integer $m \neq 0$ such that $md \in E$ so that $c(md) = m1_E \in I$. Then E/I is bounded by m and since E is an rtffr ring E/I is finite.

LEMMA 3.1.2 Let E be a semi-prime rtffr ring, let $\text{center}(E) = S$, and let $I \subset S$ be an ideal. Then $E_I I_I \subset \mathcal{J}(E_I)$.

Proof: We know that $I_I \subset \mathcal{J}(S_I)$ and we know from Lemma 1.6.3 that because E is semi-prime, E is a finitely generated S-module. Nakayama's Lemma 1.2.3 shows us that if J is a maximal right ideal $J \subset E_I$ such that $E_I I_I + J = E_I$ then $J = E_I$. (J is also an S_I-submodule of E_I.) Then $E_I I_I \subset \cap\{J \mid J \text{ is a maximal right ideal of } E\} = \mathcal{J}(E_I)$.

The next two results show us how to construct integrally closed rings $\overline{E}$ such that $\overline{E}I \subset E \subset \overline{E}$ for ideals $I \subset S$. Lemma 1.6.1 states that if I is a right ideal of finite index in $\overline{E}$ then

$$\mathcal{O}(I) = \{q \in \mathbb{Q}\overline{E} \mid qI \subset I\}$$

is an integrally closed ring.

LEMMA 3.1.3 *Let E be a semi-prime rtffr ring with center$(E) = S$. Suppose that $\overline{E}' \subset \mathbb{Q}E$ is an integrally closed ring such that $I\overline{E}' \subset E$ for some ideal $I \subset S$ of finite index in S. There is an integrally closed ring $\overline{E} \subset \mathbb{Q}E$ such that*

$$I\overline{E} \subset E \subset \overline{E}.$$

Proof: Let

$$J = \{q \in \mathbb{Q}E \mid q\overline{E}' \subset E\} \subset E.$$

Since $0 \neq I\overline{E}' \subset E$ and since S/I is finite $I \subset J$ is a right $\overline{E}'$-module of finite index in E. Write

$$\overline{E}' = \overline{E}'_1 \times \cdots \times \overline{E}'_t$$

for some classical maximal orders $\overline{E}'_i$. Then

$$J = J_1 \oplus \cdots \oplus J_t$$

for some finitely generated right $\overline{E}'_i$-modules J_i. Lemma 1.6.1 then implies that the ring

$$\overline{E}_i = \{q \in \mathbb{Q}\overline{E}_i \mid qJ_i \subset J_i\}$$

is a classical maximal order in $\mathbb{Q}\overline{E}_i$ so that

$$\overline{E} = \overline{E}_1 \times \cdots \times \overline{E}_t = \mathcal{O}(J)$$

is an integrally closed subring in $\mathbb{Q}E$. Since J is a left ideal in E, $E \subset \overline{E}$, and since $I \subset I\overline{E}' \subset J$,

$$I\overline{E} = \overline{E}I \subset \overline{E}J = J \subset E$$

which completes the proof.

EXAMPLE 3.1.4 The following is an example of an rtffr Dedekind domain $\overline{S}$ and a subring $S \subset \overline{S}$ such that $\overline{S}/S$ is *a finite group* and the minimum number of generators for $\overline{S}$ is rank$(\overline{S})$.

Specify $n > 1 \in \mathbb{Z}$, choose an algebraic number field $\mathbf{k} \neq \mathbb{Q}$, and let $\overline{S}$ denote the ring of algebraic integers in $\mathbf{k}$. Then $\overline{S}$ is a Dedekind domain. A result from number theory (see, e.g., [58]) states that $\overline{S}$ is a free abelian group on $[\mathbf{k} : \mathbb{Q}] = \text{rank}(\overline{S})$ generators. It follows that if $p \in \mathbb{Z}$ is a prime then

$$\mathbb{Z}/p\mathbb{Z}\text{-}\dim(\overline{S}/p\overline{S}) = \text{rank}(\overline{S}) \neq 1$$

so that there is a proper subring

$$S = \mathbb{Z} + p\overline{S}.$$

Note that $p\overline{S} \subset S \subset \overline{S}$ and that these inclusions are proper. Moreover, since $S/p\overline{S} \cong \mathbb{Z}/p\mathbb{Z}$, $\overline{S}/p\overline{S}$, and, hence, $\overline{S}$ are generated by exactly rank$(\overline{S})$ elements as S-modules.

EXAMPLE 3.1.5 The ring S need not decompose as a ring. Let S_1 and S_2 be rtffr Dedekind domains. For $i = 1, 2$ choose nonzero proper ideals $I_i \subset S_i$ and construct a ring S as

$$I_1 \times I_2 \subset S = I_1 \times I_2 + (1,1)\mathbb{Z} \subset S_1 \times S_2.$$

Then $(S_1 \times S_2)/S$ is finite but S is an indecomposable ring. (Observe that S does not contain the unique idempotents $(1,0), (0,1)$ of $S_1 \times S_2$.)

The next result is fundamental to our investigations. It originally appeared in [18]. A proof can be found in [10, Theorem 14.2].

THEOREM 3.1.6 [The Beaumont-Pierce-Wedderburn Theorem] *Let E be an rtffr ring. There is a semi-prime Noetherian subring $T \subset E$ such that $E \doteq T \oplus \mathcal{N}(E)$.*

Using The Beaumont-Pierce-Wedderburn Theorem we can classify a number of modules up to quasi-isomorphism.

EXAMPLE 3.1.7 This example shows that there can be little hope of classifying the additive structure of $\mathcal{N}(E)$. Let G be *any abelian group* and let

$$E = \left\{ \begin{pmatrix} n & 0 \\ x & n \end{pmatrix} \mid n \in \mathbb{Z}, x \in G \right\}.$$

The reader will show that E is a commutative ring and that $\mathcal{N}(E) = \begin{pmatrix} 0 & 0 \\ G & 0 \end{pmatrix} \cong G$ as groups. Then E is an rtffr ring iff G is an rtffr group.

Since each semi-prime rtffr ring E has finite index in an integrally closed ring it is natural to ask if some similar kind of result is true of rtffr groups. Let us agree that G is an *integrally closed group* if $E(G)$ is an integrally closed ring.

LEMMA 3.1.8 *If G is an rtffr group then there is a G-generated rtffr group $\overline{G}$ and an integer $n = n(G) > 0$ such that $n\overline{G} \subset G \subset \overline{G}$, $E(\overline{G})$ is integrally closed, and $E(\overline{G})/E(G)$ is finite.*

Proof: Let $E = \text{End}(G)$. By The Beaumont-Pierce-Wedderburn Theorem 3.1.6 there is a semi-prime rtffr subring $T \subset E$ such that $E \doteq T \oplus \mathcal{N}(E)$ and $\mathbb{Q}E = \mathbb{Q}T \oplus \mathbb{Q}\mathcal{N}(E)$ as $\mathbb{Q}T$-modules. Let $R = E(G)$ be the projected image of E into $\mathbb{Q}T$ modulo $\mathbb{Q}\mathcal{N}(E)$. Then R is a semi-prime Noetherian ring, $T \subset R$, $R \doteq T$, and $E \doteq R \oplus \mathcal{N}(E)$. By Lemma 1.6.3 there is an integer $n \neq 0$ and an integrally closed ring $\overline{T} \subset \mathbb{Q}R$ such that

$$n\overline{T} \subset T \subset R \subset \overline{T}.$$

Let $\overline{G} = \overline{T}G$ and observe that

$$n\overline{G} = (n\overline{T})G \subset TG = G \subset \overline{T}G = \overline{G}.$$

Then $\overline{G}/G$ is finite, and

$$\overline{T} \oplus \mathcal{N}(E) \doteq R \oplus \mathcal{N}(E) \doteq E \doteq \text{End}(\overline{G}).$$

Hence $\overline{T} \subset E(\overline{G})$ and

$$\overline{T} \doteq R \doteq E(G) \doteq E(\overline{G}).$$

Since $\overline{T}$ is integrally closed

$$E(G) = R \subset \overline{T} = E(\overline{G})$$

and since $\overline{G}/G$ is finite $E(\overline{G})/E(G)$ is finite. This completes the proof.

For instance G is an integrally closed group if $\text{End}(G)$ is a Dedekind domain. Theorem 2.3.3 shows us that there are plenty of strongly indecomposable groups G that are not integrally closed groups. There is a strongly indecomposable rtffr group G such that $\text{End}(G)$ is a hereditary domain but $\text{End}(G)$ is not an integrally closed group. See Example 4.5.7.

3.2 Conductor of an Rtffr Ring

We will continue to use the notation introduced in the previous subsection. Specifically, E is a semi-prime rtffr ring and $S = \text{center}(E)$. Using Lemma 1.6.3 we fix an integrally closed ring $\overline{E} \subset \mathbb{Q}E$ such that $E \subset \overline{E}$ and $\overline{E}/E$ is finite. Let $\overline{S} = \text{center}(\overline{E})$, and choose classical maximal orders $\overline{E}_1, \cdots, \overline{E}_t$ and Dedekind domains $\overline{S}_1, \cdots, \overline{S}_t$ as in (3.1) and (3.2).

An *E-lattice* is a finitely generated right E-submodule of $\mathbb{Q}E^{(k)}$ for some integer $k > 0$. *The conductor of $\overline{E}$ into E* is the largest ideal $\tau \subset S$ such that

$$\tau \overline{E} \;\subset\; E \;\subset\; \overline{E}.$$

Then

$$\tau = \{q \in S \mid q\overline{E} \subset E\}.$$

There can be many integrally closed rings $\overline{E}$ such that $\overline{E}/E$ is finite, and for each $\overline{E}$ there is a conductor τ. Thus we say that an ideal $\tau \subset S$ is *a conductor for E* if there is an integrally closed ring $\overline{E}$ such that $\overline{E}/E$ is finite and such that τ is the conductor for $\overline{E}$ into E.

Since $\overline{E}/E$ is finite τ contains an integer $n \neq 0$, so that S/τ and $E/\tau\overline{E}$ are finite rings. Since $\overline{S} = \text{center}(\overline{E})$, $\tau\overline{S} \subset S$, so that

$$(\tau\overline{S})\overline{E} = \tau\overline{E} \subset E.$$

By the maximality of τ in S, $\tau = \tau\overline{S}$ is an ideal in $\overline{S}$.

We need to know that τ is unique to E in some sense. Recall that if I is an ideal in S then the *radical of I* is

$$\sqrt{I} = \cap\{M \mid M \text{ is a maximal right ideal such that } I \subset M \subset S\}$$

LEMMA 3.2.1 *Let E be a semi-prime rtffr ring. Let $\overline{E}$ and $\overline{E}'$ be integrally closed rtffr rings that contain E as a subring of finite index, and let τ and $\tau' > 0$ be the conductors for $\overline{E}$ and $\overline{E}'$ into E, respectively. Then*

$$\sqrt{\tau} = \sqrt{\tau'}.$$

In particular given a maximal ideal $M \subset S$ then $\tau \subset M \subset S$ iff E_M is not integrally closed.

Proof: Let $I \subset S$ be the ideal defined by the finite intersection

$$I = S \cap \bigcap\{\tau_N \mid N \text{ is a maximal ideal in } S \text{ such that } E_N \neq \overline{E}_N\}.$$

Choose a maximal ideal M of S such that $E_M \neq \overline{E}_M$. Because localization commutes with finite intersection we have

$$I_M = S_M \cap \bigcap\{(\tau_N)_M \mid E_N \neq \overline{E}_N\} = S_M \cap \tau_M = \tau_M$$

since $(\tau_N)_M = \mathbb{Q}S$ for maximal ideals $M \neq N$ in S.
Thus for each maximal ideal M in S we have

$$(\overline{IE})_M = I_M \overline{E}_M = \tau_M \overline{E}_M \subset E_M.$$

By the Local-Global Theorem 1.5.1, $\overline{IE} \subset E$, and by the maximality of
τ, $\tau = I$. Therefore $\tau \subset M \subset S$ iff E_M is not integrally closed.
It follows that if τ and τ' are conductors for $\overline{E}$ into E then $\sqrt{\tau} = \sqrt{\tau'}$.

EXAMPLE 3.2.2 Let $\overline{S}$ be the ring of algebraic integers in an alge-
braic number field $\mathbf{k} \neq \mathbb{Q}$. Then $\overline{S}/p\overline{S} \cong \mathbb{Z}/p\mathbb{Z}^{(n)}$ where $n = [\mathbf{k} : \mathbb{Q}]$.
Let $S = \mathbb{Z} + p\overline{S}$. Then $p\overline{S} \subset S \subset \overline{S}$ and since $S/p\overline{S} \cong \mathbb{Z}/p\mathbb{Z}$, $p\overline{S}$ is the
conductor of $\overline{S}$ into S.

The conductor of the semi-prime rtffr ring E gives a local condition
that shows us when an E-lattice is a projective E-module. It also yields
a product decomposition $E = E_o \times E_1$ where E_o is integrally closed.

LEMMA 3.2.3 Let E be a semi-prime rtffr ring with conductor τ and
let U be an E-lattice. Then U is a projective right E-module iff the
localization U_τ is a projective right E_τ-module.

Proof: Let U be an E-lattice. It is clear that if U is a projective right
E-module then U_τ is a projective right E_τ-module.
Conversely assume that U_τ is a projective E_τ-module. We will show
that U_M is a projective right E_M-module for each maximal ideal $M \subset S$.
Fix a maximal ideal $M \subset S$. There are two cases possible.
If $\tau \subset M$ then because U_τ is projective, $U_M = (U_M)_\tau = (U_\tau)_M$ is a
projective right E_M-module.
Otherwise $\tau \not\subset M$ so by Lemma 3.2.1, E_M is an integrally closed
ring. Then U_M is an E_M-lattice, hence a projective E_M-module.
Thus U_M is a projective E-module for each maximal ideal M in S.
Inasmuch as the E-lattice U is finitely presented over the Noetherian
ring E, the Change of Rings Theorem 1.5.2 implies that U is a finitely
generated projective right E-module. This completes the proof.

Let E be a semi-prime rtffr ring with conductor $\tau \subset S$ and let

$$\mathcal{C} = \{c \in S \mid c + \tau \text{ is a unit in } S/\tau\}. \tag{3.3}$$

Then $\mathcal{C}$ is a multiplicatively closed subset of S. Given an S-module X define

$$X(\tau) = \{x \in X \mid xc = 0 \text{ for some } c \in \mathcal{C}\}.$$

Thus $X(\tau)$ is the $\mathcal{C}$-torsion submodule of X.

THEOREM 3.2.4 [T.G. Faticoni] *Let E be a semi-prime rtffr ring with conductor τ.*

1. *There are ideals $S(\tau), S^\tau \subset S$ and $E(\tau), E^\tau \subset E$ such that $S = S(\tau) \times S^\tau$ and $E = E(\tau) \times E^\tau$,*

2. *$S_\tau = (S^\tau)_\tau$ and $E_\tau = (E^\tau)_\tau$,*

3. *$E(\tau)$ is an integrally closed ring.*

Proof: 1. and 2. One shows that $S(\tau)$ is an ideal of S such that $S/S(\tau)$ is a torsion-free group. Since $S(\tau)$ is the kernel of localization at $\mathcal{C}$, $S_\tau \cong (S/S(\tau))_\tau$. Thus $S/S(\tau)$ is a cyclic S-submodule of $\mathbb{Q}S/\mathbb{Q}S(\tau)$, so that $S/S(\tau)$ is an S-lattice such that $S_\tau = (S/S(\tau))_\tau$ is a projective S_τ-module. By Lemma 3.2.3, $S/S(\tau)$ is a projective S-module, so that

$$S = S(\tau) \times S^\tau$$

for some ideal S^τ of S such that $S_\tau = (S^\tau)_\tau$.

In a similar manner $E = E(\tau) \oplus E^\tau$ and $(E^\tau)_\tau \cong E_\tau$. Since E is a semi-prime rtffr ring and since $E(\tau)$ is an ideal,

$$E = E(\tau) \times E^\tau.$$

We will return to τ presently.

3. Let M be a maximal ideal in the ring S that contains S^τ. For each $c \in S(\tau) \setminus M$ the element $(c, 0)$ is annihilated by S^τ. Then

$$S_M \cong (S/S^\tau)_M \cong S(\tau)_M$$

and

$$E_M \cong (E/E^\tau)_M \cong E(\tau)_M.$$

Suppose for the sake of contradiction that $\tau \subset M$. Let $x \in S(\tau) \setminus M$. Then x is annihilated by some element $c \in \mathcal{C}$. See (3.3). However, since $\tau \subset M$, $x + M$, a unit in S/M, is not annihilated by any of the units in S/τ. This contradiction shows us that $\tau \not\subset M$.

Therefore, by Lemma 3.2.1, $E_M \cong E(\tau)_M$ is integrally closed for each maximal ideal $\tau \not\subset M$. For $\tau \subset M \subset S$, $E(\tau)_M = 0$ since for each element $x \in E(\tau)$ there is a $c \in \mathcal{C}$ such that $xc = 0$. Thus $E(\tau)_M$ is integrally closed for each maximal ideal $M \subset S$, and hence $E(\tau)$ is integrally closed.

This completes the proof.

For example let $E = S = 2(\mathbb{Z}_6 \oplus \mathbb{Z}_{10}) + \mathbb{Z}(1 \oplus 1)$. Then $\tau = 2(\mathbb{Z}_6 \oplus \mathbb{Z}_{10})$ and $\mathcal{C} = \{(a,b) \mid a - b \in \tau\}$. Hence $S(\tau) = 0$, $S^\tau = S$, and $S_\tau = 2(\mathbb{Z}_2 \oplus \mathbb{Z}_2) + \mathbb{Z}_2(1 \oplus 1)$.

3.3 Local Correspondence

The main result of this section shows how localization at the conductor can be used to translate local isomorphism into isomorphism. The proof of the theorem is partitioned into several lemmas that are interesting in their own right.

For the proof of this theorem let τ be a conductor for E, and write $E = E(\tau) \times E^\tau$ as in Theorem 3.2.4. Given an E-lattice U let

$$U(\tau) = US(\tau) = \{x \in U \mid xc = 0 \text{ for some } c \in \mathcal{C}\}.$$

THEOREM 3.3.1 *Let E be a semi-prime rtffr ring with conductor τ, and write $E = E(\tau) \times E^\tau$ as in Theorem 3.2.4. The localization functor $(\cdot)_\tau$ induces a bijection*

$$\lambda_\tau : \{[U] \mid U \in \mathbf{P}_o(E)\} \longrightarrow \{(W) \mid W \in \mathbf{P}_o(\mathbb{Q}E(\tau) \times E_\tau)\}$$

from the set of local isomorphism classes $[U]$ of finitely generated projective right E-modules onto the set of isomorphism classes (W) of finitely generated projective right $\mathbb{Q}E(\tau) \times E_\tau$-modules.

LEMMA 3.3.2 *Let E be a semi-prime rtffr ring with conductor τ and let U, V be E-lattices. Then U is locally isomorphic to V iff $\mathbb{Q}U(\tau) \cong \mathbb{Q}V(\tau)$ and U^τ is locally isomorphic to V^τ.*

Proof: Write $U = U(\tau) \oplus U^\tau$. It is clear that U is locally isomorphic to V iff $U(\tau)$ is locally isomorphic to $V(\tau)$ as $E(\tau)$-lattices, and U^τ is locally isomorphic to V^τ as E^τ-lattices. By Theorem 3.2.4, $E(\tau)$ is an integrally closed Noetherian semi-prime rtffr ring. Then by Lemma 1.6.1, $U(\tau)$ is locally isomorphic to $V(\tau)$ iff $\mathbb{Q}U(\tau) \cong \mathbb{Q}V(\tau)$. This completes the proof.

Begin the proof of Theorem 3.3.1: Define a functor as follows.

$$L_\tau(\cdot) : \mathbf{P}_o(E) \longrightarrow \mathbf{P}_o(\mathbb{Q}E(\tau) \times E_\tau)$$
$$L_\tau(U) = \mathbb{Q}U(\tau) \oplus U_\tau.$$

Since the localization functor is additive and exact, $L_\tau(\cdot)$ is an additive exact functor.

Define a function λ_τ by

$$\lambda_\tau([U]) = (L_\tau(U)) = (\mathbb{Q}U(\tau) \oplus U_\tau).$$

We claim that λ_τ is a well defined function. Suppose that $U, V \in \mathbf{P}_o(E)$ are locally isomorphic. By Lemma 3.3.2, $\mathbb{Q}U(\tau) \cong \mathbb{Q}V(\tau)$ and U^τ is locally isomorphic to V^τ. Since τ contains an integer $n \neq 0 \in \tau$ we can choose an integer m and maps $\phi_n : U \to V$ and $\psi_n : V \to U$ such that $\gcd(m, n) = 1$, $\phi_n\psi_n = m1_V$, and $\psi_n\phi_n = m1_U$. Since $\gcd(m, n) = 1$, m is a unit of E_n, and hence ϕ_n and ψ_n lift to isomorphisms between $U_{nS} \cong V_{nS}$. Inasmuch as $nS \subset \tau$ we have $(X_{nS})_\tau = X_\tau$ for E-lattices X. Thus

$$U_\tau \cong (U_\tau)_{nS} \cong (U_{nS})_\tau \cong (V_{nS})_\tau \cong V_\tau.$$

This proves the claim, and therefore λ_τ is a well defined function.

We must show that λ_τ is a bijection.

LEMMA 3.3.3 *Let E be a semi-prime ring with conductor τ. Then for each finitely generated projective E_τ-module W there is a finitely generated projective E^τ-module U such that $U_\tau \cong W$.*

Proof: Let $W \in \mathbf{P}_o(E_\tau)$ and write $W = w_1 E_\tau + \cdots + w_s E_\tau$ for some finite set $\{w_1, \ldots, w_s\} \subset W$. Let U be the E-lattice generated by $w_1, \cdots, w_s$.

$$U = w_1 E + \cdots + w_s E.$$

There is an E-module surjection $\pi : \oplus_{i=1}^{s} x_i E \to U$ such that $\pi(x_i) = w_i$ for each i. Because localization is exact the image of $\pi_\tau : \oplus_{i=1}^{s} x_i E_\tau \to U_\tau \subset W$ is

$$U_\tau = w_1 E_\tau + \cdots + w_s E_\tau = W.$$

Moreover $U_\tau = W$ is a finitely generated projective E_τ-module so by Lemma 3.2.3, U is a finitely generated projective E-module. The lemma is proved.

COROLLARY 3.3.4 *Let E be an rtffr ring with conductor τ. Then λ_τ is a surjection.*

Let

$$X^\tau = X S^\tau$$

for E-lattices X. Then $X = X(\tau) \oplus X^\tau$. The following are technical lemmas that we need to prove that λ_τ is an injection.

LEMMA 3.3.5 *Let E be a semi-prime rtffr ring with conductor τ, and let $n \neq 0 \in \tau$ be an integer.*

1. *There is an ideal $I \subset S$ such that*

    (a) *$n \in I$,*

    (b) *$I + \tau = S$, and such that*

    (c) *$c \in S$ is a unit modulo nS iff c is a unit modulo $I \cap \tau$.*

    Consequently, $U_{nS} = U_{I \cap \tau}$ for each finitely generated projective right E-module U.

2. *$(U^\tau)_n = (U^\tau)_{nS}$ for each finitely generated projective right E-module U.*

Proof: 1. Let $\{M_1, \ldots, M_r, T_1, \ldots, T_s\}$ be a complete set of the maximal ideals of S that contain nS, and assume that $\{T_1, \ldots, T_s\}$ is a complete set of the maximal ideals of S that contain τ.
Let

$$I = M_1 \cap \cdots \cap M_r$$

and let $I \subset N \subset S$ be a maximal ideal. Using the Chinese Remainder Theorem 1.5.3 we see that

$$\frac{S}{I} \cong \frac{S}{M_1} \times \cdots \times \frac{S}{M_r}$$

and each S/M_i is a simple module. Thus $S/N \cong S/M_i$ for some $i = 1, \cdots, r$. Since S is commutative $N = \mathrm{ann}_S(1 + N) = M_i$, so that $M_1, \ldots, M_r$ is a complete list of the maximal ideals of S that contain I.

In particular I is not contained in any of the maximal ideals $T_1, \ldots, T_s$ that contain τ, hence $I + \tau$ is not contained in any maximal ideal, whence $I + \tau = S$. Another application of The Chinese Remainder Theorem 1.5.3 shows us that

$$\frac{S}{I \cap \tau} \cong \frac{S}{I} \times \frac{S}{\tau}$$

as rings.

We claim that $\{c \in S \mid c + nS \text{ is a unit in } S/nS\} = \{c \in S \mid c + (I \cap \tau) \text{ is a unit in } S/(I \cap \tau)\}$.

Since $nS \subset I \cap \tau$, and since units map to units, the elements c that are units modulo nS are units modulo $I \cap \tau$. Conversely, if $c \in S$ maps to a unit in the finite ring $S/(I \cap \tau)$ then $c \notin M_i$ or T_j for any $M_i, T_j \in \{M_1, \ldots, M_r, T_1, \ldots, T_s\}$. We have shown that $\{M_1, \ldots, M_r, T_1, \ldots, T_s\}$ is a complete list of the maximal ideals that contain $I \cap \tau$. Thus $c + nS$ is not in any of the maximal ideals of S/nS, and so $c + nS$ is a unit of S/nS. As claimed, the set of units modulo nS is the set of units modulo $I \cap \tau$. Consequently, $U_{nS} = U_{I \cap \tau}$ for each finitely generated projective right E-module. This proves part 1.

2. Since $nS \subset \tau$, the ideal $\{x \in S \mid xc = 0 \text{ for some } c \in S \text{ such that } c + nS \in S/nS \text{ is a unit}\}$ is contained in the ideal $S(\tau)$. Thus S^τ is torsion-free relative to the units modulo nS. Hence there is an inclusion $S^\tau \subset (S^\tau)_{nS}$ that lifts to one

$$(S^\tau)_n \subset (S^\tau)_{nS}.$$

To see that $(S^\tau)_n = (S^\tau)_{nS}$ let $N_1, \ldots, N_r$ be a complete list of the maximal ideals of the commutative ring S that contain nS. For each $N \in \{N_1, \ldots, N_r\}$

$$((S^\tau)_n)_N = (S^\tau)_N = ((S^\tau)_{nS})_N$$

so that

$$(S^\tau)_n = (S^\tau)_{nS}$$

by the Local Global Theorem. (Prove this as an exercise, reader.) Then $U_n^\tau \cong (U^\tau)_{nS}$ for each finitely generated projective S-module. This completes the proof.

COROLLARY 3.3.6 *Let E be a semi-prime rtffr ring with conductor τ, and let $n \neq 0 \in \tau$ be an integer. Let U and V be finitely generated projective right E-modules such that $\mathbb{Q}U(\tau) \cong \mathbb{Q}V(\tau)$ and $U_\tau \cong V_\tau$. Then $U_n \cong V_n$.*

Proof: By hypothesis, $U(\tau)$ and $V(\tau)$ are lattices over the integrally closed ring $E(\tau)$ (Theorem 3.2.4), such that $\mathbb{Q}U(\tau) \cong \mathbb{Q}V(\tau)$. Then $U(\tau)$ and $V(\tau)$ are locally isomorphic (Lemma 1.6.1(4)). Consequently, $U(\tau)_n \cong V(\tau)_n$, (Lemma 2.5.4(3)).
By Lemma 3.3.5(1)

$$U_{nS} \cong U_{I \cap \tau} \cong V_{I \cap \tau} \cong V_{nS}$$

so

$$(U^\tau)_{nS} \cong (US^\tau)_{nS} \cong (VS^\tau)_{nS} \cong (V^\tau)_{nS}.$$

Lemma 3.3.5(2) then shows us that

$$(U^\tau)_n \cong (U^\tau)_{nS} \cong (V^\tau)_{nS} \cong (V^\tau)_n.$$

Hence

$$U_n \cong U(\tau)_n \oplus (U^\tau)_n \cong V(\tau)_n \oplus (V^\tau)_n \cong V_n,$$

which proves the Corollary.

THEOREM 3.3.7 [T.G. Faticoni] *Let E be a semi-prime rtffr ring with conductor τ. Then λ_τ is an injection. Specifically, if U, $V \in \mathbf{P}_o(E)$ then U and V are locally isomorphic E-lattices iff $\mathbb{Q}U(\tau) \cong \mathbb{Q}V(\tau)$ and $U_\tau \cong V_\tau$ as E_τ-lattices.*

Proof: Suppose that $\mathbb{Q}U(\tau) \cong \mathbb{Q}V(\tau)$ and that $U_\tau \cong V_\tau$. By Lemma 3.3.6, U_n and V_n are isomorphic right E-modules for each integer $n > 0$. Thus U is locally isomorphic to V (Lemma 2.5.4), which completes the proof of the lemma.

Proof of Theorem 3.3.1 completed: By Corollary 3.3.4 and Theorem 3.3.7, λ_τ is a bijection, which completes the proof.

3.4 Canonical Decomposition

Let us extend λ_τ to include rtffr groups G and the associated semi-prime rtffr ring

$$A(G) = E(G) = \mathrm{End}(G)/\mathcal{N}(\mathrm{End}(G)).$$

Suppose that $E(G)$ has conductor τ, and let $L_\tau(\cdot)$ be the localization functor. Let $E(G)(\tau)$ denote $\{x \in E(G) \mid xc = 0$ for some $c \in S$ such that c is a unit modulo $\tau\}$. The reader will show that

$$E(G)(\tau) = E(G)S(\tau).$$

Recall Arnold's Functor $A(\cdot)$ from Section 2.4.2. Let us define

$$E_\tau(G) = \mathbb{Q}E(G)(\tau) \times (E(G))_\tau$$
$$A_\tau(\cdot) = L_\tau(\cdot) \circ A(\cdot) : \mathbf{P}_o(G) \longrightarrow \mathbf{P}_o(E_\tau(G)).$$

We say that

G *has conductor* τ if τ is a conductor for $E(G)$.

THEOREM 3.4.1 *Let G be an rtffr group with conductor τ. The functor $A_\tau(\cdot)$ induces a bijection*

$$\alpha_G(\cdot) : \{[H] \mid H \in \mathbf{P}_o(G)\} \longrightarrow \{(W) \mid W \in \mathbf{P}_o(E_\tau(G))\}$$

from the set of local isomorphism classes $[H]$ of $H \in \mathbf{P}_o(G)$ onto the set of isomorphism classes (W) of $W \in \mathbf{P}_o(E_\tau(G))$.

Proof: By [10, Corollary 12.7], $A(\cdot)$ induces a bijection

$$\{[H] \mid H \in \mathbf{P}_o(G)\} \to \{[U] \mid U \in \mathbf{P}_o(E(G))\}$$

and Theorem 3.3.1 states that the functor $L_\tau(\cdot)$ induces a bijection

$$\lambda_\tau : \{[U] \mid U \in \mathbf{P}_o(E(G))\} \to \{(W) \mid W \in \mathbf{P}_o(E_\tau(G))\}.$$

The assignment $[H] \mapsto \lambda_\tau[A(H)]$ defines the required bijection α_G.

The reader can prove the next two results as exercises.

THEOREM 3.4.2 *Let G be an rtffr group and assume that $E(G)$ is an integrally closed ring. The functor $\mathbb{Q} \otimes_{\mathbb{Z}} A(\cdot)$ induces a bijection*

$$\alpha_G(\cdot) : \{[H] \,|\, H \in \mathbf{P}_o(G)\} \longrightarrow \{(W) \,|\, W \in \mathbf{P}_o(\mathbb{Q}E(G))\}$$

from the set of local isomorphism classes $[H]$ of $H \in \mathbf{P}_o(G)$ onto the set of isomorphism classes (W) of semi-simple right $\mathbb{Q}E(G)$-modules W.

THEOREM 3.4.3 *Let G be an rtffr group with conductor $\tau \neq S$, and assume that $\mathbb{Q}E(G)$ is a simple Artinian ring. The functor $A_\tau(\cdot)$ induces a bijection*

$$\alpha_G(\cdot) : \{[H] \,|\, H \in \mathbf{P}_o(G)\} \longrightarrow \{(W) \,|\, W \in \mathbf{P}_o((E(G))_\tau)\}$$

from the set of local isomorphism classes $[H]$ of $H \in \mathbf{P}_o(G)$ onto the set of isomorphism classes (W) of finitely generated projective right $(E(G))_\tau$-modules W.

Proof: Since $\mathbb{Q}E(G)$ is simple, S is an integral domain and so $S^\tau = S$. Thus $\mathbb{Q}E(G)(\tau) = 0$. The reader can take it from here.

These results lead us to a canonical direct sum decomposition of an rtffr group G.

THEOREM 3.4.4 *Let G be an rtffr group with conductor τ. Then*

$$G \;=\; G(\tau) \oplus G^\tau$$

where $E(G(\tau)) = E(G)(\tau)$ is an integrally closed ring.

Proof: By Theorem 3.2.4, $E(G) = E(G)(\tau) \times E(G)^\tau$ where $E(G)(\tau)$ is integrally closed and $E(G)^\tau(\tau) = 0$. Let $e^2 = e \in S$ be such that $E(G)e = E(G)(\tau)$ and $E(G)(1 - e) = E(G)^\tau$. Let $G(\tau) = eG$ and $G^\tau = (1 - e)G$. Thus $G = G(\tau) \oplus G^\tau$.
By Theorem 2.4.4, $E(G)(\tau) = A(G(\tau))$ and $E(G)^\tau = A(G^\tau)$. Moreover, $E(G(\tau)) = E(G)(\tau)$ is integrally closed. This completes the proof.

If we combine the above theorem with Theorem 3.4.2 then we have shown that G can be directly decomposed as $G = G(\tau) \oplus G^\tau$ where the direct sum decompositions of $G(\tau)$ behave like modules over a semi-simple Artinian ring, and where G^τ holds the interesting perhaps badly behaved direct sum decompositions.

THEOREM 3.4.5 [T.G. Faticoni] *Let G be an rtffr group with conductor τ. Then $G = G(\tau) \oplus G^\tau$ where*

1. *Each $H \in \mathbf{P}_o(G)$ can be written uniquely as $H = H(\tau) \oplus H^\tau$ where $H(\tau) \in \mathbf{P}_o(G(\tau))$ and $H^\tau \in \mathbf{P}_o(G^\tau)$.*

2. *Direct sum decompositions of $H(\tau)$ are functorially equivalent to direct sum decompositions of finitely generated semi-simple modules over a semi-simple Artinian ring.*

3. *Direct sum decompositions of H^τ are functorially equivalent to direct sum decompositions of finitely generated projective modules over a semi-local semi-prime rtffr ring.*

Proof: Exercise.

3.5   Arnold's Theorem

The power of Theorem 3.4.1 is that it translates problems on direct sum decompositions of an rtffr group G (which are not finitely generated) into problems on direct sum decompositions of finitely generated projective right modules over the semi-local semi-prime ring

$$E_\tau(G) = \mathbb{Q}E(G)(\tau) \times (E(G))_\tau.$$

Moreover the bijection translates local isomorphism into isomorphism and conversely. For this reason when discussing the local refinement property or locally unique decompositions of G we can instead study the refinement property or unique decompositions of the pleasantly structured ring $E_\tau(G)$.

In this section we will put the above strategy to work. To begin with we will show that Theorem 3.3.1 can be used to give a short proof of an important result in the study of direct sum decompositions of rtffr groups. Part 2 of the following result is referred to in the literature as Arnold's Theorem.

THEOREM 3.5.1 *Let G be an rtffr group with conductor τ, let $A_\tau(\cdot)$ be the functor defined on page 60, and let $H \in \mathbf{P}_o(G)$.*

1. *$A_\tau(H) \cong W_1 \oplus W_2$ as right $E_\tau(G)$-modules iff H is locally isomorphic to $H_1 \oplus H_2$ for some rtffr groups H_j such that $A_\tau(H_j) \cong W_j$ for $j = 1, 2$.*

2. *[10, D. M. Arnold] H is locally isomorphic to $G_1 \oplus G_2$ as rtffr groups iff $H = H_1 \oplus H_2$ for some rtffr groups H_j such that H_j and G_j are locally isomorphic for $j = 1, 2$.*

Proof: 1. Suppose that $A_\tau(H) = W \cong W_1 \oplus W_2$. By Theorem 2.4.4, $A(H) = U$ is a finitely generated projective right $E(G)$-module such that $U_\tau \cong W$. By the proof of Lemma 3.3.3, $U \subset W$. Let $U_1 = W_1 \cap U$ and let $U_2 = U/U_1$. Because localization is exact

$$(U_2)_\tau = (U/U_1)_\tau \cong U_\tau/(U_1)_\tau \cong W/W_1 \cong W_2$$

so by Lemma 3.2.3, U_2 is a projective right $A(G)$-module. Hence $U \cong U_1 \oplus U_2$.

Furthermore, by Theorem 2.4.4 there are groups $H_1, H_2 \in \mathbf{P}_o(G)$ such that $A(H_j) \cong U_j$ for $j = 1, 2$, and $H = H_1 \oplus H_2$. Then $A_\tau(H_j) \cong (U_j)_\tau \cong W_j$ for $j = 1, 2$. The converse is clear.

2. Suppose that H is locally isomorphic to $G_1 \oplus G_2$. By Theorem 3.4.1,

$$A_\tau(H) \cong A_\tau(G_1 \oplus G_2) \cong A_\tau(G_1) \oplus A_\tau(G_2).$$

By part 1, $H \cong H_1 \oplus H_2$ for some rtffr groups H_1 and H_2 such that $A_\tau(G_i) \cong A_\tau(H_i)$ for $i = 1, 2$. Theorem 3.4.1 implies that G_i is locally isomorphic to H_i for $i = 1, 2$. The converse is evident so the proof is complete.

COROLLARY 3.5.2 *Let G be an rtffr group with conductor τ. The following are equivalent.*

1. *G is indecomposable.*

2. *If G is locally isomorphic to H then H is indecomposable.*

3. *$E(G)$ is indecomposable as a right $E(G)$-module.*

4. *$E_\tau(G)$ is indecomposable as a right $E_\tau(G)$-module.*

Recall *locally unique decompositions* and the *local refinement property* from Section 2.5. It follows from Theorem 3.4.1 and the AKS Theorem 2.1.6 that the rtffr group G has a locally unique decomposition if $E_\tau(G)$ is a *semi-perfect ring*. By the Baer-Kulikov-Kaplansky Theorem 2.1.11 completely decomposable groups have a locally unique decomposition and the reader should convince themselves that $E_\tau(G)$ is a semi-perfect ring in this case. The next two results show that the local refinement property can be treated as the refinement property.

THEOREM 3.5.3 [T.G. Faticoni] *Let G be an rtffr group with conductor τ. The following are equivalent.*

1. *G has a locally unique decomposition.*

2. *$E(G)$ has a locally unique decomposition.*

3. *$E_\tau(G)$ has a unique decomposition.*

Proof: $1 \Leftrightarrow 2$ follows immediately from Theorems 2.4.1, 2.4.4 while $1 \Leftrightarrow 3$ follows immediately from Theorem 3.5.1.

The next result shows that a locally unique decomposition can be treated as a unique decomposition.

THEOREM 3.5.4 [T.G. Faticoni] *Let G be an rtffr group with conductor τ. The following are equivalent.*

1. *G has the local refinement property.*

2. *$E(G)$ has the local refinement property.*

3. *$E_\tau(G)$ has the refinement property.*

Proof: Apply Theorems 2.4.4 and 3.5.1.

3.6   Exercises

Let E be a semi-prime rtffr ring with center S with conductor τ. Let $\overline{E} \subset \mathbb{Q}E$ be an integrally closed ring such that $\overline{E}\tau \subset E \subset \overline{E}$. Let $\overline{S} = \text{center}(\overline{E})$.

1. Let $\overline{S}$ be an Artinian ring. Show that $\overline{S}$ is local iff $e^2 = 2 \in \overline{S}$ implies that $e = 0$ or 1.

2. Show that any *group* endomorphism of $\widehat{\mathbb{Z}}_p$ is actually left multiplication by some element of $\widehat{\mathbb{Z}}_p$. (Hint: Given a map f consider $f(1)$.)

3. Let G be an rtffr. There is a bijection from the set of isomorphism classes of $e\mathbb{Q}\text{End}(G)$ where $e^2 = e \in \mathbb{Q}\text{End}(G)$ onto the set of quasi-isomorphism classes $[H]$ of quasi-summands H of G. The bijection is given by $e\mathbb{Q}\text{End}(G) \mapsto [eG]$.

4. The group G is p-*pure* in H if $G \subset H$ and if G/H is p-torsion-free. Show that

   (a) G is p-pure in H iff $pG = pH \cap G$.

   (b) a p-pure subgroup G of $\widehat{\mathbb{Z}}_p$ is strongly indecomposable.

5. Let G be a p-pure subgroup of $\widehat{\mathbb{Z}}_p$. Show that

   (a) $G/pG \cong \mathbb{Z}/p\mathbb{Z}$.

   (b) $\mathrm{End}(G)$ is a pure subring of $\widehat{\mathbb{Z}}_p$.

   (c) $\mathrm{End}(G)$ is an integral domain.

6. Let $x \in \overline{S}$ and choose maximal ideals $M_1, \ldots, M_t$ of $\overline{S}$ such that $M_1^{n_1} \cap \cdots \cap M_t^{(n_t)} \subset x\overline{S} \subset M_1 \cap \cdots \cap M_t$. Prove that *each* of the maximal ideals of $\overline{S}/x\overline{S}$ are of the form $M_i/x\overline{S}$.

7. Let $H, K \in \mathbf{P}_o(G)$. Show that the following are equivalent.

   (a) H is locally isomorphic to H as groups.

   (b) $A(H)$ is locally isomorphic to $A(K)$ as right $E(G)$-modules.

8. Let $\{G_1, \ldots, G_t\}$ be a nilpotent set of rtffr groups. Then $G = G_1 \oplus \cdots \oplus G_t$ has the refinement property iff each G_i has the refinement property.

9. (a) If E is a Noetherian semi-prime ring then $\mathbb{Q}E$ is semi-simple.

   (b) If $E \subset U \subset \mathbb{Q}E$ then $\mathrm{End}_E(U) \subset \mathbb{Q}E$.

10. If U is a generator for the ring E then the ring of $\mathrm{End}_E(U)$-module endomorphisms of U is E.

11. Let G and H be rtffr groups. Show that $H \stackrel{.}{\cong} G$ iff $A(H) \stackrel{.}{\cong} A(G)$.

12. Let $\overline{S}$ be the ring of algebraic integers in the algebraic number field **k**. Let $n \in \mathbb{Z}$ and let $E = n\mathbb{Z} + 1\mathbb{Z}$. Then $\overline{S}$ is generated as an E-module by exactly g elements where g is the composition length of the group $\overline{S}/n\overline{S}$.

13. Let I, J be ideals in the commutative ring S.

   (a) $S_I \otimes_S S_I \cong S_J \otimes_S S_I$ as S-modules.

   (b) $(X_I)_J \cong (X_J)_I$ naturally.

   (c) If $I \subset J$ then the natural map $S/I \to S/J$ takes units to units.

(d) $(S_I)_J \cong S_J$.

14. Let $I \subset S$ be an ideal and let M be an S-module such that each element of M is annihilated by some finite power of I. Then $M = 0$ iff $M_N = 0$ for each maximal ideal $I \subset N \subset S$.

15. Suppose that U and V are finitely generated projective right E-modules such that $U/xU \cong V/xV$. Show that there is an *injection* $\phi : U \to V$ such that $V = \phi(U) + V$.

16. Suppose that $U \subset W$ and that $U_I = W = W_1 \oplus W_2$. Show that $(W_1 \cap U)_I = W_1$.

17. Show that $\mathcal{J}(S_\tau) \subset \mathcal{J}(E_\tau)$.

18. Let E be an rtffr ring and suppose that E is locally isomorphic to $U_1 \oplus U_2$ for some finitely generated E-modules U_1 and U_2. Then $E = V_1 \oplus V_2$ for some cyclic right ideals V_1 and V_2 such that U_i is locally isomorphic to V_i for $i = 1, 2$.

19. If E is integrally closed then $\mathbb{Q}E(\tau) = E_\tau$. Hint: 0 is a unit in the zero ring $\{0\}$.

20. Let G be an rtffr group. Show that the following are equivalent.

    (a) G is indecomposable.
    (b) G is locally isomorphic to some indecomposable group.
    (c) If G is locally isomorphic to H then H is indecomposable.
    (d) $E(G)$ is indecomposable as a right $E(G)$-module.
    (e) $E_\tau(G)$ is indecomposable as a right $E_\tau(G)$-module.

3.7   Questions for Future Research

Let G be an rtffr group, let E be an rtffr ring, and let $S = \text{center}(E)$ be the center of E. Let τ be the conductor of G. There is no loss of generality in assuming that G is strongly indecomposable.

1. Is there some natural way of constructing the ring $T \oplus \mathcal{N}(E)$ in the Beaumont-Pierce-Wedderburn Theorem? See Theorem 3.1.6.

2. Investigate the structure of the rtffr ring E as a module over $S = \text{center}(E)$. How are the ideals of $\mathbb{Z}$ and S distributed in the ideal lattice of E?

3. Let G be an rtffr group. It is known that S is the ring of left End(G)-module homomorphisms $\phi : G \longrightarrow G$. Describe functorially S in terms of G. See [32, 35].

4. Describe a localization at (maximal) ideals in the rtffr ring E in a noncommutative setting. There is quite a large literature for this in ring theory.

5. Investigate the significance of the conductor τ of an rtffr ring. We know that $E = E(\tau) \times E_\tau$. See Theorem 3.2.4. Can we say more about τ or E_τ?

6. Describe the rings E such that $\tau = \sqrt{\tau}$.

7. Describe the rings E that have exactly one conductor τ.

8. Someone out there please come up with a better proof of Theorem 3.3.1.

9. Discuss the ideal theory of commutative rtffr rings.

10. Let G be an rtffr group. Realize the conductor τ of End(G) in terms of G.

11. Find applications for the correspondence given in Theorem 3.4.3.

12. Describe the direct sum decomposition $G = G(\tau) \oplus G^\tau$ in group theoretic terms. For example, what is G^τ?

13. Give a first principles proof of Arnold's Theorem 3.5.1.

14. Classify the Jacobson radical of S_τ. See Theorem 3.2.4.

15. Investigate the existence of commutativity in rtffr rings E.

16. Let E be an rtffr ring. Investigate the existence of central idempotents in $E/\mathcal{N}(E)$. See [40].

Chapter 4

Commuting Endomorphisms

Fix an *rtffr group* G with endomorphism ring $\mathrm{End}(G)$, and let

$$E(G) = \mathrm{End}(G)/\mathcal{N}(\mathrm{End}(G)).$$

We will study direct sum decompositions of G in the presence of partial commutative conditions on rings associated with G.

4.1 Nilpotent Sets

Let G be an rtffr group. We begin with a short discussion of idempotents in $\mathrm{End}(G)$. The element $e \in \mathrm{End}(G)$ is an *idempotent* if $e^2 = e$. We will use both $e^2 = e$ and *idempotent* to refer to the same property of e as the language permits.

Let E be a ring and let $\{e_1, \ldots, e_t\} \subset E$ be a set of idempotents. We say that $\{e_1, \ldots, e_t\}$ is *complete* if

$$e_1 + \cdots + e_t = 1_E$$

and we say that $\{e_1, \ldots, e_t\}$ is a *set of orthogonal idempotents* if

$$e_i e_j = 0 \quad \text{for each } 1 \le i \ne j \le t.$$

If $\{e_1, e_2\}$ is a set of orthogonal idempotents then we say that e_1 *and* e_2 *are orthogonal idempotents*. The reader should show that the sum of orthogonal idempotents is again an idempotent. The idempotent $e \in E$ is *primitive* if given orthogonal idempotents e_1 and e_2 such that $e = e_1 + e_2$

then either $e_1 = 0$ or $e_2 = 0$. The idempotent e is *central* if $ex = xe$ for each $x \in E$.

Idempotents in $\text{End}(G)$ characterize the direct sum decompositions of G as follows.

LEMMA 4.1.1 *Let G be a group.*

1. *If $e \in \text{End}(G)$ is an idempotent then $G = eG \oplus (1 - e)G$.*

2. *If $G = G_1 \oplus G_2$ then there are orthogonal idempotents $e_1, e_2 \in \text{End}(G)$ such that $e_1 + e_2 = 1_G$, $e_1G = G_1$, and $e_2G = G_2$.*

3. *If $e \in \text{End}(G)$ is an idempotent then e is primitive iff eG is indecomposable.*

4. *If $\{e_1, \ldots, e_t\} \subset \text{End}(G)$ is a complete set of primitive orthogonal idempotents then $G = e_1G \oplus \cdots \oplus e_tG$ is an indecomposable direct sum decomposition of G.*

5. *If $G = G_1 \oplus \cdots \oplus G_t$ is an indecomposable decomposition of G then there is a complete set of primitive orthogonal idempotents $\{e_1, \ldots, e_t\} \subset \text{End}(G)$ such that $G_i = e_iG$ for each $i = 1, \ldots, t$.*

Proof: 1. If $e^2 = e$ then each $x \in G$ can be written as $x = ex + (1-e)x$ so that $G = eG + (1-e)G$. If $x \in eG \cap (1-e)G$ then $x = ex = (1-e)ex = 0$ so that $G = eG \oplus (1 - e)G$.

2. If $G = G_1 \oplus G_2$ then let $e_i : G \to G_i$ be the unique map defined as follows. Given $x \in G$ there are unique $x_1 \in G_1$ and $x_2 \in G_2$ such that $x = x_1 \oplus x_2$. Then define $e_i \in \text{End}(G)$ by

$$e_i(x_1 \oplus x_2) = x_i \quad \text{for } i = 1, 2.$$

The reader will verify that $e_i^2 = e_i$, that $e_ie_j = 0$ for $i \neq j$, and that $e_1 + e_2 = 1_G$.

3. Let $e^2 = e \in \text{End}(G)$. If $e = e_1 + e_2$ for some nonzero orthogonal idempotents e_1 and $e_2 \in \text{End}(G)$ then as in part 1 we can write

$$eG = e_1G \oplus e_2G$$

for some nonzero groups e_1G and $(e - e_1)G$. Thus eG is decomposable if e is not primitive.

Conversely if $eG = G_1 \oplus G_2$ for some nonzero groups then as in part 2 the canonical projections $e_i : G \to G_i$ are nonzero orthogonal idempotents that satisfy $e = e_1 + e_2$. Thus e is not primitive. This completes the proof of part 3.

The proof of part 4 uses parts 1, 2, and 3 in a simple induction on t. This completes the proof of the lemma.

If $e^2 = e$ and if $H = eG$ then we say that H is *the direct summand corresponding to e*. If $G = G_1 \oplus G_2$ then the natural projection $e_1 : G \to G_1$ is *an idempotent corresponding to G_1*.

The reader can prove the following lemma.

LEMMA 4.1.2 *Let $G = G_1 \oplus G_2$ and for $i = 1, 2$ let $e_i^2 = e_i \in \mathrm{End}(G)$ correspond to G_i.*

1. $\mathrm{Hom}(G_i, G_j) = e_j \mathrm{End}(G) e_i$ *for each $i, j \in \{1, 2\}$. That is, each $f : G_i \to G_j$ satisfies $f = e_j f e_i$.*

2. $\mathrm{End}(G_i) = e_i \mathrm{End}(G) e_i$.

Each indecomposable direct summand H of G corresponds to a primitive idempotent $e \in \mathrm{End}(G)$. We are thus motivated to consider complete sets of primitive orthogonal idempotents in $\mathrm{End}(G)$. Central indecomposable idempotents are of particular importance.

LEMMA 4.1.3 *Let E be any ring, let $\{e_1, \ldots, e_t\} \subset E$ be a complete set of primitive orthogonal idempotents, and suppose that e_i is central for each $i = 1, \ldots, t$. If $e^2 = e \in E$ is primitive then $e \in \{e_1, \ldots, e_t\}$.*

Proof: Let $e^2 = e \in E$ be primitive. Because $\{e_1, \ldots, e_t\}$ is a complete set of central orthogonal idempotents $(e_i e)(e_j e) = e_i e_j e = 0$ for $i \neq j$. Then

$$e = e_1 e + \cdots + e_t e$$

is a sum of orthogonal idempotents. Since e is primitive there is a subscript, 1 say, such that $e = e_1 e$. Since e_1 is primitive and since $e_1 = e_1 e + e_1 (1 - e)$ is a sum of orthogonal idempotents we see that $e_1 (1 - e) = 0$. That is $e = e_1 e = e_1$. This completes the proof.

The set $\{G_1, \ldots, G_t\}$ of groups is *rigid* if $\mathrm{Hom}_R(G_i, G_j) = 0$ for each $1 \leq i \neq j \leq t$. B. Charles [22] calls $\{G_1, \ldots, G_t\}$ a *semi-rigid* set if $\mathrm{Hom}_R(G_i, G_j) = 0$ or $\mathrm{Hom}_R(G_j, G_i) = 0$ for each $1 \leq i \neq j \leq t$. D. Arnold, R. Hunter, F. Richman [13] call $\{G_1, \ldots, G_t\}$ a *pseudo-rigid* set if for each $1 \leq i \neq j \leq t$, each composition

$$G_i \to G_j \to G_i \tag{4.1}$$

of maps is zero. We will call $\{G_1, \ldots, G_t\}$ a *nilpotent set* if for each $1 \leq i \neq j \leq t$ each composition (4.1) of group maps is a nilpotent endomorphism of G_i.

For instance $\{\mathbb{Z}_p \mid \text{primes } p \in \mathbb{Z}\}$ is rigid. If $G_1 \subset G_2$ are rank one groups such that $\text{Hom}(G_2, G_1) = 0$ then $\{G_1, G_2\}$ is a semi-rigid set. If G_1 and G_2 are strongly indecomposable groups that are not quasi-isomorphic then we will show in Lemma 4.1.5 that $\{G_1, G_2\}$ is a nilpotent set. For instance, let $p \in \mathbb{Z}$ be a prime and let G_1, G_2 be pure subgroups of $\widehat{\mathbb{Z}}_p$ of differing ranks. Then $\{G_1, G_2\}$ is a nilpotent set.

The following classifies nilpotents sets in terms of a commutativity property.

LEMMA 4.1.4 [34, T.G. Faticoni] *Let $G = G_1 \oplus \cdots \oplus G_t$ and for each $i = 1, \ldots, t$ let $e_i \in \text{End}(G)$ be the idempotent corresponding to G_i. The following are equivalent.*

1. $\{G_1, \ldots, G_t\}$ *is a nilpotent set.*

2. $\text{Hom}(G_i, G_j) \subset \mathcal{N}(\text{End}(G))$ *for each $1 \leq i \neq j \leq t$.*

3. e_i *is central modulo $\mathcal{N}(\text{End}(G))$ for each $i = 1, \ldots, t$.*

PROOF: $1 \Rightarrow 2$ Let $i \neq j$, let $f : G_i \rightarrow G_j$, and let $x : G \rightarrow G$. Then $f = e_j f e_i$. Since $\{G_1, \ldots, G_t\}$ is a nilpotent set $e_j f e_i x e_j$ is a nilpotent endomorphism of G_j. For large $m \in \mathbb{Z}$,

$$
\begin{aligned}
(fx)^m &= \left(\sum_{k=1}^{t} e_j f e_i x e_k \right)^m \\
&= (e_j f e_i x e_j)^m \\
&= 0.
\end{aligned}
$$

The last expression is from the fact that $(xe_k)(e_j f e_i) = 0$ for $j \neq k$ and that $e_j f e_i x e_j$ is nilpotent. Then f generates a nil right ideal of $\text{End}(G)$, as required by part 2.

$2 \Rightarrow 3$ Let $x \in \text{End}(G)$. Then for fixed $i = 1, \ldots, t$

$$
\begin{aligned}
e_i x &= e_i x e_i + \sum_{i \neq j} e_i x e_j \quad \text{and} \\
x e_i &= e_i x e_i + \sum_{i \neq j} e_j x e_i.
\end{aligned}
$$

By part 2, $e_i x e_j \in \mathcal{N}(\text{End}(G))$ for each $1 \leq i \neq j \leq t$, so that

$$
e_i x - x e_i = \sum_{i \neq j} e_i x e_j - \sum_{i \neq j} e_j x e_i \in \mathcal{N}(\text{End}(G)).
$$

Thus e_i is central modulo $\mathcal{N}(\text{End}(G))$, which proves part 3.

3 $\Rightarrow$ 1 Let $i \neq j$ and let

$$G_i \xrightarrow{f} G_j \xrightarrow{g} G_i.$$

Then $gf = e_i g(e_j f e_i)$ and by part 3

$$0 = e_j e_i f \equiv e_j f e_i (\text{mod } \mathcal{N}(\text{End}(G))).$$

Thus $gf \equiv 0 (\text{mod } \mathcal{N}(\text{End}(G)))$, hence gf is nilpotent, whence $\{G_1, \ldots, G_t\}$ is a nilpotent set. This completes the logical cycle.

The following two results show us that it should be easy to determine when a set $\{G_1, \ldots, G_t\}$ of rtffr groups is nilpotent.

LEMMA 4.1.5 Let G_1 and G_2 be strongly indecomposable rtffr groups. Then $\{G_1, G_2\}$ is not a nilpotent set iff $G_1 \stackrel{.}{\cong} G_2$.

Proof: Clearly $\{G_1, G_2\}$ is not a nilpotent set if $G_1 \stackrel{.}{\cong} G_2$.

Conversely suppose that $\{G_1, G_2\}$ is not nilpotent. There is a composition

$$G_1 \xrightarrow{f} G_2 \xrightarrow{g} G_1$$

such that gf is not nilpotent. By Lemma 1.3.3, $\mathbb{Q}\text{End}(G_1)$ is a local ring and since $\mathcal{J}(\mathbb{Q}\text{End}(G_1))$ is a nilpotent ideal gf is a unit in $\mathbb{Q}\text{End}(G_1)$. Thus there is a map $h \in \text{End}(G_1)$ and an integer $n \neq 0$ such that $(hg)f = n1_{G_1}$. Then $G_2 \stackrel{.}{\cong} G_1 \oplus \ker hg$ and since G_2 is strongly indecomposable $G_2 \stackrel{.}{\cong} G_1$. This proves the lemma.

LEMMA 4.1.6 Let $\{G_1, G_2\}$ be a nilpotent set and for $i = 1, 2$ let H_i be a quasi-summand of G_i. Then $\{H_1, H_2\}$ is a nilpotent set.

Proof: Suppose that $\{G_1, G_2\}$ is a nilpotent set and let $H_1 \subset G_1$ and $H_2 \subset G_2$ be quasi-summands. Consider a composition

$$H_1 \xrightarrow{f} H_2 \xrightarrow{g} H_1$$

of group maps. Since H_2 is a quasi-summand of G_2, $f : H_1 \to G_2$. There is a map $\rho : G_2 \to H_2$ and an integer $m \neq 0$ such that $\rho(x) = mx$ for each $x \in H_2$. Let $\jmath : H_1 \to G_1$ and $\pi : G_1 \to H_1$ be the canonical injection/quasi-projection associated with the quasi-summand H_1 such that $\pi\jmath = n1_{H_1}$ for some integer n. Because H_2 is a quasi-summand of G_2

$$\jmath(g\rho f)\pi : G_1 \to G_1$$

is an endomorphism of G_1 that factors through G_2. Since $\{G_1, G_2\}$ is a nilpotent set $\jmath(g\rho f)\pi$ is a nilpotent endomorphism. Thus there is an integer $k > 0$ such that

$$
\begin{aligned}
0 &= (\jmath(g\rho f)\pi)^{k+1} \\
&= \jmath(g\rho f\pi\jmath)^k(g\rho f)\pi \\
&= \jmath(g\rho fn)^k(g\rho f)\pi \\
&= n^k\jmath(g\rho f)^{k+1}\pi.
\end{aligned}
$$

Inasmuch as π is a surjection and $n^k\jmath$ is an injection, $(g\rho f)^k = 0$. Finally for each $x \in H_1$, $g\rho f(x) = g(mf(x)) = gf(mx)$, so that

$$
(gf)^k(m^k x) = (gf)^{k-1}m^{k-1}(g\rho f)(x) = (g\rho f)^k(x) = 0.
$$

Because H_1 is torsion-free

$$
0 = (gf)^k(m^k H_1) = (gf)^k(H_1)
$$

which implies that gf is nilpotent. Hence $\{H_1, H_2\}$ is a nilpotent set.

One of the themes of this chapter is to study the direct sum decompositions of rtffr groups. With this in mind, Fitting's Lemma leads us to an important result for rtffr groups.

LEMMA 4.1.7 *Let $f : G \to G$ be an endomorphism of the rtffr group G. There is an integer $k > 0$ such that*

$$
G \doteq \ker f^k \oplus \text{image } f^k,
$$

f is nilpotent on $\ker f^k$, and f is idempotent on $\text{image } f^k$.

Proof: The given map f lifts to an endomorphism of the finite dimensional vector space $\mathbb{Q}G$ so that by Fitting's Lemma $\mathbb{Q}G = \ker f^k \oplus \text{image } f^k$ where f is nilpotent on $\ker f^k$ and f is idempotent on $\text{image } f^k$. (See [30].) Then f^k is an idempotent endomorphism of G, and hence $G \doteq \ker f^k \oplus \text{image } f^k$, as required by the lemma.

The following result is anticipated by D.M. Arnold and E.L. Lady in [14] where they show that an rtffr group $G = H \oplus K$ enjoys an exchange property if H and K satisfy a property that is equivalent to assuming that $\{H, K\}$ is a nilpotent set.

THEOREM 4.1.8 *Let $\{G_1, \ldots, G_t\}$ be a set of rtffr groups. The following are equivalent.*

1. $\{G_1, \ldots, G_t\}$ is a nilpotent set.

2. If $1 \leq i \neq j \leq t$ and if H is a quasi-summand of G_i and G_j then $H = 0$.

3. If $1 \leq i \neq j \leq t$ and if H is a strongly indecomposable quasi-summand of G_i and G_j then $H = 0$.

Proof: From the definition of nilpotent set it suffices to prove the theorem for $t = 2$.

$1 \Rightarrow 2$ is Lemma 4.1.6.

$2 \Rightarrow 3$ is clear.

$3 \Rightarrow 1$ Suppose that $\{G_1, G_2\}$ is not a nilpotent set. There is a composition

$$G_1 \xrightarrow{f} G_2 \xrightarrow{g} G_1$$

such that gf is not nilpotent. By Fitting's Lemma 4.1.7 there is an integer $k > 0$ such that

$$G_1 \doteq \ker(gf)^k \oplus \text{image } (gf)^k$$

and $(gf)^{k+1} = (gf)^k$. Then

$$G_1 \xrightarrow{f(gf)^k} G_2 \xrightarrow{g} G_1$$

are maps such that $gf(gf)^k = (gf)^{k+1}$ is idempotent on image $(gf)^k \neq 0$. But then

$$\text{image } (gf)^k \xrightarrow{f} G_2 \xrightarrow{g} \text{image } (gf)^k$$

is an integer multiple of the identity map so that image $(gf)^k$ is a quasi-summand of G_2. Thus part 3 is false and the proof is complete.

THEOREM 4.1.9 Let $G_1, \ldots, G_t$ be strongly indecomposable groups such that $G_i \cong G_j \Rightarrow i = j$. If $G \cong G_1 \oplus \cdots \oplus G_t$ then $G = G'_1 \oplus \cdots \oplus G'_s$ for some nilpotent set $\{G'_1, \ldots, G'_s\}$ of indecomposable rtffr groups.

Proof: Since G has finite rank we can write $G = G'_1 \oplus \cdots \oplus G'_s$ for some indecomposable groups $G'_1, \ldots, G'_s$. Suppose that $i \neq j$ and let H be a strongly indecomposable quasi-summand of G'_i and of G'_j. Rewriting the summands of G we can write $H^{(2)}$ as a quasi-summand of G. But by hypothesis $G_i \cong G_j \Rightarrow i = j$ for each $1 \leq i, j \leq t$. Jónsson's Theorem 2.1.10 then implies that if $H \neq 0$ then the multiplicity of H in $G_1 \oplus \cdots \oplus G_t$ is 1. Thus $H = 0$ so that $\{G'_1, \ldots, G'_s\}$ is a nilpotent set by Theorem 4.1.8.

COROLLARY 4.1.10 Let $\{G_1, \ldots, G_t\}$ be a nilpotent set of rtffr groups. Then $\{G_1^{(n_1)}, \ldots, G_t^{(n_t)}\}$ is a nilpotent set for any t-tuple $(n_1, \ldots, n_t)$ of positive integers.

EXAMPLE 4.1.11 Let G be an rtffr group. Jónsson's Theorem 2.1.10 states that there is a finite set of strongly indecomposable groups $\{G_1, \ldots, G_t\}$ and integers $n_1, \ldots, n_t > 0$ such that $G_i \cong G_j \implies i = j$ and

$$G \cong G_1^{(n_1)} \oplus \cdots \oplus G_t^{(n_t)}.$$

In this case Theorem 4.1.8 shows us that $\{G_1, \ldots, G_t\}$ is a nilpotent set. Thus each rtffr group G is quasi-isomorphic to a direct sum

$$G \cong G' \oplus H'$$

where $G' = G_1 \oplus \cdots \oplus G_t$. In particular Theorem 4.1.9 states that $G_1' \oplus \cdots \oplus G_t'$ is the direct sum of a nilpotent set $\{G_1', \ldots, G_t'\}$.

Our interest in nilpotent sets stems from the fact that they enjoy a refinement and uniqueness property. Recall the functor $A(\cdot)$ from Section 2.4.2.

THEOREM 4.1.12 Let $\{G_1, \ldots, G_t\}$ be a nilpotent set of indecomposable rtffr groups, and let $G = G_1 \oplus \cdots \oplus G_t$. Then

1. $E(G) = E(G_1) \times \cdots \times E(G_t)$ as rings.

2. G has a unique decomposition.

Proof: 1. Let $\{e_1, \ldots, e_t\} \subset \mathrm{End}(G)$ be a complete set of primitive orthogonal idempotents such that

$$e_i(G) = G_i \quad \text{and} \quad \ker e_i = \bigoplus_{i \neq j} G_j.$$

By Lemma 4.1.4 the idempotents e_i map to central idempotents $\bar{e}_i$ in $E(G)$ and by Theorem 2.4.4, $E(G)\bar{e}_i = A(G_i)$ is a cyclic projective E-module. Thus

$$\begin{aligned} E(G) &= E(G)\bar{e}_1 \oplus \cdots \oplus E(G)\bar{e}_t \\ &= A(G_1) \oplus \cdots \oplus A(G_t) \end{aligned}$$

as right ideals. Since the $\bar{e}_i$ are central

$$E(G) = A(G_1) \times \cdots \times A(G_t).$$

Thus $A(G_i)$ is a ring. The canonical map $\pi : \text{End}(G_i) \to E(G_i)$ taking $f \mapsto \bar{e}_i f \bar{e}_i$ has image in $A(G_i)$. Because $\{G_1, \ldots, G_t\}$ is a nilpotent set, each $f \in A(G_i)$ satisfies

$$f \equiv e_i f e_i (\text{mod } \mathcal{N}(\text{End}(G))).$$

Thus π has image $A(G_i)$ and so $E(G_i) \cong A(G_i)$.

2. Suppose that

$$G = G'_1 \oplus \cdots \oplus G'_s$$

for some indecomposable rtffr groups G'_i. Let $\{e'_1, \ldots, e'_s\} \subset \text{End}(G)$ be a complete set of primitive orthogonal idempotents corresponding to $G'_1 \oplus \cdots \oplus G'_s$. The e'_i map onto idempotents $\bar{e}'_i \in E(G)$. Because $\bar{e}_1, \ldots, \bar{e}_t$ is a complete set of primitive orthogonal central idempotents in $E(G)$

$$\bar{e}'_i = \bar{e}_1 \bar{e}'_i + \cdots + \bar{e}_t \bar{e}'_i$$

is a sum of orthogonal idempotents. Since $\bar{e}'_1$ is primitive

$$\bar{e}'_1 = \bar{e}_i \bar{e}'_1.$$

We may assume without loss of generality that $i = 1$. Since

$$\bar{e}_1 = \bar{e}_1 \bar{e}'_1 + \bar{e}_1 (1 - \bar{e}'_1)$$

and since $\bar{e}_1$ is primitive we see that $\bar{e}'_1 = \bar{e}_1$. Iterating this process, we produce the identities

$$\bar{e}'_i = \bar{e}_i \quad \text{for } i = 1, \ldots, s.$$

Since

$$\bar{e}'_1 + \ldots + \bar{e}'_s = 1_G,$$

$s = t$. Therefore

$$A(G_i) \cong E(G)\bar{e}_i = E(G)\bar{e}'_i = A(G'_i)$$

for each $i = 1, \ldots, t$ so that $G_i \cong G'_i$ by Theorem 2.4.4. This completes the proof.

THEOREM 4.1.13 *Let* $\{G_1, \ldots, G_t\}$ *be a nilpotent set of rtffr groups and let* $G = G_1 \oplus \cdots \oplus G_t$.

1. *If* $H \in \mathbf{P}_o(G)$ *then* $H \cong H_1 \oplus \cdots \oplus H_t$ *where for each,* $i = 1, \ldots, t$, $H_i \in \mathbf{P}_o(G_i)$.

2. *If* $G = H \oplus K \cong H \oplus K'$ *then* $K \cong K'$.

Proof: 1. Given $H \in \mathbf{P}_o(G)$ then $A(H)$ is a finitely generated projective right $E(G)$-module. Since the $\bar{e}_i$ are central in $A(G)$ we can write

$$A(H) \cong A(H)\bar{e}_1 \oplus \cdots \oplus A(H)\bar{e}_t$$

as right $E(G)$-modules. By Theorem 2.4.4 there are $H_1, \ldots, H_t$ such that

$$A(H_i) \cong A(H)\bar{e}_i$$

for each $i = 1, \ldots, t$ so that

$$\begin{aligned} A(H) &\cong A(H_1) \oplus \cdots \oplus A(H_t) \\ &\cong A(H_1 \oplus \cdots \oplus H_t). \end{aligned}$$

By Theorem 2.4.4, $H \cong H_1 \oplus \cdots \oplus H_t$.

Part 2 follows from Theorem 4.1.12 in the usual way, so the proof is complete.

COROLLARY 4.1.14 *Suppose that* $\{G_1, \ldots, G_t\}$ *is a nilpotent set of rtffr groups and suppose that*

$$H \oplus H' = G_1^{(n_1)} \oplus \cdots \oplus G_t^{(n_t)}$$

for some integers $n_1, \ldots, n_t > 0$. *For each* $i = 1, \ldots, t$ *there are rtffr groups* H_i *and* H_i' *such that* $H_i \oplus H_i' \cong G_i^{(n_i)}$ *and such that* $H \cong H_1 \oplus \cdots \oplus H_t$ *and* $H' \cong H_1' \oplus \cdots \oplus H_t'$.

The following result is somewhat of a surprise.

THEOREM 4.1.15 *Let* $G_1, \ldots, G_t$ *be strongly indecomposable rtffr groups such that* $G_i \cong G_j \Rightarrow i = j$, *and let* $G \doteq G_1 \oplus \cdots \oplus G_t$.

1. G *has a unique decomposition* $G = G_1' \oplus \cdots \oplus G_r'$.

2. *If* $H \in \mathbf{P}_o(G)$ *then* $H = H_1 \oplus \cdots \oplus H_s$ *where for each* $i = 1, \ldots, s$, $H_i \in \mathbf{P}_o(G_i')$.

Proof: Apply Theorems 4.1.9 and 4.1.12.

4.2 Commutative Rtffr Rings

In this section we begin our investigation of the $\text{End}(G)$-module structure of rtffr groups G such that $E(G)$ is a *commutative ring*. We will consider the finitely generated projective modules over rtffr commutative rings. Our techniques will produce good results on these rings.

4.2.1 Modules over Commutative Rings

Our first result gives us an easily verified condition that implies that the ring $E(G)$ is commutative.

LEMMA 4.2.1 [T.G. Faticoni] *Let G be a strongly indecomposable rtffr group, let $E = \text{End}(G)$, let $\mathcal{N} = \mathcal{N}(\text{End}(G))$, and let $\mathbf{k} = \mathbb{Q}E/\mathcal{N}$.*

1. *The degree $[\mathbf{k} : \mathbb{Q}]$ divides the integers*

   (a) *$\mathbb{Q}$-dim$(\mathbb{Q}E)$,*

   (b) *$\mathbb{Q}$-dim$(\mathbb{Q}\mathcal{N})$,*

   (c) *$\mathbb{Q}$-dim$(\mathbb{Q}\mathcal{N}G)$,*

   (d) *$\mathbb{Q}$-dim$(\mathbb{Q}G)$, and*

   (e) *$\mathbb{Q}$-dim$(\mathbb{Q}\mathcal{N}^k G/\mathbb{Q}\mathcal{N}^{k+1}G)$ for each integer $k \geq 0$.*

2. *If $\mathbb{Q}$-dim$(\mathbb{Q}E)$ or $\mathbb{Q}$-dim$(\mathbb{Q}G)$ is a square-free integer then $\mathbf{k}$ is a commutative field.*

Proof: 1. By the Beaumont-Pierce-Wedderburn Theorem 3.1.6 we can identify $\mathbf{k}$ with a subring of $\mathbb{Q}E$. Since $\mathbb{Q}E$ and $\mathbb{Q}G$ are then $\mathbf{k}$-vector spaces and since $\mathbb{Q}\mathcal{N}$ and $\mathbb{Q}\mathcal{N}G$ are $\mathbf{k}$-vector subspaces of $\mathbb{Q}E$ and $\mathbb{Q}G$ it follows that $[\mathbf{k} : \mathbb{Q}]$ divides each of the integers in (a)-(e).

2. Since $\mathbb{Q}E$ is a local ring $\mathbb{Q}E/\mathbb{Q}\mathcal{N} = \mathbf{k}$ is a division ring. By part 1, the $\mathbf{k}$-dimension of $\mathbb{Q}E$ is a divisor of $\mathbb{Q}$-dim$(\mathbb{Q}E)$. Since the dimension of a division ring over its center is a perfect square (see [69]), and since $\mathbb{Q}$-dim$(\mathbb{Q}E)$ is assumed to be square-free $\mathbf{k}$ is commutative. A similar proof applies if $\mathbb{Q}$-dim$\mathbb{Q}G$ is square-free. This completes the proof.

LEMMA 4.2.2 *Let S be an Artinian commutative ring and let U be a finitely generated projective S-module. If $\text{ann}_S(U) = 0$ then $U \cong S \oplus U'$ for some S-module U'.*

Proof: Let U be a finitely generated projective S-module. Since S is an Artinian commutative ring S is a finite product of local commutative rings say

$$S = S_1 \times \cdots \times S_t.$$

Since $\mathrm{ann}_S(U) = 0$

$$U = US_1 \oplus \cdots \oplus US_t$$

is a direct sum of nonzero projective modules. Since projective modules over local rings are free $US_i \cong S_i^{(n_i)}$ for some integers $n_1, \ldots, n_t > 0$. Inasmuch as $S = S_1 \times \cdots \times S_t$, it follows that $U \cong S \oplus U'$ for some S-module U'. This completes the proof.

The next result is of independent interest since it deals with the structure of right ideals in (not necessarily commutative) rtffr rings. A simple consequence of part 1 that we will not use in our deliberations is that *each module of finite composition length over an rtffr ring is finitely presented.*

LEMMA 4.2.3 *Let I be a right ideal in the rtffr ring E.*

1. *If I is a maximal right ideal in E then E/I is a finite p-group for some prime $p \in \mathbb{Z}$.*

2. *If $I \oplus J$ has finite index in E for some right ideal J in E then I is finitely presented.*

Proof: 1. Let I be a maximal right ideal in E. Then $\mathcal{N}(E) \subset I$ and by the Beaumont-Pierce-Wedderburn Theorem 3.1.6 we have a group decomposition $E \cong E/\mathcal{N}(E) \oplus \mathcal{N}(E)$. Hence $E/\mathcal{N}(E)$ is a reduced group. Thus we can assume without loss of generality that $\mathcal{N}(E) = 0$. That is, $\mathbb{Q}E$ is a semi-simple Artinian ring. Thus $\mathbb{Q}I = e\mathbb{Q}E$ for some $e^2 = e \in \mathbb{Q}E$. There is an integer $m \neq 0$ such that $m(1-e)E \subset E$. Left multiplication by $m(1-e)$ induces an embedding of right E-modules

$$E/(\mathbb{Q}I \cap E) = E/(e\mathbb{Q}E \cap E) \cong m(1-e)E \subset E.$$

Evidently, $E/(\mathbb{Q}I \cap E)$ is a right rtffr E-module. Furthermore since E/I is simple, either E/I is an elementary p-group for some prime $p \in \mathbb{Z}$ or E/I is divisible. Since E/I maps onto $E/(\mathbb{Q}I \cap E)$, E/I is an elementary p-group, and since E has finite rank, E/I is finite.

2. There is an integer $n \neq 0$ such that $n \in I \oplus J$. Thus $(I \oplus J)/nE$ is a finite right ideal in E/nE so that $I \oplus J$ is finitely generated. Suppose that $\pi : P \to I$ is a surjection where P is a finitely generated projective

right E-module. Then π lifts to a surjection $\pi : \mathbb{Q}P \to \mathbb{Q}I$ which must split. Let $f : \mathbb{Q}I \to \mathbb{Q}P$ be the map such that $\pi f = 1_{\mathbb{Q}I}$. Since I and P are finitely generated we may assume that $f : I \to P$ is such that $\pi f = m1_I$ for some integer $m \neq 0$. Then $P \doteq f(I) \oplus \ker \pi$ and since P is finitely generated $\ker \pi$ is finitely generated. Hence $I = \pi(P)$ is a finitely presented right ideal of E which completes the proof.

The following result anticipates Chapter 6 where we link the condition $IG \neq G$ to the splitting of exact sequences involving G. The next theorem is proved in [10, Theorem 5.9] for ideals of finite index in commutative rings. We give a different proof.

THEOREM 4.2.4 [10, D.M. Arnold, Theorem 5.9] *Let G be an rtffr group with commutative endomorphism ring $E = \mathrm{End}(G)$. Then $IG \neq G$ for each proper ideal $I \subset E$.*

Proof: It suffices to show that $IG \neq G$ for each maximal proper ideal $I \subset E$. So choose a ideal $I \subset E$ such that E/I is simple and such that $IG = G$. By Lemma 4.2.3(1) there is a prime $p \in \mathbb{Z}$ such that E/I is a finite p-group so that $pE \subset I$. Thus I/pE is a maximal ideal in the commutative Artinian ring E/pE. There is an idempotent $\bar{e} \in E/pE$ such that $I/pE = (\bar{e}E + \mathcal{J})/pE$ where $\mathcal{J}/pE = \mathcal{J}(E/pE)$. Then

$$\frac{G}{pG} = \frac{IG}{pG} = \frac{\bar{e}G + \mathcal{J}G}{pG} = \frac{\bar{e}G + pG}{pG}$$

by Nakayama's Lemma 1.2.3. Since G/pG has annihilator pE in E, and since $\bar{1} - \bar{e} \notin pE$ annihilates $\bar{e}$, $\bar{e} = \bar{1}$. Thus $I/pE = E/pE$ and $I = E$. This completes the proof.

The above result implies that if $\mathrm{End}(G)$ is commutative then $\mathrm{Hom}(G, IG) = I$ for each maximal ideal $I \subset S$. It *does not* state that $\mathrm{Hom}(G, IG) = I$ for each nonzero ideal $I \subset S$. It would be interesting to investigate the set $\{$ideals $I \subset S \mid \mathrm{Hom}(G, IG) = I\}$.

LEMMA 4.2.5 *Let S be an Artinian commutative ring and let M be an S-module such that $\mathrm{ann}_S(M) = 0$, let $\mathrm{length}(X)$ denote the composition length of X as a module over some subring R of S, and assume that $\mathrm{length}(S) < \infty$. The following are equivalent.*

1. $\mathrm{ann}_S(M/MI) = I$ *for each ideal* $I \subset S$ *and* $\mathrm{length}(M) \leq \mathrm{length}(S)$.

2. $S \cong M$ *as S-modules.*

Proof: Part 1 readily follows from part 2, so assume part 1 is true. Suppose that S is a local commutative Artinian ring. Since S is Artinian there is a simple ideal $I \neq 0$ such that $\text{length}(I) = k = \text{length}(S/\mathcal{J}(S))$. By hypothesis $\text{ann}_S(M/MI) = I$. Furthermore, because $MI \neq 0$, $k \leq \text{length}(IM)$, so that

$$\begin{aligned}\text{length}(S/I) &= \text{length}(S) - k \\ &\geq \text{length}(M) - \text{length}(MI) \\ &= \text{length}(M/MI).\end{aligned}$$

Induction on $\text{length}(S)$ implies that $S/I \cong M/MI$. Since S is local $I \subset \mathcal{J}(S)$ so that $M = xS + IM$ implies that $M = xS$ by Nakayama's Lemma 1.2.3. By hypothesis $\text{ann}_S(x) = \text{ann}_S(M) = 0$ and so $M \cong S$.

To complete the proof the Artinian commutative ring S satisfies

$$S = S_1 \times \cdots \times S_t$$

for some local rings $S_1, \ldots, S_t$. Then

$$M = MS_1 \oplus \cdots \oplus MS_t$$

where $\text{ann}_{S_i}(MS_i) = 0$. If $\text{length}(M) = \text{length}(S)$ then

$$\text{length}(M) = \text{length}(MS_1) + \cdots + \text{length}(MS_t).$$

By the above argument $\text{length}(MS_i) \leq \text{length}(S_i)$ implies that $MS_i \cong S_i$. Then $M \cong S$. This completes the proof.

4.2.2 Projectives over Commutative Rings

Although the following is true for general commutative rings we will prove it for rtffr rings. We lift this proof from [75, Lemma VI.8.6].

LEMMA 4.2.6 Let S be a commutative rtffr ring and let $I^2 = I$ be an ideal of S. Then $I = eS$ for some $e^2 = e \in S$.

Proof: Suppose that $I = \sum_{i=1}^n x_i S$. Since $I^2 = I$ for each x_i there are $y_{ij} \in I$ such that

$$x_i = \sum_j y_{ij} x_j,$$

which leads to the following system of equations.

$$\begin{cases} 0 = (1-y_{11})x_1 - y_{12}x_2 - \cdots - y_{1n}x_n \\ 0 = y_{21}x_1 - (1-y_{22})x_2 - \cdots - y_{2n}x_n \\ \vdots \\ 0 = y_{n1}x_1 - y_{n2}x_2 - \cdots - (1-y_{nn})x_n \end{cases}$$

Thus the determinant $\det(M)$ of the coefficient matrix

$$M = \begin{pmatrix} (1 - y_{11}) & -y_{12} & \cdots & -y_{1n} \\ y_{21} & -(1 - y_{22}) & \cdots & -y_{2n} \\ \vdots & & & \vdots \\ y_{n1} & -y_{n2} & \cdots & -(1 - y_{nn}) \end{pmatrix}$$

satisfies $x_i \det(M) = 0$ for each $i = 1, \ldots, n$. (See, e.g., [58, page 334-335].) Thus $I \det(M) = 0$. However if we think of the determinant as a homogeneous polynomial of degree n then we see that it is of the form $1 - e$ for some $e \in I$. Then $(1 - e)e \in (1 - e)I = 0$ implies that $e^2 = e$ and so $eS = I$, as required to prove the lemma.

Since the trace ideal I of a finitely generated projective module P is an idempotent ideal such that $IP = P$ we have the following result.

COROLLARY 4.2.7 *Let S be a commutative ring and let P be a finitely generated projective S-module. There is an $e^2 = e \in S$ such that $P = eP$ and $(1 - e)P = 0$.*

Proof: Let I be the trace ideal of the projective S-module P. Since $I^2 = I$, Lemma 4.2.6 implies that $I = eS$ for some $e^2 = e \in S$, and since $IP = P$, $eP = eSP = IP = P$.

COROLLARY 4.2.8 *Let S be a commutative ring and let P be a finitely generated projective S-module. If $\text{ann}_S(P) = 0$ then P is a generator over S.*

Proof: By Lemma 4.2.6, $I = \text{trace}_S(P) = eS$, and because $\text{ann}_S(P) = 0$, $(1 - e)P = 0$ implies that $e = 1$. Thus $I = S$ and hence P generates S.

LEMMA 4.2.9 *Let S be an rtffr commutative ring and let $S \subset U \subset \mathbb{Q}S$ be a finitely generated S-module. Then U is a projective S-module iff U is an invertible fractional ideal for S.*

Proof: Suppose that $S \subset U \subset \mathbb{Q}S$ is a projective S-module. By Lemma 4.2.6 there is an $e^2 = e \in S$ such that $\text{trace}_S(U) = eS$ and $(1 - e)U = 0$. Since $S \subset U$, $e = 1$. That is, U generates S. Since $\text{Hom}_S(U, S) = \{q \in \mathbb{Q}S \mid qU \subset S\} = U^*$, $UU^* = S$, whence U is invertible.

Conversely suppose that $S \subset U \subset \mathbb{Q}S$ and that $UU^* = S$. Then U generates S. Because S is commutative $S \subset \text{End}_S(U) \subset \mathbb{Q}S$ and it is known that U is then a finitely generated projective $\text{End}_S(U)$-module. Furthermore since $S = UU^*$

$$\text{End}_S(U) = \text{End}_S(U)S = \text{End}_S(U)UU^* = UU^* = S$$

and hence U is a finitely generated projective S-module. This completes the proof.

We will use the following result to construct a direct sum decomposition of a projective module.

LEMMA 4.2.10 *Suppose that S is an rtffr commutative ring and let U be a finitely generated projective S-module such that $\text{ann}_S(U) = 0$.*

1. *Given a nonzero ideal $T \doteq S$ there is a map $f : U \to S$ such that $f(U) + T = S$.*

2. *Suppose that S is a semi-prime ring and let $\tau \subset S$ be a conductor for S. If $I \doteq S$ is an ideal such that $I + \tau = S$ then I is an invertible ideal in S.*

Proof: 1. Since $\text{ann}_S(U) = 0$, Lemma 4.2.9 implies that $U^{(k)} \cong S \oplus U'$ for some integer $k \neq 0$ so that $\text{ann}_{S/T}(U/UT) = 0$. Since U/UT is a finitely generated projective S/T-module Lemma 4.2.2 implies that $U/UT \cong S/T \oplus W$ for some S/T-module W. Thus there is a surjection

$$\bar{f} : U \to S/T.$$

Since U is projective $\bar{f}$ lifts to a map $f : U \to S$ such that $f(U) + T = S$.

2. This is [10, Corollary 12.14]. We give a different proof. It suffices to prove this under the additional assumption that S is indecomposable as a ring. Localize at τ. Then $I_\tau + \tau_\tau = S_\tau$. Inasmuch as $\tau_\tau \subset \mathcal{J}(S_\tau)$, $I_\tau = S_\tau$. Hence $I_M \cong S_M$ for each maximal ideal $\tau \subset M \subset S$. For maximal ideals $M \subset S$ such that $\tau \not\subset M$, S_M is an integrally closed ring by Theorem 3.2.1. That is, S_M is a finite product of local Dedekind domains. Such a local ring is a *pid*. Since $I_M \doteq S_M$, we see that $I_M \cong S_M$. Hence I is locally isomorphic to S so that I is a progenerator = invertible ideal in S. This proves the lemma.

4.2.3   Direct Sums over Commutative Rings

Now that we have disposed of the more general lemmas we will demonstrate that finitely generated projectives over commutative rtffr rings enjoy a natural direct sum decomposition.

THEOREM 4.2.11 *Suppose that S is an rtffr commutative semi-prime ring, and let U be a finitely generated projective S-module. Then*

$$U \cong U_1 \oplus \ldots \oplus U_s$$

for some finitely generated projective ideals $U_1, \ldots, U_s$ in S.

Moreover, if $\mathrm{ann}_S(U) = 0$ then U_1 can be chosen to be an invertible ideal in S.

Proof: Let $\sigma = \mathrm{trace}_S(U)$. Then $\sigma^2 = \sigma$ so by Lemma 4.2.6, $\sigma = eS$ for some central $e^2 = e \in S$ and hence U is a projective generator over the ring eS. Thus we may assume without loss of generality that $e = 1$ and that $\mathrm{ann}_S(U) = 0$. Let R be an indecomposable factor of S and consider UR. Since U generates S, UR generates R.

By Lemma 1.6.3 there is an integrally closed $\overline{R}$ and the associated conductor $\tau \subset R$ such that

$$\tau \overline{R} \subset R \subset \overline{R} \subset \mathbb{Q}R.$$

By Theorem 3.2.4, $R = R(\tau) \times R^\tau$, and since R is indecomposable $R(\tau) = 0$ and $R = R^\tau$. Because R is an rtffr ring we can choose n so large that R_n is a reduced group. Then Lemma 3.3.5(2) implies that $R_n = R_{nR}$ and subsequently by Lemma 1.2.1, $n \in \mathcal{J}(R_n)$.

By Lemma 4.2.10 there is a map $f : U \to R$ such that $f(U) + nR = R$. Localize at n. By Lemma 1.2.1

$$f(U)_n + nR_n = R_n = f(U)_n \supset R$$

by Nakayama's Lemma 1.2.3. Then $mR \subset f(U) \subset R$ for some integer m such that $\gcd(m, n) = 1$. By Lemma 4.2.10(2), $f(U)$ is an invertible ideal in the semi-prime rtffr ring R. Thus

$$U = f(U) \oplus U' = U'' \oplus U'$$

for some invertible ideal U'' in R.

Since $S = R_1 \times \cdots \times R_r$ for indecomposable factors R_i we see that

$$U = U_1 \oplus \cdots \oplus U_r \oplus U''$$

for some projective S-module U'', where for each $i = 1, \ldots, t$, U_i is an invertible ideal in R_i. An induction on the rank of U completes the proof.

THEOREM 4.2.12 *Let S be an indecomposable commutative rtffr ring and let $U \neq 0$ be a finitely generated projective S-module. Then $U = U_1 \oplus \cdots \oplus U_t$ for some invertible ideals $U_1, \ldots, U_t$ in S. In this case U is locally isomorphic to $S^{(r)}$ for some integer $r > 0$.*

4.2.4   Commutative Endomorphism Rings

Now that we have the ring theoretic results out of the way we can apply them to structure theorems for rtffr groups with commutative endomorphism rings. Our approach will be to fix an rtffr group G and consider the semi-prime rtffr ring

$$E(G) = \text{End}(G)/\mathcal{N}(\text{End}(G))$$

giving general conditions under which $E(G)$ is commutative.

Recall the *unique decomposition* from Section 2.1 and the *local refinement property* from Section 2.5.

THEOREM 4.2.13 [T.G. Faticoni] *Let* $\{G_1, \ldots, G_t\}$ *be a nilpotent set of indecomposable rtffr groups such that* $E(G_i)$ *is a commutative ring for each* $i = 1, \ldots, t$, *and let* $G = G_1 \oplus \cdots \oplus G_t$. *Then* G *has the local refinement property and* G *has a unique decomposition.*

Proof: Since $\{G_1, \ldots, G_t\}$ is a nilpotent set of indecomposable groups, Theorem 4.1.12(2) states that $G = G_1 \oplus \cdots \oplus G_t$ has a unique decomposition.

For each $i = 1, \ldots, t$, $A(G_i) = E(G_i)$ is indecomposable projective module. Then

$$E(G) = E(G_1) \times \cdots \times E(G_t)$$

as rings by Theorem 4.1.12(1). Let $H \in \mathbf{P}_o(G)$. By Theorem 2.4.4, $A(H)$ is a finitely generated projective $E(G)$-module, and by Theorem 4.2.11 there are indecomposable projective ideals $U_1, \ldots, U_s$ in $E(G)$ such that

$$A(H) \cong U_1 \oplus \cdots \oplus U_s.$$

Since $U_1 = U_1 E(G_1) \oplus \cdots \oplus U_1 E(G_t)$ is indecomposable there is an index, say 1, such that $U_1 = U_1 E(G_1)$. Moreover, each U_i is locally isomorphic to $E(G_i)$. By Theorem 2.4.4, there are $H_i \in \mathbf{P}_o(G_i)$ such that $U_i = A(H_i)$, H_i is locally isomorphic to G_i, and $H = H_1 \oplus \cdots \oplus H_s$. Thus G has the local refinement property. This completes the proof.

COROLLARY 4.2.14 *Let* $G_1, \ldots, G_t$ *be strongly indecomposable rtffr groups such that* $G_i \cong G_j \Rightarrow i = j$. *Suppose that* $G \doteq G_1 \oplus \cdots \oplus G_t$.

1. *If* $H \in \mathbf{P}_o(G)$ *then there are groups* $H_i \in \mathbf{P}_o(G_i)$ *for each* $i = 1, \ldots, t$ *such that* $H = H_1 \oplus \cdots \oplus H_t$.

2. G *has a unique decomposition.*

THEOREM 4.2.15 *Let $G_1, \ldots, G_t$ be strongly indecomposable rtffr groups such that $G_i \cong G_j \Rightarrow i = j$. Let $G \doteq G_1 \oplus \cdots \oplus G_t$ and assume that $E(G_i)$ is a commutative ring for each $i = 1, \ldots, t$. Then G has the local refinement property and a unique decomposition.*

Proof: By Theorem 4.1.15, $G = G'_1 \oplus \cdots \oplus G'_s$ for some nilpotent set $\{G'_1, \ldots, G'_s\}$ of indecomposable groups. By hypothesis and by Theorem 4.1.12(1),

$$\mathbb{Q}E(G) = \mathbb{Q}E(G_1) \times \cdots \times \mathbb{Q}E(G_t)$$

is a commutative ring, so $E(G'_i)$ is an indecomposable commutative ring for each $i = 1, \ldots, t$. Theorem 4.2.13 then states that G has the local refinement property and a unique decomposition.

A good source of commutative rings that occur naturally is the class of rtffr *E-rings*. The ring E is an *E-ring* if the map $\lambda : E \to \text{End}_{\mathbb{Z}}(E, +)$ sending $r \in E$ to left multiplication by r, $[\lambda(r)](x) = rx$, is an isomorphism. Little is known about groups G such that $\text{End}(G)$ is an *E-ring*.

LEMMA 4.2.16 *If E is an rtffr E-ring then E is commutative.*

Proof: The map $\lambda : E \to \text{End}(E, +)$ is an isomorphism so given a $y \in E$, the map $x \mapsto xy$ is also $\lambda(z)$ for some $z \in E$. Then $1y = \lambda(z)(1) = z$ and so $xy = yx$ for each $x \in E$.

LEMMA 4.2.17 *Let G be an rtffr group and let $E = \text{End}(G)$. Suppose that $G \cong E$. Then E is an E-ring.*

Proof: Since $G \doteq E$ we have $\text{End}(E) \doteq \text{End}(G) = E$. The reader can show that E is an *E-ring*.

LEMMA 4.2.18 *Let G be an rtffr group with semi-prime commutative endomorphism ring $E = \text{End}(G)$ such that $E \subset G \subset \mathbb{Q}E$. Then $G_p \doteq E_p$ for each prime $p \in \mathbb{Z}$. In particular $G/nG \cong E/nE$ as groups for each integer $n \neq 0$.*

Proof: Since E is semi-prime Lemma 1.6.3 states that there is an integrally closed ring $\overline{E}$ such that $E \subset \overline{E} \subset \mathbb{Q}E$ and $\overline{E}/E$ is finite. Let $\overline{G} = \overline{E}G$. Then $G \doteq \overline{G}$ so that

$$\overline{E} \subset \text{End}(\overline{G}) \doteq E \doteq \overline{E}.$$

Because $\overline{E}$ is integrally closed, $\overline{E} = \text{End}(\overline{G})$ and $\overline{G}$ is a flat $\overline{E}$-module. Evidently we can assume without loss of generality that $E = \overline{E}$ and that

$G = \overline{G}$. Furthermore, since the integrally closed ring $\overline{E}$ is a finite product of Dedekind domains, we may further assume that E is a Dedekind domain.

Let $p \in \mathbb{Z}$ be a prime and consider $E_p \subset G_p \subset \mathbb{Q}E$. Fix a maximal ideal $pE \subset M \subset E$. Since $p \in M$ we see that $(E_p)_M = E_M$, and hence

$$E_M \subset G_M \subset \mathbb{Q}E.$$

Since E is a Dedekind domain, E_M is a discrete valuation domain, so either $G_M = \mathbb{Q}E$ or $G_M = xE_M$ for some $x \in \mathbb{Q}E$.

Assume for the sake of contradiction that $G_M = \mathbb{Q}E$. Since M is a maximal ideal in E

$$G/MG \cong G_M/M_M G_M = \mathbb{Q}E/M_M\mathbb{Q}E = 0$$

so that $G = MG$. But E is commutative so by Theorem 4.2.4, $G \neq MG$ for maximal ideals $M \subset E$. This contradiction shows us that $G_M \neq \mathbb{Q}E$.

Thus $G_M = xE_M$ for some $x \in \mathbb{Q}E$ so there is an integer $k(M)$ such that $p^{k(M)}G_M \subset E_M$. Since $\mathcal{M} = \{\text{maximal ideals } M \mid p \in M \subset E\}$ is a finite set we can let

$$k = \sup\{k(M) \mid M \in \mathcal{M}\} < \infty.$$

Then

$$p^k G_M \subset E_M = (E_p)_M \quad \text{for each } M \in \mathcal{M}$$

so that

$$\begin{aligned}
p^k G_p \ &\subset \ \cap\{p^k G_M \mid p \in M \subset E \text{ is a maximal ideal}\} \\
&\subset \ \cap\{E_M \mid p \in M \subset E \text{ is a maximal ideal}\} \\
&= \ E_p \\
&\subset \ G_p
\end{aligned}$$

by the Local-Global Theorem 1.5.1. Consequently, $G_p \doteq E_p$. In particular $G/p^k G \cong E/p^k E$ as groups for each integer $k > 0$.

Let $n = p_1^{k_1} \cdots p_t^{k_t}$ be a product of powers of distinct primes. Then

$$G/nG \cong G/p_1^{k_1}G \oplus \cdots \oplus G/p_t^{k_t}G \cong E/nE$$

as groups. This completes the proof.

4.3   *E*-Properties

From these preliminary results we begin our investigation of the properties of G as a left E-module. The first *E-property* that we will discuss is *E-flat*. The group G is E-flat if G is a flat left E-module. Given a semi-prime commutative endomorphism ring we give a necessary and sufficient condition that the rtffr group G be E-flat.

THEOREM 4.3.1 [T.G. Faticoni] *Let G be an rtffr group such that $E = \text{End}(G)$ is commutative and assume that $E \subset G \subset \mathbb{Q}E$. The following are equivalent.*

1. $\text{Hom}(G, IG) = I$ *for each ideal I of finite index in E.*

2. *G is an E-flat group.*

Proof: $1 \Rightarrow 2$ Fix a prime p. By hypothesis $\text{ann}_E(G/IG) = I$ for each ideal I of finite index in E, and by Lemma 4.2.18

$$G_p/p^k G_p \cong G/p^k G \cong E/p^k E \cong E_p/p^k E_p$$

as groups for each integer $k > 0$. Then by Lemma 4.2.5, $G/p^k G \cong E/p^k E$ as E-modules. So there is an element $x \in G_p$ such that $E_p x + p^k G_p = G_p$. By Lemma 4.2.18, $G_p \doteq E_p$ is a finitely generated E_p-module so that $G_p \cong E_p x$ by Nakayama's Lemma 1.2.3. Since E_p is commutative $G_p = E_p x \cong E_p$. Thus G is locally flat and hence G is E-flat. This proves part 2.

$2 \Rightarrow 1$ Suppose that G is E-flat and fix a prime $p \in \mathbb{Z}$. By Lemma 4.2.18, G_p is an ideal of finite index in E_p. By hypothesis E_p is Noetherian and G_p is a flat left E_p-module, so G_p is a projective ideal of finite index in the commutative ring E_p. But then by Lemma 4.2.9, G_p is invertible, say $U_p G_p = E_p$.

Let $I \subset E$ be an ideal of finite index in E and let $J = \text{Hom}(G, IG)$. Then for each prime $p \in \mathbb{Z}$

$$J_p = JE_p = JU_p G_p = JG_p U_p = IG_p U_p = IE_p = I_p$$

so that $J = I$ by the Local-Global Theorem 1.5.1. This proves part 1 and completes the proof.

The examples will help illustrate the conditions in the above result.

EXAMPLE 4.3.2 Let **k** be a field extension of $\mathbb{Q}$ such that $[\mathbf{k} : \mathbb{Q}] = 3$, and let $\mathcal{O}$ denote the ring of algebraic integers in **k**. There are rational primes p, q such that $\mathcal{O}/p\mathcal{O}$ is a field of degree 3 over $\mathbb{Z}/p\mathbb{Z}$ and $q\mathcal{O} = MM'M''$ for distinct maximal ideals M, M', M'' in $\mathcal{O}$.

Since the extension $\mathbf{k}/\mathbb{Q}$ is minimal $\mathcal{O}_M \cap \mathcal{O}_p = \overline{E}$ is an E-ring. (See Appendix K.) Let

$$E = \mathbb{Z} + p\overline{E} \subset G \subset \overline{E}$$

where G is an E-submodule of $\overline{E}$ such that $G/p\overline{E}$ has dimension 2 over $\mathbb{Z}/p\mathbb{Z}$. Since $E \subset G \subset \overline{E}$, $E \subset \operatorname{End}(G) \doteq \overline{E}$, and since $\overline{E}$ is a Dedekind domain

$$E \subset \operatorname{End}(G) \subset \overline{E}.$$

Because $[\operatorname{End}(G)/p\overline{E} : E/p\overline{E}]$ is then a proper divisor of $[\overline{E}/p\overline{E} : E/p\overline{E}]$ $= 3$, $\operatorname{End}(G) = E$.

However G is not flat as follows. Since we chose G such that $G/p\overline{E}$ has $\mathbb{Z}/p\mathbb{Z}$-dimension 2, G is generated by at least 2 elements as an E-module. Thus $G_p/p\overline{E}_p \cong G/p\overline{E}$ is generated by two elements, so that G_p requires at least two generators over E_p. By our choice of E and p, E_p is a local ring with $\mathcal{J}(E_p) = p\overline{E}_p$. Finitely generated flat modules over the Noetherian domain E_p are projective, hence free E_p-modules. Thus any projective module $U \doteq E_p$ satisfies $U \cong E_p$. In particular U is cyclic. Since G_p is not cyclic, it is not flat over E_p, whence G is not E-flat.

EXAMPLE 4.3.3 Let E be a commutative semi-prime rtffr ring that is not hereditary. (There are plenty of these. Try to construct one as we did in the previous example.) There is an ideal $I \subset E$ that is not projective, hence not flat, so that $I \oplus E$ is not a flat left E-module. Evidently

$$E = \mathcal{O}(I \oplus E) = \{q \in \mathbb{Q}E \,|\, q(I \oplus E) \subset I \oplus E\}.$$

Then by Theorem 2.3.4 there is a short exact sequence

$$0 \to I \oplus E \longrightarrow G \longrightarrow \mathbb{Q}E \oplus \mathbb{Q}E \to 0$$

such that $E \cong \operatorname{End}(G)$. The fact that $\mathbb{Q}E$ is a flat left E-module implies that there is an isomorphism of functors

$$\operatorname{Tor}^1_E(I \oplus E, \cdot) \cong \operatorname{Tor}^1_E(G, \cdot).$$

Since $I \oplus E$ is not flat $\operatorname{Tor}^1_E(I \oplus E, \cdot) \neq 0$ so that G is not flat. Observe that $I \oplus E \subset G \not\subset \mathbb{Q}E$.

EXAMPLE 4.3.4 Let E be any noncommutative rtffr ring whose additive structure is a free group. Butler's Construction 2.3.1 shows us that there is a group G such that $\mathbb{Q}G = \mathbb{Q}E = \mathbb{Q}\mathrm{End}(G)$. Obviously $\mathrm{End}(G)$ is not commutative.

The next theorem illustrates a theme that we will come back to several times in this book. We show that a number of the *E-properties* that we are examining agree on rtffr groups that have semi-prime commutative endomorphism rings. Notice that the endomorphism ring in the next result need not be commutative or semi-prime.

THEOREM 4.3.5 *Let G be an rtffr group with endomorphism ring $E = \mathrm{End}(G)$. If $E \subset G \subset \mathbb{Q}E$ then the following are equivalent.*

1. $G \overset{.}{\cong} E$ as groups.

2. G is an E-finitely generated group.

3. G is an E-finitely presented group.

In this case $\mathrm{End}(G)$ is an E-ring.

Proof: 3 $\Rightarrow$ 2 is clear.

2 $\Rightarrow$ 1 Say $x_1, \ldots, x_n$ generate G as an E-module. Since $E \subset G \subset \mathbb{Q}E$ there is an integer $m \neq 0$ such that $mx_i \in E$ so that $mG \subset E \subset G$.

1 $\Rightarrow$ 3 Since $G \overset{.}{=} E$, Lemma 4.2.3 states that G is finitely presented. This proves 3 and completes the logical cycle.

We say that G is *locally E-free* iff $G_p \cong E_p$ as E-modules for each $p \neq 0 \in \mathbb{Z}$. The Local-Global Theorem 1.5.1 can be used to show that G is E-flat if G is locally E-free. The converse holds in the present situation.

THEOREM 4.3.6 *Let G be an rtffr group such that $E = \mathrm{End}(G)$ is commutative, and suppose that $E \subset G \subset \mathbb{Q}E$. Then G is an E-flat group iff G is locally E-free.*

Proof: If G is locally E-free then the above comment shows that G is E-flat.

Conversely suppose that G is E-flat. Fix a prime $p \in \mathbb{Z}$ and choose by Lemma 3.3.5 a multiple n of p so large that E_n is a reduced group. By Lemma 4.2.18, $G_n \overset{.}{=} E_n$, so by Lemma 4.2.3, G_n is a finitely presented ideal in E_n. Lemma 4.2.10 states that there is a map $\pi : G_n \to E_n$ such that $\pi(G_n) + nE_n = E_n$. By Lemma 1.2.1, $n \in \mathcal{J}(E_n)$, so Nakayama's Lemma 1.2.3 implies that $\pi(G_n) = E_n$. Inasmuch as G_n has finite rank, π is an isomorphism. Thus G is locally E-free.

The group G is E-*projective* if G is a projective left E-module, G is an E-*generator* if G generates the category of left E-modules, and G is an E-*progenerator* if G is E-finitely generated, E-projective, and an E-generator. The right ideal $I \subset E$ is *invertible* if $II^* = I^*I = E$ where $I^* = \{q \in \mathbb{Q}E \mid qI \subset E\}$. These E-*properties* coincide when $\text{End}(G)$ is a commutative semi-prime ring.

THEOREM 4.3.7 *Let G be an rtffr group with commutative semi-prime endomorphism ring $E = \text{End}(G)$. If $E \subset G \subset \mathbb{Q}E$ then the following are equivalent.*

1. *G is an E-projective group.*

2. *G is an E-generator group.*

3. *G is an E-progenerator group.*

4. *G is an invertible ideal in $\text{End}(G)$.*

In this case G is E-finitely presented and $\text{End}(G)$ is an E-ring.

Proof: $4 \Rightarrow 3$ An invertible ideal in a commutative ring is a progenerator.

$3 \Rightarrow 2$ is clear.

$2 \Rightarrow 1$ If G is a generator over E then G is projective as a module over $\text{End}_E(G)$. But $\text{End}_E(G) = \text{center}(E) = E$ since E is commutative. Thus G is E-projective.

$1 \Rightarrow 4$ By hypothesis G is a projective E-module such that $E \subset G \subset \mathbb{Q}E$. There is a cardinal c and a direct sum $E^{(c)} \cong G \oplus K$ of E-modules. Let $\{x_1, \ldots, x_n\} \subset G$ be a maximal linearly independent subset of G. There is a finite direct sum $E^{(m)}$ that contains the x_i so $G \subset E^{(m)}$. Since E is semi-prime, E is Noetherian whence G is finitely generated. Since $E \subset G \subset \mathbb{Q}E$, G/E is a finitely generated torsion E-module, so that G/E is finite. Hence G is a finitely generated projective ideal of finite index in E. Thus G is invertible which proves part 4 and completes the logical cycle.

EXAMPLE 4.3.8 Let G be the groups constructed in Example 4.3.2. Then $\text{End}(G)$ is a commutative nonhereditary integral domain. Observe that $G \doteq \text{End}(G)$ is finitely generated but that G is neither E-projective nor an E-generator.

EXAMPLE 4.3.9 (See [43].) Let E be any ring whose additive structure $(E, +)$ is a (locally) free group. Then $E = \text{End}(G)$ for some group G such that $E \subset G \subset \mathbb{Q}E$. In fact the reader can prove that $G_p \cong E_p$ for each prime $p \in \mathbb{Z}$ so that G is an E-flat group. But G is not E-finitely generated and not E-projective.

4.4 Square-Free Ranks

We investigate the strongly indecomposable rtffr groups whose rank is a square-free integer. We show that many E-*properties* coincide for these groups and that direct sums of nilpotent sets of these groups have the local refinement property.

4.4.1 Rank Two Groups

By a Theorem of J.D. Reid's below [10, Theorem 3.3], a strongly indecomposable torsion-free group of rank two has a commutative endomorphism ring. Here is a ready-made interesting class of rtffr groups to which our techniques will provide good results. We will refine the work in the previous section by proving the following result. Given a subgroup $X \subset Y$, let X_* be *the pure subgroup of Y* generated by X.

THEOREM 4.4.1 [43, T.G. Faticoni and H.P. Goeters] *A strongly indecomposable torsion-free group of rank two is E-flat.*

The proof of this theorem depends on the following classification of strongly indecomposable rank two groups.

THEOREM 4.4.2 [J.D. Reid] *Let G be a strongly indecomposable torsion-free group of rank two. Then $\mathbb{Q}\text{End}(G)$ satisfies one of the following conditions.*

1. $\mathbb{Q}\text{End}(G) = \mathbb{Q}$.

2. $\mathbb{Q}\text{End}(G)$ *is a field extension of $\mathbb{Q}$ of degree 2.*

3. $\mathbb{Q}\text{End}(G) = \mathbb{Q}[x]/(x^2)$ *where x is an indeterminant and where (x^2) is the ideal generated by x^2.*

In any case $\text{End}(G)$ is a commutative ring.

Proof of Theorem 4.4.1: Let $E = \text{End}(G)$. By Theorem 4.4.2, E is a commutative ring, so that $\mathbb{Q}E$ is a local commutative ring and $\mathbb{Q}G$ is a $\mathbb{Q}E$-module.

If $\mathbb{Q}E = \mathbb{Q}$ then E is a *pid* and hence G is a flat left E-module.

If $\mathbb{Q}E$ is a field extension of $\mathbb{Q}$ of degree 2 then $\mathbb{Q}E \cong \mathbb{Q}G$ as $\mathbb{Q}E$-modules since G has rank two. Thus we can assume without loss of generality that $E \subset G \subset \mathbb{Q}E$ as E-modules. We will show that G is locally flat.

Let $M \subset E$ be a maximal ideal in E and let p be the unique prime integer in M. Since $pE_M \subset M_M \subset E_M$ and since E_M/pE_M has $\mathbb{Z}_p/p\mathbb{Z}_p$-dimension at most 2, either $pE_M = M_M$ or $pE_M \neq M_M \subset E_M$.

Suppose that $pE_M = M_M$. Then

$$E_M \supset pE_M \supset p^2 E_M \supset \cdots$$

is a composition series for E_M so that E_M is a discrete valuation domain. Hence G_M is flat as an E_M-module.

Otherwise $pE_M \neq M_M \subset E_M$. By Lemma 4.2.18, $G_M \doteq E_M$, so that $G_M/pG_M \cong E_M/pE_M$ as groups. Since M_M/pE_M is the only proper nonzero ideal in the commutative ring E_M/pE_M, and since $G_M \neq M_M G_M$ (Nakayama's Lemma 1.2.3), we see that

$$M_M/pE_M = \text{ann}_{E_M/pE_M}(G_M/M_M G_M).$$

Thus Lemma 4.2.5 states that $G_M/M_M G_M = E_M(x + M_M G_M)$ for some $x \in G_M$. Another application of Nakayama's Lemma 1.2.3 shows us that

$$G_M = E_M x + M_M G_M = E_M x.$$

Since $\text{ann}_{E_M}(x) = \text{ann}_{E_M}(G_M) = 0$, $G_M \cong E_M$ is a flat left E-module.

The last case to consider is $\mathbb{Q}E = \mathbb{Q}[x]/(x^2)$. Observe that $\mathbb{Q}G$ is a $\mathbb{Q}E$-module of composition length 2. Since $\text{ann}_{\mathbb{Q}E}(\mathbb{Q}G) = 0$ Lemma 4.2.5 shows us that $\mathbb{Q}E = \mathbb{Q}G$ so that $E \subset G \subset \mathbb{Q}E$. Let $N = \mathcal{N}(E)G$ and notice that since $\mathcal{N}(E)^2 = 0$, $\mathcal{N}(E) = \text{Hom}(G/N_*, G)$. Since G/N_* and N are rank one groups $N = \text{Hom}(G/N_*, G)G/N_*$ is a pure subgroup of G. Thus given a prime $p \in \mathbb{Z}$ such that $pG \neq G$, we have $G_p/N_p \cong \mathbb{Z}_p$, so that $G_p = E_p x + N_p = E_p x + \mathcal{N}(E)_p G_p$. In the by now familiar manner $G_p = E_p x$ is a flat E-module.

In any case G is a locally flat E-module so that G is E-flat. This completes the proof.

An appeal to the above theorem and Theorems 4.3.5 and 4.3.7 will prove the following result.

THEOREM 4.4.3 *Let G be a strongly indecomposable torsion-free group of rank two. The following are equivalent.*

1. *$G \stackrel{.}{\cong} \text{End}(G)$ as groups.*

2. *G is an E-finitely generated group.*

3. *G is an E-finitely presented group.*

4. *G is an E-projective group.*

5. *G is an E-generator group.*

6. *G is an E-progenerator group.*

7. *G is an invertible ideal in $\text{End}(G)$.*

If one and hence all of the above conditions is true then $\text{End}(G)$ is an E-ring and $\mathbb{Q}\text{End}(G)$ is a field extension of $\mathbb{Q}$ of degree 2.

The next example addresses the hypotheses of the above theorem.

EXAMPLE 4.4.4 *Let $\mathbf{k} = \mathbb{Q}[\alpha]$ be a field extension of $\mathbb{Q}$ of degree 2 and let E denote the ring of algebraic integers in $\mathbf{k}$. Then $(E, +)$ is a free group of rank 2. By a construction of M.C.R. Butler 2.3.1 there is a group $E \subset G \subset \mathbf{k} = \mathbb{Q}E$ such that $E = \text{End}(G)$. We observe that E is not an E-ring, G and E are not quasi-isomorphic, but G is an E-flat group.*

4.4.2 Groups of Rank ≥ 3

By Lemma 1.3.3, G is strongly indecomposable iff $\mathbb{Q}\text{End}(G)$ is a local ring. In this case $\mathbb{Q}E(G) = \mathbb{Q}\text{End}(G)/\mathcal{N}(\mathbb{Q}\text{End}(G))$ is a division ring. Because $\mathbb{Q}E(G)$ is commutative if $\text{rank}(G)$ is square-free (Lemma 4.2.1) and since we can pass from G to $E(G)$ using the functor $A(\cdot)$ and Theorem 2.4.4, we can study G by applying some commutative ring theory to $E(G)$.

Our theorems can be applied to strongly indecomposable rtffr groups of rank three. Compare the next result to Reid's Theorem 4.4.2.

THEOREM 4.4.5 *Let G be a strongly indecomposable torsion-free group of rank 3, let $E = \text{End}(G)$, and let $\mathcal{N} = \mathcal{N}(\mathbb{Q}\text{End}(G))$. Then $\mathbb{Q}E$ falls into one of the following classes.*

1. *$\mathcal{N} = 0$ and $\mathbb{Q}E$ is a field extension of $\mathbb{Q}$ of degree 1 or 3.*

2. $\mathcal{N} \neq 0$ and $\mathbb{Q}$-dim$(\mathbb{Q}E) = 3$. Then $\mathbb{Q}E/\mathcal{N} = \mathbb{Q}$. Moreover

(a) If $\mathbb{Q}$-dim$(\mathbb{Q}E) = 3$ and $\mathcal{N}^2 \neq 0$ then $\mathbb{Q}E \cong \mathbb{Q}[x]/(x^3)$.

(b) If $\mathbb{Q}$-dim$(\mathbb{Q}E) = 3$ and $\mathcal{N}^2 = 0$ then $\mathcal{N} = \mathbb{Q} \oplus \mathbb{Q}$ as ideals in $\mathbb{Q}E$.

3. $\mathcal{N}^2 \neq 0$, $\mathbb{Q}$-dim$(\mathbb{Q}E) \geq 4$, and $\mathbb{Q}E/\mathcal{N} = \mathbb{Q}$.

In any case $\mathbb{Q}E/\mathcal{N}$ is a field extension of $\mathbb{Q}$ of degree 1 or 3.

Proof: Suppose that rank$(G) = 3$. Because G is strongly indecomposable of square-free rank, Theorem 1.3.3 states that $\mathbb{Q}E/\mathcal{N} = \mathbf{k}$ is a field extension of $\mathbb{Q}$ whose degree $[\mathbf{k} : \mathbb{Q}]$ divides $\mathbb{Q}G$ and $\mathcal{N}\mathbb{Q}G$.

1. Suppose that $\mathcal{N} = 0$. Then $\mathbf{k} = \mathbb{Q}E$ is a field extension of $\mathbb{Q}$ whose degree $[\mathbf{k} : \mathbb{Q}]$ divides $\mathbb{Q}$-dim$(\mathbb{Q}G) = 3$. Thus $[\mathbf{k} : \mathbb{Q}] = 1$ or 3.

2. Suppose that $\mathcal{N} \neq 0$ and that $\mathbb{Q}$-dim$(\mathbb{Q}E) = 3$. Then $[\mathbf{k} : \mathbb{Q}] < 3$ is a proper divisor of $\mathbb{Q}$-dim$(\mathbb{Q}G) = 3$ (Lemma 4.2.1), so that $[\mathbf{k} : \mathbb{Q}] = 1$. Thus $\mathbf{k} = \mathbb{Q}$. Consider the following cases.

2(a). Suppose that $\mathcal{N}^2 \neq 0$. Then

$$0 \subset \mathcal{N}^2 \subset \mathcal{N} \subset \mathbb{Q}E$$

is a chain of distinct subspaces of $\mathbb{Q}E$ which shows that $\mathcal{N}/\mathcal{N}^2 \cong \mathbb{Q}$ as $\mathbb{Q}$-vector spaces. By Nakayama's Lemma 1.2.3, $\mathcal{N} = \mathbb{Q}Ey$ for some $y \in \mathbb{Q}E$ such that $y^2 \neq 0$ and $y^3 = 0$. Evidently there is a surjection $f : \mathbb{Q}[x] \to \mathbb{Q}[y]$ such that $f(x) = y$ with ker $f = (x^3) =$ the ideal generated by x^3. Then a comparison of dimensions shows us that $\mathbb{Q}[x]/(x^3) \cong \mathbb{Q}[y]$.

2(b). Assume that $\mathbb{Q}$-dim$(\mathbb{Q}E) = 3$ and that $\mathcal{N}^2 = 0$. Then $\mathbf{k} \cong \mathbb{Q}$ implies that the ideal structure of $\mathcal{N}$ is that of a two dimensional $\mathbb{Q}$-vector space, say $\mathcal{N} \cong \mathbb{Q}x \oplus \mathbb{Q}y$ as ideals in $\mathbb{Q}E$. (Note: $\mathbb{Q}E$ need not be commutative. See the example below.)

3. Assume that $\mathcal{N}^2 \neq 0$ and that $\mathbb{Q}$-dim$(\mathbb{Q}E) \geq 4$. By Lemma 4.2.1, $[\mathbf{k} : \mathbb{Q}]$ divides $\mathbb{Q}$-dim$(\mathbb{Q}G) = 3$ and $\mathbb{Q}$-dim$(\mathcal{N}\mathbb{Q}G) < 3$, so $[\mathbf{k} : \mathbb{Q}] = 1$. This completes the proof.

EXAMPLE 4.4.6 Choose a commutative ring E whose additive structure is a free abelian group of rank 3 and such that $\mathbb{Q}E$ is a local ring. For example, let E be the algebraic integers in a field extension $\mathbf{k}$ of degree 3 over $\mathbb{Q}$. By Butler's Construction in Theorem 2.3.1, $E = \mathrm{End}(G)$ for some group G such that $E \subset G \subset \mathbb{Q}E$. G is strongly indecomposable because $\mathbb{Q}E$ is a local ring.

EXAMPLE 4.4.7 We construct a strongly indecomposable rank 3 group G such that $\text{End}(G)$ is a *noncommutative* ring of rank 3.

Let $E = R/I$ where

$$R = \left\{ \begin{pmatrix} a & 0 & 0 \\ b & a & 0 \\ d & c & a \end{pmatrix} \;\middle|\; a,b,c,d \in \mathbb{Z} \right\} \text{ and } I = \begin{pmatrix} 0 & 0 & 0 \\ 0 & 0 & 0 \\ \mathbb{Z} & 0 & 0 \end{pmatrix}.$$

Then E is a ring such that $(E, +)$ is a free group of rank 3. By Butler's Construction in Theorem 2.3.1, there is a group G such that $\mathbb{Q}E = \mathbb{Q}\text{End}(G)$ and $E \subset G \subset \mathbb{Q}E$. Since $\mathbb{Q}E$ is a local ring G is strongly indecomposable.

EXAMPLE 4.4.8 We construct a strongly indecomposable rank 3 group G such that $E = \text{End}(G)$ is a noncommutative ring of rank 4.

Let $M = R/I$ where

$$R = \left\{ \begin{pmatrix} a & 0 & 0 \\ b & a & 0 \\ d & c & a \end{pmatrix} \;\middle|\; a,b,c,d \in \mathbb{Z} \right\} \text{ and } I = \begin{pmatrix} 0 & 0 & 0 \\ 0 & 0 & 0 \\ 0 & \mathbb{Z} & 0 \end{pmatrix}.$$

Then $M = R/I$ is a left R-module such that $\text{ann}_R(M) = 0$, and whose additive structure is a free group of rank 3. A slight variation of Butler's Construction 2.3.1 produces a group G such that $M \subset G \subset \mathbb{Q}M$ and such that $\mathbb{Q}R \cong \mathbb{Q}\text{End}(G)$. Thus $\text{rank}(R) = 4$, $\text{rank}(G) = 3$, and because $\mathbb{Q}R$ is a local ring, G is strongly indecomposable.

For groups of rank 4 or 5 we have the following theorems.

The ring of *Hamiltonian quaternions over* $\mathbb{Q}$ is the $\mathbb{Q}$-vector space $\mathbb{H}$ with basis $\{1, i, j, k\}$ where we define $i^2 = j^2 = k^2 = ijk = -1$. Extend the multiplication linearly to give $\mathbb{H}$ a ring structure. Since $ij = -ji$ the Hamiltonian Quaternions form a *noncommutative domain*.

THEOREM 4.4.9 *Let G be a strongly indecomposable torsion-free group of rank 4, let $E = \text{End}(G)$ and let $\mathbf{k} = \mathbb{Q}E/\mathcal{N}(\mathbb{Q}E)$. Then $\mathbf{k}$ is a field extension of $\mathbb{Q}$ of degree 1, 2, or 4.*

1. *If $[\mathbf{k} : \mathbb{Q}] \leq 2$ then $\mathbf{k}$ is a commutative field.*

2. *If $[\mathbf{k} : \mathbb{Q}] = 4$ then either $\mathbf{k}$ is a commutative field or $\mathbf{k}$ is $\mathbb{H} =$ the noncommutative ring of Hamiltonian Quaternions over $\mathbb{Q}$.*

Proof: The proof is a repeated application of Lemma 4.2.1 and is left as an exercise for the reader.

EXAMPLE 4.4.10 Let E be a maximal $\mathbb{Z}$-order in the ring of Hamiltonian Quaternions. Then $(E, +)$ is a free group of rank 4 and E is a noncommutative domain. By Butler's Construction in Theorem 2.3.1, $E = \text{End}(G)$ for some group G such that $E \subset G \subset \mathbb{Q}E$. Compare to item 2 in the above theorem.

THEOREM 4.4.11 *Let G be a strongly indecomposable torsion-free group of rank 5, let $E = \text{End}(G)$, and let $\mathbf{k} = \mathbb{Q}E/\mathcal{N}(\mathbb{Q}E)$.*

1. $\mathbf{k}$ *is a commutative field of degree 1 or 5 over $\mathbb{Q}$.*

2. *If $\mathcal{N} \neq 0$ then $\mathbf{k} = \mathbb{Q}$ is commutative.*

THEOREM 4.4.12 *Let G be a strongly indecomposable torsion-free group of rank p for some prime $p \in \mathbb{Z}$, let $E = \text{End}(G)$, and let $\mathbf{k} = \mathbb{Q}E/\mathcal{N}(\mathbb{Q}E)$. Then*

1. $\mathbf{k}$ *is a commutative field of degree 1 or p over $\mathbb{Q}$.*

2. *If $\mathcal{N} \neq 0$ then $\mathbf{k} = \mathbb{Q}$ is commutative.*

EXAMPLE 4.4.13 An example of $\text{End}(G)$ for a strongly indecomposable rank 5 group.

Let R be a noncommutative ring such that $(R, +)$ is a free abelian group of rank 4 and let E be the ring $\{(z, r) \mid z \in \mathbb{Z}, r \in R\}$ where $(z, r)(z', r') = (zz', rz' + zr' + rr')$. Then E is an associative ring with identity $(1, 0)$. By Butler's Construction in Theorem 2.3.1 there is a group $E \subset G \subset \mathbb{Q}E$ such that $\mathbb{Q}E \cong \mathbb{Q}\text{End}(G)$. Since $\mathbb{Q}E$ is a local ring of $\mathbb{Q}$-dimension 5, G is a strongly indecomposable group of rank 5.

The indecomposable group G has *the refinement property* if given $H \in \mathbf{P}_o(G)$ then $H \cong G^m$ for some integer $m > 0$.

THEOREM 4.4.14 *A strongly indecomposable rtffr group having prime rank and having nonzero nilradical has the refinement property.*

Proof: Let $H \in \mathbf{P}_o(G)$. Recall the functor $A(\cdot)$ from Theorem 2.4.4. Then $A(H)$ is a finitely generated projective $E(G)$-module. By the above Theorem, $\mathbb{Q}E(G) = \mathbb{Q}$ when $\text{rank}(G) = p$ is prime and $\mathcal{N}(\text{End}(G)) \neq 0$, so that $E(G)$ is a pid. Hence $A(H)$ is a free $E(G)$-module of rank, say, m, whence $H \cong G^{(m)}$. This completes the proof.

4.5 Refinement and Square-Free Rank

We have shown in Lemma 4.2.1 that $E(G)$ is a commutative ring for strongly indecomposable groups of square-free rank. Then we can use the machinery that we have developed for direct sum decompositions of projective modules over commutative rtffr rings. The following results extend the Baer-Kulikov-Kaplansky Theorem 2.1.11 to direct sums of strongly indecomposable groups with square-free ranks. Recall *unique decomposition* from page 25 and the *local refinement property* from page 45. We give an extensive class of rtffr groups that have a locally unique decomposition.

THEOREM 4.5.1 [T.G. Faticoni] *Let $G_1, \ldots, G_t$ be strongly indecomposable rtffr groups of square-free rank such that $G_i \cong G_j \Rightarrow i = j$. If $G \doteq G_1 \oplus \cdots \oplus G_t$ then G has the local refinement property and G has a unique decomposition.*

Proof: By Lemma 4.2.1, $E(G_i)$ is commutative for each $i = 1, \ldots, t$, so by Theorem 4.2.15, G has the local refinement property and a unique decomposition.

Of course primes are square-free.

COROLLARY 4.5.2 *Let $G_1, \ldots, G_t$ be strongly indecomposable rtffr groups of prime ranks $p_1, \ldots, p_t$ such that $G_i \cong G_j \Rightarrow i = j$. If $G \doteq G_1 \oplus \cdots \oplus G_t$ then G has the local refinement property and G has a unique decomposition.*

COROLLARY 4.5.3 *Let $G_1, \ldots, G_t$ be strongly indecomposable rtffr groups of differing prime ranks. If $G \doteq G_1 \oplus \cdots \oplus G_t$ then G has the local refinement property and G has a unique decomposition.*

Notice in the next result that although we cannot conclude that G has a unique decomposition we can say that G has a *locally* unique decomposition.

THEOREM 4.5.4 *Let $G_1, \ldots, G_t$ be strongly indecomposable rtffr groups of square-free rank such that $G_i \cong G_j \Rightarrow i = j$, and let*

$$G = G_1 \oplus \cdots \oplus G_t. \tag{4.2}$$

(Note the equation here.) Then G has the local refinement property and a locally unique decomposition.

Proof: Suppose that G has a direct sum decomposition (4.2). By Theorem 4.5.1, $G_1 \oplus \cdots \oplus G_t$ has the local refinement property. The local refinement property implies a locally unique decomposition by Lemma 2.5.10. This completes the proof.

THEOREM 4.5.5 Let H be a strongly indecomposable rtffr group of square-free rank, let $n > 0$ be an integer, and let $G = H^{(n)}$.

1. Each $K \in \mathbf{P}_o(G)$ is a direct sum $K = K_1 \oplus \cdots \oplus K_s$ where each K_i is locally isomorphic to H.

2. $H^{(n)}$ is a locally unique decomposition of G.

The following examples will show that in some sense our results are the best possible.

EXAMPLE 4.5.6 There is a set $\{G_1, G_2\}$ of quasi-isomorphic rank two groups such that $G = G_1 \oplus G_2$ does not have the local refinement property. Theorem 2.2.1 states that there are acd groups G_1, G_2, G_3, G_4, such that

1. $G = G_1 \oplus G_2 \cong G_3 \oplus G_4$

2. $\operatorname{rank}(G_1) = \operatorname{rank}(G_2) = 2$, $\operatorname{rank}(G_3) = 1$, $\operatorname{rank}(G_4) = 3$.

Evidently $\{G_1, G_2\}$ is not a nilpotent set. Such a group G fails to have the local refinement property. Examination of the construction of G_1 and G_2 shows that $E(G_i) = \mathbb{Z}$ is commutative for $i = 1, 2$.

EXAMPLE 4.5.7 There is a strongly indecomposable torsion-free group of rank 4 that does not have the local refinement property.
     Proof: Let $\mathcal{O}$ denote a classical maximal $\mathbb{Z}$-order in the ring of Hamiltonian Quaterions over $\mathbb{Q}$. Then $\mathcal{O}/3\mathcal{O} \cong M_2(\mathbb{Z}/3\mathbb{Z})$. Let $3\mathcal{O} \subset E \subset \mathcal{O}$ be such that $E/3\mathcal{O} \cong \begin{pmatrix} \mathbb{Z}/3\mathbb{Z} & 0 \\ \mathbb{Z}/3\mathbb{Z} & \mathbb{Z}/3\mathbb{Z} \end{pmatrix}$. It is known (see, e.g., [69]) that E is a hereditary Noetherian prime ring. Let I be the ideal $3\mathcal{O} \subset I \subset E$ such that $I/3\mathcal{O} = \begin{pmatrix} 0 & 0 \\ \mathbb{Z}/3\mathbb{Z} & \mathbb{Z}/3\mathbb{Z} \end{pmatrix}$. Then $I^2 = I$ is a projective ideal. Since $I/3\mathcal{O}$ does not generate $\begin{pmatrix} \mathbb{Z}/3\mathbb{Z} & 0 \\ 0 & 0 \end{pmatrix}$, I is not a generator for E. Therefore I is not locally isomorphic to E. It is clear from the above construction that I is generated by two elements so we can write $E \oplus E \cong I \oplus J$.

By Corner's Theorem 2.3.3, there is a group G such that $E = \mathrm{End}(G)$. Then by the Arnold-Lady-Murley Theorem 2.4.1, $I \otimes_E G \cong IG \in \mathbf{P}_o(G)$, $G \oplus G \cong IG \oplus JG$, and IG is not locally isomorphic to G. Thus G does not have a locally unique decomposition. The reader can show that $\mathrm{End}(IG) = \mathrm{End}_{\mathcal{O}}(I) = \mathcal{O}$.

The above example illustrates the type of failure of local refinement that occurs whenever we are given an rtffr ring E and a finitely generated projective right ideal I that does not generate E. Such a projective right ideal I always exists in a hereditary rtffr ring E that is not integrally closed.

4.6 Hereditary Endomorphism Rings

There is another interesting class of rtffr groups G for which each $H \in \mathbf{P}_o(G)$ is a direct sum of strongly indecomposables.

THEOREM 4.6.1 Let $\{G_1, \ldots, G_t\}$ be a set of strongly indecomposable rtffr groups such that $G_i \cong G_j \Longrightarrow i = j$. Assume that $E(G_i)$ is a hereditary domain for each $i = 1, \ldots, t$ and let

$$G = G_1 \oplus \cdots \oplus G_t.$$

Given $H \in \mathbf{P}_o(G)$ there are integers $m_1, \ldots, m_t \geq 0$ and indecomposable groups $H_{ij} \in \mathbf{P}_o(G_i)$ such that

$$H_{ij} \stackrel{\cdot}{\cong} G_i \quad \text{for each } j = 1, \ldots, m_i$$

and such that

$$H = [\bigoplus_{j=1}^{m_1} H_{1j}] \oplus \cdots \oplus [\bigoplus_{j=1}^{m_t} H_{tj}].$$

Proof: Let $H \in \mathbf{P}_o(G)$. Since $\{G_1, \ldots, G_t\}$ is a nilpotent set there are $H_i \in \mathbf{P}_o(G_i)$ such that

$$H = H_1 \oplus \cdots \oplus H_t$$

so we may as well assume that $G = G_1$ and that $H = H_1$.

Under these reductions $H \in \mathbf{P}_o(G)$ and G is strongly indecomposable. Since $H \oplus H' \cong G^{(n)}$ for some integer n, Jónsson's Theorem 2.1.10 states that $H \stackrel{\cdot}{\cong} G^{(m)}$ for some integer $m > 0$. Then

$$A(H) \stackrel{\cdot}{\cong} E(G)^{(m)}$$

as right $E(G)$-modules. Since $E(G)$ is a hereditary domain

$$A(H) \cong U_1 \oplus \cdots \oplus U_m$$

where $U_i \doteq E(G)$ for each $i = 1, \ldots, m$. By Theorem 2.4.4 we can write

$$H \cong H_1 \oplus \cdots \oplus H_m$$

where $A(H_i) \cong U_i$ for each $i = 1, \ldots, m$. Finally since $A(\cdot)$ preserves quasi-isomorphism classes, $A(H_i) \doteq E(G)$ implies that $H_i \doteq G$. This completes the proof.

4.7   Exercises

Let E be an rtffr ring, let G, H be rtffr groups.

1. Let E be a ring and let P be a right E-module. Then P is finitely generated projective over E iff P is a generator in the category of left $\mathrm{End}_E(P)$-modules.

2. Let M be a finitely presented right E-module. Show that M is a projective right E-module iff M_I is a projective right E_I-module for each maximal ideal $I \subset S = \mathrm{center}(E)$.

3. Let $\mathcal{O}$ denote the ring of algebraic integers in the field $\mathbb{Q}[\sqrt{-5}]$.

   (a) Show by brute force that the ideal $I = \langle 2, 1 + \sqrt{-5} \rangle$ in $\mathcal{O}$ is a maximal nonprincipal ideal.

   (b) Show that $2\mathcal{O} = I_1^2$.

   (c) Conclude that $I \oplus I \cong \mathcal{O} \oplus \mathcal{O}$ so that $\mathcal{O}$ has the local refinement property but not the refinement property.

4. Let E be the ring constructed in Example 4.5.7 and let I be the maximal ideal.

   (a) Show that E is hereditary.

   (b) Show that I is an ideal generated by two elements.

   (c) Show that I is not a generator.

5. Let E be a hereditary prime rtffr ring and let $\overline{E}$ be a classical maximal order such that $E \subset \overline{E}$ and $\overline{E}/E$ is finite.

   (a) Show that $\overline{E}$ is an ideal over E that is projective but not a generator over E.

(b) Show that there is an idempotent ideal Δ of finite index in E such that $\mathcal{O}(\Delta) = \overline{E}$.

(c) Show that the hereditary prime rtffr ring E is a classical maximal order iff E and 0 are the only idempotent ideals in E.

6. Show that if E is an integral domain then E and 0 are the only idempotent ideals in E.

7. Construct an indecomposable rtffr subring $E \subset \mathbb{Q} \times \mathbb{Q}$ such that each finitely generated projective E-module is a generator for E and $E \cong \mathbb{Z} \oplus \mathbb{Z}$ as group.

8. Let S be a commutative rtffr ring and let U be a finitely generated ideal of finite index in S. Show that U is projective iff U is invertible iff $U_p \cong S_p$ for each prime p.

9. Let S be a commutative rtffr ring and let U be a finitely generated ideal of finite index in S. Show that U is invertible iff U is locally isomorphic to S.

10. Let S be a discrete valuation domain with field of fractions $\mathbb{Q}S$. If $S \subset M \subset \mathbb{Q}S$ and if $M \neq \mathbb{Q}S$ then $M = Sx$ for some $x \in \mathbb{Q}S$.

11. Prove Fitting's Lemma. If G is a left E-module of finite composition length and if $f \in \mathrm{End}(G)$ then there is an integer $k > 0$ such that $G = \ker f^k \oplus \mathrm{image}\, f^k$ where f is a nilpotent endomorphism of $\ker f^k$ and f is an idempotent endomorphism of image f^k.

12. If G is a torsion-free reduced p-local group of rank n and if $r_p(G) = n$ then $G \cong \mathbb{Z}_p^{(n)}$.

13. Let E be a commutative ring and let M be a projective E-module. Then M generates E.

4.8 Questions for Future Research

Let G be an rtffr group, let E be an rtffr ring, and let $S = \mathrm{center}(E)$ be the center of E. Let τ be the conductor of G. There is no loss of generality in assuming that G is strongly indecomposable.

1. Give a general theory of nilpotent sets for infinite sets of groups or modules. See [34].

2. Extend the theory in chapter 4 on the direct sum properties of nilpotent sets $G = G_1 \oplus \cdots \oplus G_t$ to direct sums in which $G_i \doteq G_j$ for each $i \neq j$. See [13].

3. Find a condition on G, more general than square-free rank, that implies commutativity in $\mathbb{Q}\mathrm{End}(G)$ or $\mathbb{Q}\mathrm{E}(G)$.

4. Let S be a commutative rtffr ring. Discuss the localization of S at its maximal ideals and at multiplicatively closed sets contained in S.

5. Characterize the rtffr groups G that possess the (local) refinement property.

6. Characterize the rtffr groups G that possess the (locally) unique decomposition property.

7. Discuss the direct sum of strongly indecomposable rank two groups. These rank two groups have commutative endomorphism rings.

8. Just what is an E-flat group?

9. Discuss the structure of strongly indecomposable groups G of square-free rank. Extend that discussion to a large class of rtffr groups.

10. Discuss the local-global theory of (*almost completely decomposable*) rtffr groups.

11. Give a general ideal lattice for the nilradical of a strongly indecomposable rtffr group. Specifically, generalize Theorems 4.4.2, 4.4.5, and 4.4.9 to larger ranks.

12. Find some kind of localization theory between the group G and S, the center of $\mathrm{End}(G)$.

13. Discuss the decomposition of groups G such that $E(G)$ is hereditary.

14. Let G be an rtffr group. If $E(G)$ is a hereditary ring then how does the ideal structure of $E(G)$ affect G. For example, each right ideal is projective in $E(G)$. There are at most finitely many isomorphism classes of right ideals in $E(G)$. At most finitely many maximal right ideals are not progenerators for $E(G)$. Discuss how these properties affect the properties of direct summands of G.

15. Extend the results in chapter 4 to include rtffr groups G such that $E(G)$ is integrally closed. That is, $E(G)$ is a product of classical maximal orders.

Chapter 5

Refinement Revisited

Recall that an indecomposable rtffr group G has the *refinement property* if given a direct sum $G^{(n)} \cong H \oplus H'$ of groups then $H \cong G^{(m)}$ for some integer $m > 0$. We showed in Theorem 4.2.13 that if $E(G)$ is a commutative ring then the indecomposable rtffr group G possesses the *local refinement property*. That is, given a direct sum $G^{(n)} \cong H \oplus H'$ of groups then $H \cong G_1 \oplus \cdots \oplus G_m$ for some groups $G_1, \ldots, G_m$ that are locally isomorphic to G. This brings two questions to mind.

1. If G has the local refinement property under what conditions does G have the refinement property?

2. If G has the local refinement property under what conditions does a subgroup H of finite index in G have the local refinement property?

We will investigate these problems in this chapter. For example, let G be a group such that $E(G)$ has finite index in a Dedekind domain $\overline{E}$. Then G has the local refinement property. But if $\overline{E}$ is not a *pid* then we will show that G has the local refinement property but it does not have the refinement property. By the Baer-Kulikov-Kaplansky Theorem 2.1.11, if $\overline{G}$ is a completely decomposable group then $\overline{G}$ has the refinement property. However Example 2.2.1 shows that it is not uncommon for a subgroup G of finite index in $\overline{G}$ to fail to possess the local refinement property. We will give conditions on $\overline{G}$ under which subgroups of finite index in $\overline{G}$ have the refinement property when $\overline{G}$ has the refinement property.

5.1   Counting Isomorphism Classes

5.1.1   Class Groups and Class Numbers

The *integral closure* $\overline{E}$ of the integral domain E is the largest subring $\overline{E} \subset \mathbb{Q}E$ such that $E \subset \overline{E}$ and such that $\overline{E}/E$ is finite. If $E = \overline{E}$ then E is an *integrally closed ring*. For example, each subring of $\mathbb{Q}$ is integrally closed. If $i^2 = -1$ and if $E = \{a + 2bi \mid a, b \in \mathbb{Z}\}$ then the reader can show that $\overline{E} = \mathbb{Z}[i]$. Given a Dedekind domain $\overline{E}$ and a nonzero ideal $I \subset E$ then the integral closure of $E = I + \mathbb{Z}$ is $\overline{E}$. It is known that the integral closure $\overline{E}$ of an rtffr integral domain E is a Dedekind domain, [10, Corollaries 10.12, 11.4].

For the remainder of this section we let E
be an rtffr integral domain with integral closure $\overline{E}$.

The nonzero ideal $I \subset E$ is *invertible* if $I^*I = II^* = E$ where

$$I^* \;=\; \{q \in \mathbb{Q}E \mid qI \subset E\}.$$

By Lemma 4.2.9 a nonzero ideal I in the rtffr integral domain E is invertible iff it is projective. By Theorem 4.2.12, each finitely generated projective module over the commutative Noetherian integral domain E is a direct sum of invertible ideals.

Fix a *strongly indecomposable rtffr group* G such that $E(G)$ is a commutative ring, or equivalently an *rtffr integral domain*. This is the case, e.g., when $\operatorname{rank}(G)$ is a square-free integer. Theorem 4.2.13 shows us that G has the local refinement property. We define

---

$\Gamma(G) = \{\text{isomorphism classes } (H) \mid G \text{ is locally isomorphic to } H\}$

---

and we define *the class number of G* as

$$h(G) = \operatorname{card}(\Gamma(G)).$$

If E is an integral domain then a *fractional ideal of E* is a nonzero E-submodule of the field of fractions of E. If we define $IJ = \{\sum_{\text{finite}} x_i y_i \mid x_i \in I \text{ and } y_i \in J\}$ for any fractional ideals I, J of E then

$$\gamma(E) = \text{the set of invertible fractional ideals of E}$$

is a group with $I^{-1} = I^*$ and identity E.

Let $\rho(E) \subset \gamma(E)$ denote the set of principal fractional ideals of E. Evidently $\rho(E)$ is a subgroup of $\gamma(E)$. We define the *class group of E* to be

$$\Gamma(E) = \gamma(E)/\rho(E).$$

It is readily shown that fractional ideals I and J are isomorphic iff there is a $q \in \mathbb{Q}E$ such that $qI = J$. Then the coset $I\rho(E)$ in $\Gamma(E)$ is the isomorphism class (I) of I. Thus

$$\Gamma(E) = \{\text{isomorphism classes } (I) \mid I \text{ is an invertible ideal in } E\}$$

and we define *the class number of E* to be

$$h(E) = \text{card}(\Gamma(E)).$$

The integral domain $\overline{E}$ is a Dedekind domain so each ideal in $\overline{E}$ is invertible. Thus $\Gamma(\overline{E})$ is the set of isomorphism classes of nonzero ideals in $\overline{E}$, and $h(\overline{E})$ is the number of isomorphism classes of nonzero ideals in $\overline{E}$.

The next result shows us that $h(G)$ is finite. The proof is due to E.L. Lady and can be found in [10, Theorem 11.11, page 137].

THEOREM 5.1.1 [E.L. Lady] *Let G be an rtffr group. There are at most finitely many isomorphism classes of groups H that are locally isomorphic to G. Consequently $h(G) < \infty$.*

THEOREM 5.1.2 [Jordan-Zassenhaus] *(See [10, Theorem 13.13, page 155].) Let E be an rtffr integral domain. There are at most finitely many isomorphism classes of nonzero ideals in E. Consequently $h(E) < \infty$.*

Recall that the *conductor of $\overline{E}$ into E* is

$$\tau = \text{ the largest ideal } \tau \text{ in } E \text{ such that } \tau\overline{E} \subset E \subset \overline{E} .$$

From section 3.2, τ is an ideal of $\overline{E}$ contained in E. Our investigations will relate the local refinement property in strongly indecomposable rtffr groups G to the class number of the integral closure $\overline{E}$ of $E(G)$.

LEMMA 5.1.3 *Let G be an rtffr group. Then $h(G) = h(E(G))$.*

Proof: Given a group H that is locally isomorphic to G, Lemma 2.5.6 implies that $G \oplus G \cong H \oplus H'$ for some group H'. Then $H \in \mathbf{P}_o(G)$. By Lemma 2.5.1, $A(H)$ is locally isomorphic to $A(G)$. Then by Lemma 2.4.4, $A(\cdot)$ induces a functorial bijection $\alpha : \Gamma(G) \to \Gamma(E(G))$. Thus $h(G) = h(E(G))$. The reader will fill in the elementary details.

Our investigation of $h(G)$ for the rtffr group G is then translated into an investigation of $h(E)$ for an rtffr integral domain E.

5.1.2   Class Group of the Integral Closure

The integral domain E is an *rtffr integral domain* if its additive structure $(E, +)$ is an rtffr group. In this section E denotes an rtffr integral domain with integral closure $\overline{E}$. We will prove the following result.

THEOREM 5.1.4 [T.G. Faticoni] *Let E be an rtffr integral domain. Then $\gamma(\overline{E})$ is a free abelian group, and there is a group surjection $\alpha : \gamma(E) \longrightarrow \gamma(\overline{E})$. Thus $\gamma(E) \cong \gamma(\overline{E}) \oplus \ker \alpha$.*

Proof: Each fractional ideal I is uniquely the product of powers of finitely many maximal ideals of $\overline{E}$, [69, page 47, Theorem 4.8]. Thus $\gamma(\overline{E})$ is a free group on the set of maximal ideals of $\overline{E}$.

Define $\alpha(J) = J\overline{E}$ for each $J \in \gamma(E)$. Suppose that J is an invertible ideal in E. Because $\mathbb{Q}E$ is commutative and because $J^*J = E$, $J^*(J\overline{E}) = \overline{E}$. Thus $J\overline{E}$ is an invertible ideal in $\overline{E}$, hence α is well defined.

Inasmuch as $\mathbb{Q}E$ is commutative, we have $IJ\overline{E} = I\overline{E} \cdot J\overline{E}$, for any ideals $I, J \subset E$, so that α is a group homomorphism.

It remains to show that α is a surjection. The proof is a series of lemmas.

LEMMA 5.1.5 *Let $T \neq 0$ be an ideal in E and let I be an invertible ideal in E. There is an ideal $I' \subset E$ such that $I' \cong I$ and $I' + T = E$.*

Proof: Since E is an integral domain $T \doteq E$, and by Lemma 4.2.9, I is a projective E-module. Then by part 1 of Lemma 4.2.10 there is a map $f : I \to E$ such that $f(I) + T = E$. Since E is an integral domain f is an injection so letting $I' = f(I) \cong I$ completes the proof.

LEMMA 5.1.6 *Let E be an rtffr integral domain with conductor τ and integral closure $\overline{E}$. Let $J \subset E$ be an ideal such that $J + \tau = E$.*

1. *If $\tau \subset M \subset E$ is a maximal ideal then $J_M = E_M$.*

2. *If $\tau \not\subset M \subset E$ is a maximal ideal then $\tau_M = E_M = \overline{E}_M$. Thus J_M is a principal ideal in $\overline{E}_M$.*

3. *J is an invertible ideal in E.*

Proof: 1. Given a maximal ideal $\tau \subset M \subset E$ then $J_M + \tau_M = E_M$ and $\tau_M \subset M_M = \mathcal{J}(E_M)$. By Nakayama's Lemma 1.2.3, $E_M = J_M$.

2. Given a maximal ideal $M \subset E$ such that $\tau \not\subset M$ then $\tau + M = E$ so that $\tau_M + M_M = E_M$. Since $M_M = \mathcal{J}(E_M)$, $\tau_M = E_M$, and so the inclusion $\tau\overline{E} \subset E \subset \overline{E}$ becomes

$$\overline{E}_M = E_M\overline{E}_M = \tau_M\overline{E}_M \subset E_M \subset \overline{E}_M.$$

Thus $E_M = \overline{E}_M$. It follows that J_M is a projective ideal in the Dedekind domain $\overline{E}_M$. Since $\overline{E}_M$ is then a local domain, J_M is free, and since $\mathbb{Q}J_M \subset \mathbb{Q}\overline{E}_M$, $\mathbb{Q}J_M \oplus V = \mathbb{Q}\overline{E}_M$ for some vector space V over $\mathbb{Q}\overline{E}_M$. Inasmuch as the field $\mathbb{Q}\overline{E}_M$ is indecomposable $J_M \cong \overline{E}_M$.

3. In parts 1 and 2 we have shown that J_M is a projective ideal in E_M for each maximal ideal $M \subset E$. Then by the Local-Global Theorem 1.5.1, J is a projective ideal in E. By Lemma 4.2.9, J is an invertible ideal in E. This completes the proof of the lemma.

LEMMA 5.1.7 *Let E be an rtffr integral domain with integral closure $\overline{E}$. Let $\overline{J} \subset \overline{E}$ be an ideal such that $\overline{J} + \tau = \overline{E}$. Then $\overline{J} \cap E$ is an invertible ideal in E such that $(\overline{J} \cap E) + \tau = E$.*

Proof: Because $\tau \subset E$ the modular law implies that $(\overline{J} + \tau) \cap E = (\overline{J} \cap E) + \tau = E$. By part 3 of Lemma 5.1.6, $\overline{J} \cap E$ is then an invertible ideal in E.

LEMMA 5.1.8 *Let E be an rtffr integral domain with conductor τ and with integral closure $\overline{E}$. Let $\overline{J} \subset \overline{E}$ be an ideal such that $\overline{J} + \tau = \overline{E}$. Then $[\overline{J} \cap E]\overline{E} = \overline{J}$.*

Proof: Let $\overline{J} \subset \overline{E}$ be an ideal such that $\overline{J} + \tau = \overline{E}$ and let $J = \overline{J} \cap E$. By the previous Lemma, $J + \tau = E$. Given a maximal ideal $\tau \subset M \subset E$ we have $J_M = E_M$ by Lemma 5.1.6(1). Furthermore because $\tau_M \subset M_M = \mathcal{J}(E_M)$ we have

$$\overline{E}_M = \overline{J}_M + \tau_M = \overline{J}_M + \tau_M \overline{E}_M = \overline{J}_M$$

by Nakayama's Lemma 1.2.3. Then

$$(J\overline{E})_M = J_M \overline{E}_M = E_M \overline{E}_M = \overline{E}_M = \overline{J}_M.$$

Given a maximal ideal $\tau \not\subset M \subset E$ then Lemma 5.1.6(2) implies that J_M is an ideal in $E_M = \overline{E}_M$. Then

$$(J\overline{E})_M = J_M \overline{E}_M = J_M = \overline{J}_M \cap E_M = \overline{J}_M \cap \overline{E}_M = \overline{J}_M$$

so that $J\overline{E} = \overline{J}$ by the Local-Global Theorem 1.5.1.

Continuation of the Proof of Theorem 5.1.4: Let $\overline{J}$ be a fractional ideal of $\overline{E}$. There is a copy $\overline{J}'$ of $\overline{J}$ such that $\overline{J}' + \tau = \overline{E}$ (Lemma 5.1.5). Thus there is a $q \in \mathbb{Q}\overline{E}$ such that $\overline{J} = q\overline{J}'$. Moreover by Lemma 5.1.8, $\overline{J}' \cap E$ is an invertible ideal of E such that $(\overline{J}' \cap E)\overline{E} = \overline{J}'$. Then $J = q(\overline{J}' \cap E)$ is an invertible ideal of $\overline{E}$ such that $J\overline{E} = q\overline{J}' = \overline{J}$. Therefore $\alpha : \gamma(E) \longrightarrow \gamma(\overline{E})$ is a surjection, and hence $\gamma(E) \cong \gamma(\overline{E}) \oplus \ker \alpha$. This completes the proof of Theorem 5.1.4.

It is clear that the principal fractional ideals $J = aE$ of E are mapped to principal fractional ideals $\alpha(J) = a\overline{E}$ of $\overline{E}$. Because $\Gamma(\overline{E})$ is formed by taking the quotient of $\gamma(\overline{E})$ modulo the subgroup of principal fractional ideals, α induces a group surjection

$$\beta : \Gamma(E) \longrightarrow \Gamma(\overline{E}).$$

THEOREM 5.1.9 Let E be an rtffr integral domain with integral closure $\overline{E}$. Then $\beta : \Gamma(E) \longrightarrow \Gamma(\overline{E})$ is a surjection of finite abelian groups.

Proof: By the Jordan-Zassenhaus Theorem 5.1.2, $\Gamma(E)$ and $\Gamma(\overline{E})$ are finite abelian groups. Thus β is a group surjection of finite abelian groups.

COROLLARY 5.1.10 [T.G. Faticoni] Let G be a strongly indecomposable rtffr group such that $E(G)$ is a commutative integral domain with integral closure $\overline{E}$. Then $h(\overline{E})$ divides $h(G)$.

Proof: By Theorem 5.1.9, there is a surjection β of finite groups. Then by Lagrange's Theorem $h(E(G)) = \mathrm{card}(\Gamma(E(G)))$ is divisible by $\mathrm{card}(\Gamma(\overline{E})) = h(\overline{E})$. This completes the proof.

5.1.3   Class Number and Refinement

We continue to investigate $h(G)$ when G is a strongly indecomposable rtffr group such that $E(G)$ is a commutative integral domain with integral closure $\overline{E}$. We say that G has the $\mathcal{L}$-property if G is isomorphic to each group that is locally isomorphic to G. The $\mathcal{L}$ stands for E.L. Lady who first used local isomorphism on rtffr groups. See [56] where local isomorphism is called *near* isomorphism.

THEOREM 5.1.11 *Let G be a strongly indecomposable rtffr group such that $E(G)$ is an integral domain. The following are equivalent.*

1. $h(G) = 1$.

2. $h(E(G)) = 1$.

3. G has the $\mathcal{L}$-property.

4. G has the refinement property.

Proof: $1 \Leftrightarrow 2$ is Lemma 5.1.3.

$1 \Leftrightarrow 3$ Let $h(G) = 1$ and suppose that G is locally isomorphic to H. Since $h(G) = 1$ implies that there is only one isomorphism class in $\Gamma(G)$, $(G) = (H)$ or equivalently $G \cong H$. The converse is proved in the same way.

$2 \Rightarrow 4$ By Theorem 2.4.7 we can prove that G has the refinement property if we prove that $E(G)$ has the refinement property. Suppose that $P \oplus P' \cong E(G)^{(n)}$ for some integer n. Since $E(G)$ is an rtffr integral domain Theorem 4.2.15 states that $E(G)$ has the local refinement property so that $P = J_1 \oplus \cdots \oplus J_r$ for some invertible ideals $J_1, \ldots, J_r$ of $E(G)$. Since $h(E(G)) = 1$, $E(G) \cong J_i$ for each $i = 1, \ldots, r$, and hence $E(G)$ has the refinement property.

$4 \Rightarrow 3$ Let G be locally isomorphic to H. Since G is strongly indecomposable Theorem 3.5.1 implies that H is indecomposable. By Lemma 2.5.6 there is a direct sum $G \oplus G \cong H \oplus H'$ of groups. Since G has the refinement property $G \cong H$, so that G has the $\mathcal{L}$-property. This proves part 3 and completes the logical cycle.

Let G be a strongly indecomposable rtffr group. We say that G has the *proper local refinement property* if G has the local refinement property but G does not have the refinement property. That is, there are isomorphic direct sums $G^{(n)} \cong H \oplus K$ such that in no direct sum decomposition of $H = G_1 \oplus \cdots \oplus G_n$ is each G_i isomorphic to G. The group G has the *proper $\mathcal{L}$-property* if there is a group that is locally isomorphic to G but that is not isomorphic to G.

THEOREM 5.1.12 *Let G be a strongly indecomposable rtffr group. Assume that $E(G)$ is an rtffr integral domain with integral closure $\overline{E(G)}$.*

1. $h(G) \neq 1$.

2. $h(\overline{E(G)}) \neq 1$.

3. G has the proper local refinement property.

4. G has the proper $\mathcal{L}$-property.

The next result shows us that the proper $\mathcal{L}$-property and the proper local refinement property are inherited by subgroups of finite index. This kind of genetic behavior is uncommon in the study of rtffr groups.

THEOREM 5.1.13 [T.G. Faticoni] *Let G be a strongly indecomposable rtffr group such that $E(G)$ is a commutative integral domain. Let $\overline{E}$ be the integral closure of $E(G)$.*

1. *If $h(\overline{E}) \neq 1$ then G has the proper local refinement property.*

2. $h(\overline{E}) = 1$ *if G has the refinement property.*

Proof: Part 2 is the contrapositive of part 1. Suppose that $h(\overline{E}) > 1$. Since $h(\overline{E})$ divides $h(G)$, $h(G) > 1$, whence G does not have the refinement property, but G has the proper refinement property. This proves the Theorem.

It is rare to find a property for an rtffr group G that is passed onto other groups in the quasi-equality class of G.

COROLLARY 5.1.14 [T.G. Faticoni] *Let $\overline{G}$ be a strongly indecomposable rtffr group such that $E(\overline{G})$ is a Dedekind domain, and assume that $\overline{G}$ has the proper local refinement property. If G is quasi-isomorphic to $\overline{G}$ then G has the proper local refinement property.*

Since $E(G)$ is a commutative integral domain when G is a strongly indecomposable rtffr group of square-free rank we have proved the following corollary.

COROLLARY 5.1.15 [T.G. Faticoni] *Let G be a strongly indecomposable rtffr group of square-free rank. Let $\overline{E}$ be the integral closure of $E(G)$.*

1. *If $h(\overline{E}) \neq 1$ then G has the proper local refinement property.*

2. *$h(\overline{E}) = 1$ if G has the refinement property.*

Recall from Lemma 3.1.8 that if G is an rtffr group then there is an rtffr group $G \subset \overline{G}$ such that $\overline{G}/G$ is finite, $E(\overline{G})$ is an integrally closed integral domain, $E(G) \subset E(\overline{G})$, and $E(\overline{G})/E(G)$ is finite.

THEOREM 5.1.16 *Let $\overline{G}$ be a strongly indecomposable integrally closed rtffr group of square-free rank. If $\overline{G}$ has the proper local refinement property and if $G \doteq \overline{G}$ then G has the proper refinement property.*

Proof: Suppose that $\overline{G}$ has the proper refinement property. By Theorem 5.1.11, $h(E(\overline{G})) \neq 1$. Now $E(\overline{G})$ is integrally closed and $E(\overline{G})/E(G)$ is finite so $E(\overline{G})$ is the integral closure of $E(G)$. Hence Corollary 5.1.15 states that G has the proper refinement property.

THEOREM 5.1.17 *Let $G_1, \ldots, G_t$ be strongly indecomposable rtffr groups of square-free ranks such that $G_i \cong G_j \Rightarrow i = j$. For each $i = 1, \ldots, t$ let $\overline{E}_i$ be the integral closure of $E(G_i)$, and let $G = G_1^{(n_1)} \oplus \cdots \oplus G_t^{(n_t)}$ for some integers $n_1, \ldots, n_t > 0$.*

1. *If some $\overline{E}_i$ is not a pid then G has the proper local refinement property.*

2. *If G has the refinement property then $\overline{E}_i$ is a pid for each $i = 1, \ldots, t$.*

Proof: Part 1 is the contrapositive of part 2.

2. Suppose that G has the refinement property. One shows that G_i has the refinement property for each $i = 1, \ldots, t$, so by Theorem 5.1.12, $\overline{E}_i$ is a *pid* for each $i = 1, \ldots, t$.

The above theorems connect the refinement property for strongly indecomposable rtffr groups to the class number problem for Dedekind domains in algebraic number fields. Thus a classification of the strongly indecomposable rtffr groups that have the refinement property would naturally lead to a classification of the algebraic number fields whose class number is 1. The determination of the class number of an algebraic number field is generally regarded as a difficult problem. See [69]. Thus it would seem that the refinement property will be a fun one but a hard one to investigate.

5.1.4   Quadratic Number Fields

It is natural to ask for $h(G)$ when $\mathbb{Q}E(G)$ is a small field. Since there is a rather complete description of quadratic number fields $\mathbf{k}$ such that $h(\mathbf{k}) = 1$ we will look there for a classification of strongly indecomposable rtffr groups G that have the refinement property.

PROPOSITION 5.1.18 *Let G be a strongly indecomposable rtffr group.*

1. *If $\mathbb{Q}E(G) = \mathbb{Q}$ then G has the refinement property.*

2. *If $\mathrm{rank}(G)$ is prime and if $\mathcal{N}(\mathrm{End}(G)) \neq 0$ then G has the refinement property.*

Proof: Part 1. follows from the fact that each subring of $\mathbb{Q}$ is a *pid*.

2. By hypothesis and Lemma 4.2.1, $\mathbb{Q}$-dim$(\mathbb{Q}E(G))$ is a proper divisor of the prime $\mathbb{Q}$-dim$(\mathbb{Q}G)$ so that $\mathbb{Q}E(G) = \mathbb{Q}$. Now use part 1.

LEMMA 5.1.19 *Let $\mathcal{O}$ be the ring of algebraic integers in a quadratic number field $\mathbb{Q}[\sqrt{d}]$ for some square-free integer $d \neq 0$. Let E be a subring of finite index in $\mathcal{O}$.*

1. *Suppose that $d < 0$. Then $h(E) = 1$ if $d \in \{-1, -2, -3, -7, -11, -19, -43, -67, -163\}$.*

2. *Suppose that $1 < d < 100$ is a square-free integer. Then $h(E) \neq 1$ iff $d \in X = \{10, 15, 26, 30, 34, 35, 39, 42, 51, 55, 58, 65, 66, 70, 74, 78, 79, 82, 85, 87, 91, 95\}$.*

Proof: 1. By [76, Theorem 10.5], the negative integers $d \in \{-1, -2, -3, -7, -11, -19, -43, -67, -163\}$ are the only negative integers for which the ring $\mathcal{O}$ of algebraic integers in $\mathbb{Q}[\sqrt{d}]$ is a *pid*. Since $\mathcal{O}$ is the integral closure of E, Theorem 5.1.9 implies that $h(\mathcal{O}) \neq 1$ divides $h(E) \neq 1$.

2. By [76, Theorem 10.5], the 22 values in X are exactly the square-free integers $1 < d < 100$ such that $h(\mathcal{O}) \neq 1$. Now apply Theorem 5.1.12. This completes the proof.

THEOREM 5.1.20 [T.G. Faticoni] *Let G be a strongly indecomposable rtffr group of rank 2.*

1. *If $\mathbb{Q}\text{End}(G) = \mathbb{Q}$ or $\mathcal{N}(\text{End}(G)) \neq 0$ then G has the refinement property.*

2. *Otherwise, $\mathbb{Q}\text{End}(G) = \mathbb{Q}[\sqrt{d}]$ for some square-free integer $d \neq 0$. Assume that $\text{End}(G) \doteq \mathcal{O}$ where $\mathcal{O}$ is the ring of algebraic integers in $\mathbb{Q}\text{End}(G)$. Then G has the proper refinement property provided that either*

    (a) *$d \notin \{-1, -2, -3, -7, -11, -19, -43, -67, -163\}$ or*

    (b) *$1 < d < 100$ and $d \in \{10, 15, 26, 30, 34, 35, 39, 42, 51, 55, 58, 65, 66, 70, 74, 78, 79, 82, 85, 87, 91, 95\}$.*

Proof: 1. Our assumptions amount to assuming that $\mathbb{Q}E(G) = \mathbb{Q}$ so that $E(G)$ is a *pid* in this case. Thus the groups G in this case have the refinement property.

2. Since $E(G) \doteq \mathcal{O}$, $\mathcal{O}$ is the integral closure of $E(G)$. By the previous theorem $h(\mathcal{O}) \neq 1$ in case d satisfies parts (a) or (b). By Corollary 5.1.15, G has the proper refinement property.

Reader be warned: Suppose $\mathcal{O}$ is the ring of algebraic integers in *any* algebraic number field. Choose any nonzero ideal $I \subset \mathcal{O}$. Then regardless of the class number of $\mathcal{O}$, $\mathcal{O}_I$ is a *pid*.

5.1.5 Counting Theorems

The following result is central to our discussion.

THEOREM 5.1.21 [I. Kaplansky] (See [69]). *Let E be a Noetherian integral domain and let $I_1, \ldots, I_m, J_1, \ldots, J_n$ be invertible ideals in E. Then*

$$I_1 \oplus \cdots \oplus I_m \cong J_1 \oplus \cdots \oplus J_n \iff m = n \text{ and } I_1 \cdots I_m \cong J_1 \cdots J_n.$$

Proof: ($\iff$) follows from [69, Theorem 38.13].

Given an integer $n > 0$, a strongly indecomposable rtffr group G with square free rank, and an integral domain E let

$$
\begin{aligned}
\nu_n(G) \;=\; & \text{card}\{\text{unordered } n\text{-tuples } \{H_1, \ldots, H_n\} \mid \\
& \text{each } H_i \text{ is an indecomposable rtffr group,} \\
& \text{and } G^{(n)} \cong H_1 \oplus \cdots \oplus H_n\} \\
\nu_n(E) \;=\; & \text{card}\{\text{unordered } n\text{-tuples } \{I_1, \ldots, I_n\} \mid \\
& \text{each } I_i \text{ is an invertible ideal in } E, \\
& \text{and } E^{(n)} \cong I_1 \oplus \cdots \oplus I_n\}
\end{aligned}
$$

and let

$$
\begin{aligned}
e(G) \;=\; & \text{least integer } e > 0 \text{ such that } x^e = 1 \\
& \text{for all } x \in \Gamma(E(G)) \\
\ell(G) \;=\; & \text{least integer } t > 0 \text{ such that } \Gamma(E(G)) \\
& \text{is a direct sum of } t \text{ cyclic groups, say} \\
& \Gamma(E(G)) = \langle (J_1) \rangle \times \cdots \times \langle (J_t) \rangle
\end{aligned}
$$

Observe that $e(G), \ell(G) \le h(G) \le e(G)\ell(G)$. If $\Gamma(E(G))$ is a cyclic group then $\ell(G) = 1$ and $e(G) = h(E(G)) = h(G)$. We will prove the following theorem.

THEOREM 5.1.22 [T.G. Faticoni] *Let G be a strongly indecomposable rtffr group of square free rank, and let $n > 0$ be an integer.*

1. *If $\Gamma(E(G))$ is a cyclic group then $\nu_n(G) \le h(G)^{n-1}$.*

2. *In general, $\nu_n(G) \le e(G)^{n\ell(G)} \le h(G)^{n\ell(G)}$.*

The proof is a series of lemmas. Let

$$
\begin{aligned}
K_n(G) \;=\; & \{\text{ordered } m\text{-tuples } (I_1, \cdots, I_n) \mid I_1, \ldots, I_n \\
& \text{are invertible ideals in } E \text{ such that} \\
& E^{(n)} \cong I_1 \oplus \cdots \oplus I_n\}.
\end{aligned}
$$

LEMMA 5.1.23 *Let E be an rtffr integral domain, and let $n > 0$ be an integer. Then*

1. *There is a short exact sequence*

$$0 \longrightarrow K_n(E) \longrightarrow \Gamma(E)^{(n)} \xrightarrow{\ \mu\ } \Gamma(E) \longrightarrow 0 \qquad (5.1)$$

   of abelian groups where $\mu((I_1), \ldots, (I_n)) = (I_1 \cdots I_n)$.

2. *$\nu_n(E) \leq \operatorname{card}(K_n(E)) = h(E)^{n-1}$.*

Proof: 1. Given $(I_1, \ldots, I_n) \in \ker \mu$ then $E \cdots E = E \cong I_1 \cdots I_n$ so that

$$E^{(n)} \cong I_1 \oplus \cdots \oplus I_n$$

by Theorem 5.1.21. Thus $\ker \mu \subset K_n(E)$.

Conversely given $(I_1, \ldots, I_n) \in K_n(E)$ then

$$E^{(n)} \cong I_1 \oplus \cdots \oplus I_n$$

so that $E \cong I_1 \cdots I_n$ by Theorem 5.1.21. Thus $(I_1, \ldots, I_n) \in \ker \mu$, so that $K_n(E) = \ker \mu$.

2. By part 1 there is a short exact sequence (5.1) so that Lagrange's Theorem shows us that

$$h(E)^n = \operatorname{card}(\Gamma(E)^{(n)}) = h(E) \cdot \operatorname{card}(K_n(E)).$$

That is, $\operatorname{card}(K_n(E)) = h(E)^{n-1}$. Furthermore there is an obvious surjection from $K_n(E)$ onto the set of unordered n-tuples $\{I_1, \ldots, I_n\}$ such that $E^{(n)} \cong I_1 \oplus \cdots \oplus I_n$. Then $\nu_n(E) \leq \operatorname{card}(K_n(E))$. This completes the proof of the lemma.

Lemma 5.1.23(2) proves part 1 of **Theorem 5.1.22** once we recall that $h(G) = h(E(G))$.

LEMMA 5.1.24 *Let E be a Noetherian integral domain, and let $n > 0$ be an integer. Then*

$$\nu_n(E) \leq e(E)^{n\ell(E)}.$$

Proof: Let $t = \ell(E)$ and write the finite abelian group $\Gamma(E)$ as

$$\Gamma(E) = \langle (J_1) \rangle \times \cdots \times \langle (J_t) \rangle$$

for some fixed invertible ideals $J_1, \ldots, J_t \subset E$. Then $(J_i)^{e(G)} = (E)$ for each $i = 1, \ldots, t$ so that each $(I) \in \Gamma(E)$ can be written as

$$I = J_1^{e_1} \cdots J_t^{e_t} \quad \text{for some integers} \quad 0 \leq e_i < e(G).$$

Form the set

$$\mathrm{Mat}(E) \;=\; \{n \times t \text{ matrices } (e_{ij})_{n \times t} \;\mid\; 0 \le e_{ij} < e(G) \text{ for each}$$
$$1 \le i \le n \text{ and } 1 \le j \le t\}.$$

Since each element of $\Gamma(E)$ is a product of powers of $(J_1), \ldots, (J_t)$, for each element $(I_1, \ldots, I_n) \in K_n(E)$ there is an $(e_{ij})_{n \times t} \in \mathrm{Mat}(E)$ such that

$$I_i = J_1^{e_{i1}} \cdots J_t^{e_{it}}.$$

There is then a (set theoretic) surjection $\mathrm{Mat}(E) \longrightarrow K_n(E)$ such that

$$\nu_n(E) = \mathrm{card}(K_n(E)) \le \mathrm{card}(\mathrm{Mat}(E)) \le e(E)^{nt} = e(E)^{n\ell(E)}.$$

This completes the proof of the lemma.

LEMMA 5.1.25 *Let G be a strongly indecomposable rtffr group of square free rank. Then*

$$\nu_n(G) = \nu_n(E(G)).$$

Proof: Let G be a strongly indecomposable rtffr group of square free rank. Suppose that $G^{(n)} = H_1 \oplus \cdots \oplus H_n$ for some indecomposable rtffr groups $H_1, \ldots, H_n$. Then by Theorem 2.4.4

$$E(G)^{(n)} = A(H_1) \oplus \cdots \oplus A(H_n)$$

where $A(H_i)$ is an indecomposable projective right $E(G)$-module. Since $E(G)$ is a Noetherian integral domain, Theorem 4.2.12 states that $A(H_i)$ is an invertible ideal in $E(G)$. Thus $\nu_n(G) \le \nu_n(E(G))$.

Conversely suppose that $E(G)^{(n)} = I_1 \oplus \cdots \oplus I_n$ for some invertible ideals I_i in $E(G)$. By Theorem 5.1.21, $E(G) \cong I_1 \cdots I_n$, and Theorem 2.4.4 implies that there are indecomposable groups H_i such that $A(H_i) = I_i$ and

$$G^{(n)} \cong H_1 \oplus \cdots \oplus H_n.$$

Then $\nu_n(E(G)) \le \nu_n(G)$ and it follows that $\nu_n(G) = \nu_n(E(G))$.

Proof of Theorem 5.1.22(2): By definition and Lemma 5.1.25

$$e(G) = e(E(G)), \quad \ell(G) = \ell(E(G)), \quad \nu_n(G) = \nu_n(E(G))$$

so that by Lemma 5.1.24

$$\nu_n(G) = \nu_n(E(G)) \le e(E(G))^{n\ell(E(G))} = e(G)^{n\ell(G)}.$$

This completes the proof.

As an application of the above counting results we have a power cancellation result.

THEOREM 5.1.26 [T.G. Faticoni] *Let G be a strongly indecomposable rtffr group of square free rank. If $e(G)$ and n are relatively prime then*

$$G^{(n)} \cong H^{(n)} \implies G \cong H.$$

Proof: Suppose that $G^{(n)} \cong H^{(n)}$. By Lemma 2.5.7, G and H are locally isomorphic. Theorem 2.4.4 implies that $A(H)$ is an ideal in $E(G)$ such that $E(G)^{(n)} \cong A(H)^{(n)}$. By Theorem 5.1.21, $E(G) \cong A(H)^n$, so that the class $(A(H)) \in \Gamma(E(G))$ has order r dividing n. Since r must also divide $e(G)$, and since $e(G)$ and n are relatively prime, $r = 1$. Hence $E(G) \cong A(H)$, and therefore $G \cong H$.

THEOREM 5.1.27 *Let G be an rtffr group such that $\Gamma(E(G))$ is a p-group for some prime p. Then $G^{(n)} \cong H^{(n)} \implies G \cong H$ for each integer n not divisible by p.*

Consider now some small values of n. Specifically we will show that these estimates are very good when $n = 2$. Let

$$h_n(G) = \operatorname{card}\{x \in \Gamma(E(G)) \mid x^n = 1\}.$$

THEOREM 5.1.28 [T.G. Faticoni] *Let G be a strongly indecomposable rtffr group of square free rank.*

1. $h_n(G) =$ *the number of isomorphism classes of groups H such that $G^{(n)} \cong H^{(n)}$.*

2. $h(G) - h_2(G) =$ *the number of unordered pairs $\{(H), (K)\}$ of distinct isomorphism classes of rtffr groups such that $G \oplus G \cong H \oplus K$.*

Proof: 1. By Theorem 2.4.4 it suffices to assume that $G = E(G)$. So suppose that $E(G)^{(n)} \cong I^{(n)}$ for some invertible ideal $I \subset E(G)$. Theorem 5.1.21 states that $E(G) \cong I \cdots I$ (n factors), so that $(I) \in \{x \in \Gamma(E) \mid x^n = 1\}$. Hence the number of isomorphism classes of ideals $I \subset E(G)$ such that $E(G)^{(n)} \cong I^{(n)}$ is at most $h_n(G)$. The converse is proved by a similar argument.

2. Suppose that $G \oplus G \cong H \oplus K$ as groups and that $H \not\cong K$. By Theorem 2.4.4, $E(G) \oplus E(G) \cong A(H) \oplus A(K)$ and $A(H) \not\cong A(K)$. Then by Theorem 5.1.21, $A(H)A(K) \cong E(G)$, or in other words $(A(H)) \neq (A(K))$ and $(A(H)) = (A(K))^{-1}$. Hence

$$
\begin{aligned}
\mathrm{card}\{(x, x^{-1}) \mid x^{-1} \neq x \in \Gamma(E(G))\} &= \text{\# of } x \text{ such that } x^2 \neq 1 \\
&= h(E(G)) - h_2(G) \\
&= h(G) - h_2(G)
\end{aligned}
$$

by Lemma 5.1.25. This completes the proof.

THEOREM 5.1.29 [T.G. Faticoni] *Let G be a strongly indecomposable rtffr group of square free rank. If $h(G)$ is an odd integer then*

1. *$G \oplus G \cong H \oplus H$ implies that $G \cong H$.*

2. *The number of sets $\{H, K\}$ of nonisomorphic rtffr groups such that $G \oplus G \cong H \oplus K$ is $h(G) - 1$.*

Proof: 1. Suppose that $G \oplus G \cong H \oplus H$. The group H corresponds to an element $(I) \in \Gamma(\overline{E})$ such that $I \cdot I \cong \overline{E}$. Since $\mathrm{card}(\overline{E}) = h(\overline{E}) = h(G)$ is odd, $I \cong \overline{E}$. By Theorem 2.4.4, $H \cong G$.

2. Since $h(G)$ is odd, $h_2(G) = 1$. Apply Theorem 5.1.28 to complete the proof.

5.2 Integrally Closed Groups

The rings in this section need not be commutative and the groups need not be strongly indecomposable.

Recall that the rtffr group G is *integrally closed* group if $E(G)$ is an integrally closed ring. We show in Lemma 3.1.8 that each rtffr group has finite index in an integrally closed group. We continue our theme that groups possessing the (local) refinement property and a locally unique decomposition are more common than once believed by identifying a naturally occurring class of subgroups of finite index in integrally closed groups that possess the local refinement property and a locally unique decomposition. Our goal should be contrasted with Example 2.2.1. Said example begins with a completely decomposable group $\overline{G}$ and constructs a subgroup G of finite index in $\overline{G}$. By Theorem 2.1.11, $\overline{G}$ possesses a unique decomposition and it has the refinement property, but there are indecomposable decompositions $G = G_1 \oplus G_2 = H_1 \oplus H_2$ such that $\text{rank}(G_1) = \text{rank}(G_2) = 2$ and $\text{rank}(H_1) = 1$. This juxtaposition of uniqueness with nonuniqueness invites us to study further. We can make some sense of this seemingly counterintuitive coincidence by examining direct sum decompositions of *semi-primary groups*. Definitions follow.

5.2.1 Review of Notation and Terminology

Fix a nilpotent set $\{G_1, \cdots, G_t\}$ of indecomposable rtffr groups, fix integers $t, n_1, \cdots, n_t > 0$, and let

$$G = G_1^{(n_1)} \oplus \cdots \oplus G_t^{(n_t)}. \tag{5.2}$$

Nilpotency implies that $G_i \cong G_j \Rightarrow i = j$. We will make extensive use of the ideal structure of the semi-prime rtffr ring $E(G)$, and the functor

$$A(\cdot) \ : \ \mathbf{P}_o(G) \longrightarrow \mathbf{P}_o(E(G))$$

from Theorem 2.4.4. Then because $\{G_1, \ldots, G_t\}$ is a nilpotent set (Theorem 4.1.8) we have

$$
\begin{aligned}
E(G) &= A(G_1)^{(n_1)} \oplus \cdots \oplus A(G_t)^{(n_t)} \\
&= \text{Mat}_{n_1}(E(G_1)) \times \cdots \times \text{Mat}_{n_t}(E(G_t)).
\end{aligned}
$$

There is at least one interesting class of rtffr ring that has the refinement property.

PROPOSITION 5.2.1 *The rtffr group G with conductor τ has the local refinement property and a locally unique decomposition if $E_\tau(G)$ is a semi-perfect ring.*

Proof: Given a semi-perfect ring $E_\tau(G)$, Lemma 1.7.1 and the AKS Theorem 2.1.6 state that $E_\tau(G)$ has the refinement property. Then Theorem 3.5.4 implies that G has the local refinement property. Lastly apply Lemma 2.5.10.

Thus we are motivated to look for semi-perfect rtffr rings of the form $E_\tau(G)$. We will show that there are some fairly common rtffr groups G, called *semi-primary groups*, for which $E_\tau(G)$ is semi-perfect.

We showed in Lemma 4.2.1 that $E(G)$ is commutative if G is strongly indecomposable and if rank(G) is a square-free integer. Thus there are plenty of rtffr groups G with direct sum decomposition (5.2) such that

$$\mathbb{Q}E(G) = \mathrm{Mat}_{n_1}(\mathbf{k}_1) \times \cdots \times \mathrm{Mat}_{n_t}(\mathbf{k}_t)$$

where $\mathbf{k}_i$ is the field of fractions of the integral domain $E(G_i)$.

5.2.2   Locally Semi-Perfect Rings

Let E be a semi-prime rtffr ring with conductor τ. We give some fairly general conditions on τ and E that make E_τ semi-perfect.

5.2.2 Let $S = \mathrm{center}(E)$ and choose an integrally closed ring $\overline{E} \subset \mathbb{Q}E$ such that $\tau\overline{E} \subset E \subset \overline{E}$. Then

$$\overline{E} = \overline{E}_1 \times \cdots \times \overline{E}_t$$

where each $\overline{E}_i$ is a classical maximal order in the field $\mathbf{k}_i$. Let

$$\mathrm{center}(\overline{E}) = \overline{S} = \overline{S}_1 \times \cdots \times \overline{S}_t$$

where $\overline{S}_i = \mathrm{center}(\overline{E}_i)$ is a Dedekind domain for each $i = 1, \ldots, t$. As in section 3.2, τ is an ideal in $\overline{S}$ so we can write

$$\tau = \tau_1 \oplus \cdots \oplus \tau_t \quad \text{where } 0 \neq \tau_i \subset \overline{S}_i \text{ is an ideal of finite index.}$$

Recall that an ideal $I \neq 0$ in a commutative ring S is a *primary ideal* if S/I is a local ring with nilpotent Jacobson radical. For instance, $q\mathbb{Z}$ is a primary ideal in $\mathbb{Z}$ iff q is a prime power. The primary ideals in a Dedekind domain are powers of maximal ideals.

THEOREM 5.2.3 *Suppose that E is a semi-prime rtffr ring with conductor $\tau = \tau_1 \oplus \cdots \oplus \tau_t$ as in (5.2.2). Then E has the local refinement property and a locally unique decomposition if E satisfies the following two conditions:*

1. There are integers $t, n_1, \ldots, n_t > 0$ and fields $\mathbf{k}_1, \ldots, \mathbf{k}_t$ such that

$$\mathbb{Q}E = \mathrm{Mat}_{n_1}(\mathbf{k}_1) \times \cdots \times \mathrm{Mat}_{n_t}(\mathbf{k}_t)$$

and

2. τ_i is a primary ideal in $\overline{S}_i$ for each $i = 1, \ldots, t$.

Proof: The proof is a series of lemmas. The notations and conditions of the Theorem are used without fanfare. The first result is called the Change of Rings Theorem. I leave the proof as an exercise in the nature of long exact sequences and localizations.

THEOREM 5.2.4 *[72, page 107, Theorem 3.84] Let U be a finitely generated S-module and let I be an ideal in S. Then*

$$(\mathrm{End}_S(U))_I \cong \mathrm{End}_{S_I}(U_I)$$

as rings.

LEMMA 5.2.5 *Idempotents in $E_\tau / E_\tau \tau_\tau$ lift to idempotents of E_τ.*

Proof: We first show that $(\overline{E}_i)_\tau$ is a semi-perfect ring.
Fix $i \in \{1, \ldots, t\}$. By condition 1 of Theorem 5.2.3

$$\mathbb{Q}\overline{S}_i = \mathrm{center}(\mathbb{Q}\overline{E}_i) = \mathrm{center}(\mathrm{Mat}_{n_i}(\mathbf{k}_i)) = \mathbf{k}_i.$$

Let V_i be a simple left module over the simple Artinian ring $\mathrm{Mat}_{n_i}(\mathbf{k}_i)$ and let $U_i = \overline{E}_i v_i$ be a cyclic left $\overline{E}_i$-submodule of V_i. Then

$$\mathbb{Q}U_i = \mathbb{Q}\overline{E}_i v_i = V_i \cong \mathbf{k}_i^{(n_i)}$$

as $\mathbf{k}_i$-vector spaces. Since $\overline{S}_i$ is a Dedekind domain, since $\mathbb{Q}\overline{E}_i$ is a torsion-free S_i-module, and since $\overline{E}_i$ is a finitely generated torsion-free $\overline{S}_i$-module we see that U_i is a finitely generated projective $\overline{S}_i$-module. Thus $\mathrm{End}_{\overline{S}_i}(U_i)$ is a finitely generated torsion-free $\overline{S}_i$-module. Furthermore

$$\overline{E}_i \subset \mathrm{End}_{\overline{S}_i}(U_i) \subset \mathrm{End}_{\mathbf{k}_i}(\mathbb{Q}U_i) = \mathrm{End}_{\mathbf{k}_i}(\mathbf{k}_i^{(n_i)}) = \mathrm{Mat}_{n_i}(\mathbf{k}_i)$$

and $\overline{E}_i$ is a classical maximal order in $\mathrm{Mat}_{n_i}(\mathbf{k}_i)$. Since $\mathrm{End}_{\overline{S}_i}(U_i)/\overline{E}_i$ is then a finitely generated torsion $\overline{S}_i$-module, $\overline{E}_i = \mathrm{End}_{\overline{S}_i}(U_i)$, and since U_i is a finitely generated $\overline{S}_i$-module Theorem 5.2.4 implies that

$$(\overline{E}_i)_\tau = (\overline{E}_i)_{\tau_i} \cong \mathrm{End}_{(\overline{S}_i)_{\tau_i}}(U_{\tau_i}).$$

Condition 2 states that τ_i is a primary ideal in $\overline{S}_i$ so that

$$\frac{\overline{S}_i}{\tau_i} \cong \frac{(\overline{S}_i)_{\tau_i}}{(\tau_i)_{\tau_i}}$$

is a local ring. Since $(\tau_i)_{\tau_i} \subset \mathcal{J}((S_i)_{\tau_i})$, $(\overline{S}_i)_{\tau_i}$ is a local ring.

Moreover U is a finitely generated projective $\overline{S}_i$-module. Since U_{τ_i} is then a free module over the local ring $(\overline{S}_i)_{\tau_i}$ $U_{\tau_i} \cong (\overline{S}_i)_{\tau_i}^{(n_i)}$. Then

$$(\overline{E}_i)_\tau \cong \mathrm{End}_{(\overline{S}_i)_{\tau_i}}(U_{\tau_i}) \cong \mathrm{Mat}_{n_i}((\overline{S}_i)_{\tau_i})$$

is a semi-perfect ring. As claimed $\overline{E}_\tau$ is semi-perfect.

Now, let $e \in E_\tau$ be such that $e^2 - e \in \tau_\tau E_\tau$. Inasmuch as $\tau \overline{E} \subset E \subset \overline{E}$, setting $I = \tau_\tau^2 \overline{E}_\tau$ gives us

$$I = \tau_\tau(\tau_\tau \overline{E}_\tau) \subset \tau_\tau E_\tau \subset \mathcal{J}(\overline{E}_\tau).$$

Since idempotents lift modulo the nilpotent ideal $\tau_\tau E_\tau / I \subset E_\tau / I$, there is an $(f + I)^2 = f + I \in E_\tau / I \subset \overline{E}_\tau / I$ such that $e - f \in \tau_\tau E_\tau$. By the above claim $\overline{E}_\tau$ is semi-perfect and $I \subset \mathcal{J}(\overline{E}_\tau)$, so by Lemma 1.7.4 there is a $g^2 = g \in \overline{E}_\tau$ such that $f + I = g + I$. Hence $g \in f + I \subset E_\tau$, whence idempotents in $E_\tau / \tau_\tau E_\tau$ lift to idempotents in E_τ. This completes the proof of the lemma.

The rtffr ring E is *semi-local* iff there is some integer $n \neq 0$ such that each maximal ideal of E contains n. For example, if E is a semi-prime rtffr ring with conductor τ then E_τ is a semi-local ring.

LEMMA 5.2.6 *Let E be a semi-prime rtffr ring with conductor τ that satisfies conditions 1 and 2 of Theorem 5.2.3. Then E_τ is a semi-perfect ring.*

Proof: Suppose that E satisfies conditions 1 and 2. Since $\tau_\tau E_\tau \subset \mathcal{J}(E_\tau)$ and since τ has finite index in S, $E_\tau / \mathcal{J}(E_\tau)$ is a (finite =) semi-simple Artinian ring. Furthermore if $\overline{e}^2 = \overline{e} \in E_\tau / \mathcal{J}(E_\tau)$ then $\overline{e}$ lifts modulo the nilpotent ideal $\mathcal{J}(E_\tau) / \tau_\tau E_\tau$ to an

$$(e + \tau_\tau E_\tau)^2 = (e + \tau_\tau E_\tau) \in E_\tau / \tau_\tau E_\tau.$$

By Lemma 5.2.5, there is an $f^2 = f \in E_\tau$ such that $e - f \in \tau_\tau E_\tau$. Therefore E_τ is semi-perfect.

Proof of Theorem 5.2.3: By Lemma 5.2.6, E_τ is semi-perfect, so Lemma 1.7.1 and the AKS Theorem 2.1.6 imply that E_τ has the refinement property. Using Theorem 3.5.3 we conclude that E has the local refinement property.

In case $\mathbb{Q}E = \text{Mat}_n(\mathbf{k})$ for some field $\mathbf{k}$ then $S = \text{center}(E)$ is an integral domain with field of quotients $\mathbf{k}$. If $\mathbf{k}$ has finite degree over $\mathbb{Q}$ then there is a unique Dedekind domain $S \subset \overline{S} \subset \mathbf{k}$ called the *integral closure of S* such that $\overline{S}/S$ is finite.

COROLLARY 5.2.7 *Suppose that E is a prime rtffr ring with conductor τ such that*

$$\mathbb{Q}E \cong \text{Mat}_n(\mathbf{k})$$

for some integer $n > 0$ and algebraic number field field $\mathbf{k}$. Let $\overline{S}$ be the integral closure of S. If τ is a primary ideal in $\overline{S}$ then E has the local refinement property.

EXAMPLE 5.2.8 Let $\mathbf{k}$ be an algebraic number field, let $\mathcal{O}$ be the ring of algebraic integers in $\mathbf{k}$ with class number $h \neq 1$, and let $I \neq 0$ be an ideal in $\mathcal{O}$ such that $\mathcal{O} \not\cong I$. By Theorem 5.2.3 the finitely generated projectives over E have the local refinement property. However since the class number h of $\mathcal{O}$ is not 1, $I^{(h)} \cong \mathcal{O}^{(h)}$, while $I \not\cong \mathcal{O}$. Thus $\mathcal{O}$ has the local refinement property.

5.2.3 Semi-Primary Rtffr Groups

Let $G \subset \overline{G}$ be rtffr groups and suppose that $\overline{G}$ has the local refinement property. The results of this section will give conditions on $\overline{G}$ under which a subgroup of finite index in $\overline{G}$ has the local refinement property. This should be contrasted with examples due to Corner and Fuchs-Loonstra (see [10, 46]), that show that subgroups of finite index in completely decomposable rtffr groups can have badly behaved direct sum decompositions.

The group G is a *Dedekind group* if $E(G)$ is a Dedekind domain. If G is a Dedekind domain then $E(G)$ is a commutative ring. We assume that we are given a finite nilpotent set $\{G_1, \ldots, G_t\}$ of Dedekind rtffr groups and integers $n_1, \ldots, n_t > 0$. Let

$$\overline{G} = G_1^{(n_1)} \oplus \cdots \oplus G_t^{(n_t)}. \tag{5.3}$$

For each $i = 1, \ldots, t$ choose a primary ideal P_i such that

$$\mathcal{N}(\text{End}(G_i)) \subset P_i \subset \text{End}(G_i).$$

Notice that because P_i is primary in $\text{End}(G_i)$ then $\text{End}(G_i)/P_i$ is finite. The action $P_i G_i^{(n_i)}$ is unambiguous since $P_i \subset \text{End}(G_i)$. Then

$$(P_1 \oplus \cdots \oplus P_t)\overline{G} = P_1 G_1^{(n_1)} \oplus \cdots \oplus P_t G_t^{(n_t)}$$

is uniquely defined.

The rtffr groups we are interested in are those G such that

$$P_1 G_1^{(n_1)} \oplus \cdots \oplus P_t G_t^{(n_t)} \subset G \subset \overline{G}. \qquad (5.4)$$

In this case we say that G has *semi-primary index in* $\overline{G}$. Alternatively we say that G is a *semi-primary rtffr group* if there are integers $t, n_1, \ldots, n_t > 0$, a nilpotent set $\{G_1, \ldots, G_t\}$ of Dedekind rtffr groups, and primary ideals $\mathcal{N}(\text{End}(G_i)) \subset P_i \subset \text{End}(G_i)$ such that (5.4) is true.

Given a semi-primary rtffr group G we will use $\overline{G}$ and P_i in the way they are used in the above discussion.

Evidently each Dedekind rtffr group is a semi-primary group so semi-primary groups exist in abundance. The following examples illustrate the settings that we are trying to generalize.

EXAMPLE 5.2.9 Let $p \in \mathbb{Z}$ be a prime, let $\overline{E} = \text{Mat}_2(\mathbb{Z})$, and let $E = 1\mathbb{Z} + \text{Mat}_2(p\mathbb{Z})$. Then $e^2 = e \in E \Rightarrow e \in \{0, 1\}$. Using Corner's Theorem 2.3.3 there is an rtffr group G such that $E = \text{End}(G)$. Let $\overline{G} = \overline{E}G$. One shows that $\overline{E} = \text{End}(\overline{G})$ and that

$$p\overline{G} \subset G \subset \overline{G}.$$

Then G is a semi-primary group, G is indecomposable, and $\overline{G}$ has a nontrivial direct sum decomposition. Moreover by Theorem 5.2.3, G and $\overline{G}$ have the local refinement property and unique decompositions. This example demonstrates that we are properly extending Theorem 2.4.6.

EXAMPLE 5.2.10 Let $G_1, \ldots, G_t \subset \mathbb{Q}$ be rank 1 groups such that $G_i \cong G_j \Rightarrow i = j$. The Baer-Kulikov-Kaplansky Theorem 2.1.11 states that the group $\overline{G}$ in (5.3) has unique decomposition. If $q \in \mathbb{Z}$ is a prime power and if G is a group such that

$$q\overline{G} \subset G \subset \overline{G}$$

then G is a semi-primary group. It is traditional [60] to call G an *acd group with primary regulating quotient*.

EXAMPLE 5.2.11 Let $\{G_1, \ldots, G_t\}$ be a finite nilpotent set of rtffr groups such that $\text{End}(G_i)$ is a *pid* for each $i = 1, \ldots, t$. Let $\overline{G}$ be the group in (5.3) and for each $i = 1, \ldots, t$ choose prime powers

$$q_i \in \text{End}(G_i).$$

Then each group G such that

$$q_1 G_1^{(n_1)} \oplus \cdots \oplus q_t G_t^{(n_t)} \subset G \subset \overline{G}$$

is a semi-primary rtffr group.

EXAMPLE 5.2.12 Let $\overline{G}$ be a (strongly indecomposable) Dedekind rtffr group, fix an integer $n > 0$, and choose a primary ideal

$$\mathcal{N}(\text{End}(\overline{G})) \subset P \subset \text{End}(\overline{G}).$$

If G is any rtffr group such that

$$P\overline{G}^{(n)} \subset G \subset \overline{G}^{(n)}$$

then G is a semi-primary rtffr group. If $n = 1$ then we are looking at semi-primary rtffr groups G such that

$$P\overline{G} \subset G \subset \overline{G}.$$

The next result, the main result of this chapter, explains our interest in semi-primary rtffr groups.

THEOREM 5.2.13 [T.G. Faticoni]   *Semi-primary rtffr groups have the local refinement property and a locally unique decomposition.*

Proof: The proof of this theorem consists of showing that $E(G)$ satisfies conditions 1 and 2 in Theorem 5.2.3.

Let G be a semi-primary rtffr group. There is a nilpotent set $\{G_1, \ldots, G_t\}$ of Dedekind rtffr groups, integers $n_1, \ldots, n_t > 0$, and primary ideals

$$\mathcal{N}(\text{End}(G_i)) \subset P_i \subset \text{End}(G_i)$$

such that

$$(P_1 \oplus \cdots \oplus P_t)\overline{G} \subset G \subset \overline{G}$$

where $\overline{G}$ is given in (5.3). Let

$$P = P_1 \oplus \cdots \oplus P_t.$$

For each $i = 1, \ldots, t$, $E(G_i)$ is a Dedekind domain whose field of fractions $\mathbb{Q}E(G_i)$ is a number field. Since $\{G_1, \ldots, G_t\}$ is a nilpotent set Theorem 1.3.1 implies that

$$E(\overline{G}) = \text{Mat}_{n_1}(E(G_1)) \times \cdots \times \text{Mat}_{n_t}(E(G_t)).$$

Since $G \doteq \overline{G}$, $\text{End}(G) \doteq \text{End}(\overline{G})$, so that $\mathbb{Q}E(G) \cong \mathbb{Q}E(\overline{G})$ as rings. Then

$$\mathbb{Q}E(G) = \text{Mat}_{n_1}(\mathbb{Q}E(G_1)) \times \cdots \times \text{Mat}_{n_t}(\mathbb{Q}E(G_t))$$

satisfies condition 1 of Theorem 5.2.3.

As for condition 2, for each $i = 1, \ldots, t$, P_i maps onto a nonzero primary ideal $T_i \subset E(G_i)$. Let $T = T_1 \oplus \cdots \oplus T_t$. Then T is an ideal of finite index in $\overline{S}$. Since $P\overline{G} \subset G \subset \overline{G}$, $P\text{End}(\overline{G}) \subset \text{End}(G)$, and so

$$PE(G_i) = TE(G_i) = T_iE(G_i) \subset E(G_i).$$

Inasmuch as $T_i \subset \tau_i \subset E(G_i)$ (where $\tau = \tau_1 \oplus \cdots \oplus \tau_t$ is the conductor of $E(G)$), and since T_i is a primary ideal in $E(G_i)$, τ_i is a primary ideal in $E(G_i)$. Thus $E(G)$ satisfies condition 2.

Then by Theorem 5.2.3, $E(G)$ has the local refinement property and a locally unique decomposition. This completes the proof of the theorem.

5.2.4   Applications to Refinement

Some applications will show that the seemingly strong conditions in Theorem 5.2.3 are actually quite common.

If H is a rank one group then $\text{End}(H) \subset \mathbb{Q}$ is a *pid* and a given prime $p \in \mathbb{Z}$ is either a unit in $\text{End}(H)$ or a prime. Thus the rank one groups are Dedekind groups whose primary ideals are generated by powers of primes in $\mathbb{Z}$. We say that $\overline{G}$ is *homogeneous completely decomposable of type* σ if $\overline{G} = H^{(n)}$ for some rank one group H of type σ. If $\overline{G}$ is a completely decomposable rtffr group then

$$\overline{G} = G[\sigma_1] \oplus \cdots \oplus G[\sigma_t]$$

where $G[\sigma_i]$ is a homogeneous completely decomposable group of type σ_i, and the types $\sigma_1, \ldots, \sigma_t$ are distinct. Using this notation we can use Theorem 5.2.13 to study the uniqueness of direct sum decompositions of groups of semi-primary index in a completely decomposable rtffr group.

The next two results should be contrasted with the *acd* groups constructed in Examples 2.2.1.

THEOREM 5.2.14 [T.G. Faticoni] *Let* $\overline{G} = G[\sigma_1] \oplus \cdots \oplus G[\sigma_t]$ *be a completely decomposable rtffr group, where the groups* $G[\sigma_i]$ *are homogeneous completely decomposable of type* σ_i, *and where the types* $\sigma_1, \ldots, \sigma_t$ *are distinct. If* G *is a subgroup of finite index in* $\overline{G}$ *such that for each* $i = 1, \ldots, t$, $(G + G[\sigma_i])/G$ *is a finite primary abelian group for each* $i = 1, \ldots, t$ *then* G *has a locally unique decomposition.*

Proof: By hypothesis $G[\sigma_i] = G_i^{(n_i)}$ for some rank one group G_i, $i = 1, \ldots, t$. The reader can show that $\{G_1, \ldots, G_t\}$ is a nilpotent set. Suppose that G has finite index in $\overline{G}$ and that for each $i = 1, \ldots, t$, $(G + G[\sigma_i])/G$ is a finite p_i-group. There are integers $m_1, \ldots, m_t > 0$ such that

$$p_1^{m_1} G[\sigma_1] \oplus \cdots \oplus p_t^{m_t} G[\sigma_t] \subset G \subset G[\sigma_1] \oplus \cdots \oplus G[\sigma_t]$$

or in other words G has semi-primary index in $\overline{G}$. By Theorem 5.2.13, G has the local refinement property.

COROLLARY 5.2.15 [T.G. Faticoni and P. Schultz] *(See [45].) Let $\overline{G}$ be a completely decomposable rtffr group and let $q \in \mathbb{Z}$ be a prime power. If G is a group such that $q\overline{G} \subset G \subset \overline{G}$ then G has a locally unique decomposition.*

For the purposes of this discussion let

$$H = H_1 \oplus \cdots \oplus H_s$$

where $H_i \subset \mathbb{Q}$ for each $i = 1, \ldots, s$. A *kernel group* is a group G that fits into an exact sequence

$$0 \to G \longrightarrow H \overset{\sigma}{\longrightarrow} \mathbb{Q}$$

where σ is the unique map such that $\sigma(x) = x$ for each $x \in H_i$ and for each $i = 1, \ldots, s$. Dually a *cokernel group* is a group G that fits into an exact sequence

$$0 \to X \longrightarrow H \overset{\pi}{\longrightarrow} G \to 0$$

where X is a pure rank one subgroup of H. We call G a *bracket group* if G is either a kernel group of a cokernel group. If G is a strongly indecomposable bracket group then it is known that $\text{End}(G) \subset \mathbb{Q}$ is a *pid*. See [10]. Thus the strongly indecomposable bracket groups are Dedekind groups.

A *strongly homogeneous group* is a group G such that for any pure rank one subgroups $X, Y \subset G$ there is an automorphism $\alpha : G \to G$ such that $\alpha(X) = Y$. It is known that if G is a strongly indecomposable strongly homogeneous rtffr group then $\text{End}(G)$ is a *pid* in which the primes $p \in \mathbb{Z}$ such that $pG \neq G$ are primes in $\text{End}(G)$. See [10]. Thus the strongly indecomposable strongly homogeneous rtffr groups are Dedekind groups.

The group G is a *Murley group* if $\mathbb{Z}/p\mathbb{Z}$-dim$(G/pG) \leq 1$ for each prime $p \in \mathbb{Z}$. It is known that if G is a strongly indecomposable Murley group then $\text{End}(G)$ is a *pid* in which $p \in \mathbb{Z}$ is a prime in $\text{End}(G)$ if $pG \neq G$. Evidently G is a Murley group iff G is a pure subgroup of $\widehat{\mathbb{Z}}$. See [62].

Direct sums of rtffr groups G such that $\text{End}(G)$ is a *pid* are considered by D.M. Arnold, R. Hunter, and F. Richman in [13]. Certainly G is a Dedekind group if $\text{End}(G)$ is a *pid*.

We have shown that if G is a *strongly indecomposable group of square-free rank* then $E(G)$ is an rtffr integral domain. Lemma 3.1.8 implies that G has finite index in a Dedekind group, thus motivating the assumption that G is a Dedekind group.

Let

$$\mathcal{D}$$

denote the set of strongly indecomposable rtffr groups G such that G falls into at least one of the following categories.

1. Groups G such that $E(G)$ is a *pid*,

2. Strongly indecomposable Murley groups,

3. Strongly indecomposable strongly homogeneous groups,

4. Strongly indecomposable bracket groups,

5. Rank one groups.

THEOREM 5.2.16 [T.G. Faticoni] *Let $\{G_1, \ldots, G_t\}$ be a nilpotent subset of $\mathcal{D}$ and let $\overline{G}$ be the direct sum (5.3) for some integers $n_1, \ldots, n_t > 0$. If $q_i \in E(G_i)$ are powers of primes in the pid $E(G_i)$ and if G is a group such that*

$$q_1 G_1^{(n_1)} \oplus \cdots \oplus q_t G_t^{(n_t)} \subset G \subset \overline{G}$$

then G has the local refinement property.

Proof: By the above discussion for each $G_i \in \mathcal{D}$, $E(G_i)$ is a *pid* so each G_i is a Dedekind rtffr group. Thus G is a semi-primary group. An application of Theorem 5.2.13 shows us that G has the local refinement property.

In [13] it is shown that if $\text{End}(\overline{G})$ is a *pid* and if $H_1, \ldots, H_r \doteq \overline{G}$ satisfy $\text{End}(H_i) = \text{End}(\overline{G})$ then $H = H_1 \oplus \cdots \oplus H_r$ has a uniqueness of decomposition that is different from local uniqueness and that is a little too technical for this book. In general there is an integer $q \neq 0$ such that $q\overline{G} \subset H_i \subset \overline{G}$ for each $i = 1, \ldots, r$. Our techniques allow us to slightly expand the main theorem in [13] by restricting q and by relaxing the conditions on $\operatorname{End}(\overline{G})$ and $\operatorname{End}(H_i)$.

COROLLARY 5.2.17 Let $\{G_1, \ldots, G_t\}$ be a nilpotent subset of $\mathcal{D}$ such that $E(G_1) \cong E(G_i)$ for each $i = 1, \ldots, t$, and let

$$\overline{G} \;=\; G_1^{(n_1)} \oplus \cdots \oplus G_t^{(n_t)}. \tag{5.5}$$

If $q \in E(G_1)$ is a prime power then each group

$$q\overline{G} \subset G \subset \overline{G}$$

has the local refinement property.

COROLLARY 5.2.18 Let $\overline{G}$ be a rtffr group such that $E(\overline{G})$ is a pid, let $n \neq 0$ be an integer, and let $q \in E(\overline{G})$ be a power of some prime in $E(\overline{G})$. Then any group $q\overline{G}^{(n)} \subset G \subset \overline{G}^{(n)}$ has the local refinement property.

COROLLARY 5.2.19 Let $\{G_1, \ldots, G_t\}$ be a nilpotent subset of $\mathcal{D}$ such that $E(G_i) \cong \mathbb{Z}$ for each $i = 1, \ldots, t$, and let $\overline{G}$ be a direct sum (5.5). If $q \in \mathbb{Z}$ is a prime power and if $q\overline{G} \subset G \subset \overline{G}$ then G has the refinement property.

We end this section with a result that shows that direct sums of semi-primary groups G are up to local isomorphism actually direct sums of semi-simple $\operatorname{End}(G)$-modules. The reader will enjoy proving this result.

THEOREM 5.2.20 [T.G. Faticoni] Let G be a semi-primary rtffr group. Let τ be the conductor of $E(G)$ and assume that $E(G)(\tau) = 0$. Let $(\mathcal{S}, \tau)$ be the category of semi-simple $\operatorname{End}(G)$-modules annihilated by τ.

1. There is a functor

$$C(\cdot) : \mathbf{P}_o(G) \to (\mathcal{S}, \tau) : H \to A(H) \otimes_{E(G)} E(G)_\tau / \mathcal{J}(E(G)_\tau).$$

2. $C(\cdot)$ induces a bijection from the set of local isomorphism classes (H) of $H \in \mathbf{P}_o(G)$ onto the set of isomorphism classes (M) of semi-simple right $E(G)$-modules such that $M\tau = 0$.

5.3   Exercises

1. Show that $\mathcal{J}(S_\tau) \subset \mathcal{J}(E_\tau)$.

2. Let E be a rtffr ring and suppose that E is locally isomorphic to $U_1 \oplus U_2$ for some finitely generated E-modules U_1 and U_2. Then $E = V_1 \oplus V_2$ for some cyclic right ideals V_1 and V_2 such that U_i is locally isomorphic to V_i for $i = 1, 2$.

3. Let G be a rtffr group. Show that the following are equivalent.

   (a) G is indecomposable.

   (b) G is locally isomorphic to some indecomposable rtffr group.

   (c) If G is locally isomorphic to H then H is indecomposable.

   (d) $E(G)$ is indecomposable as a right $E(G)$-module.

   (e) $E_\tau(G)$ is indecomposable as a right $E_\tau(G)$-module.

4. Show that G has a locally unique decomposition if either

   (a) $E(G)$ is a semi-perfect ring or

   (b) G is a completely decomposable rtffr group or

   (c) $G = H^{(n)}$ where $E(H)$ is a Dedekind domain.

5. Suppose that S is a Noetherian integral domain and that its field of fractions $\mathbf{k}$ is a finitely generated S-module. Show that $\mathbf{k} = S$.

6. Using the notation of Theorem 5.2.13 show that there is a ring theoretic embedding $E(G) \to E(\overline{G})$ that takes regular elements to regular elements.

7. Using the notation of Theorem 5.2.13 show that $\mathbb{Q}E(\overline{G}) = \mathbb{Q}E(G)$.

8. Let $\mathcal{O}$ be the ring of algebraic integers in the number field $\mathbf{k}$. Show that

   (a) $\mathrm{Mat}_n(\mathbf{k})$ is the classical ring of quotients of $\mathrm{Mat}_n(\mathcal{O})$.

   (b) $\mathrm{Mat}_n(\mathcal{O})$ is a maximal order in $\mathrm{Mat}_n(\mathbf{k})$.

9. Let E be a semi-prime rtffr ring with conductor τ. Say $E = E_1 \times E_2$.

   (a) Show that $\tau = \tau_1 \oplus \tau_2$ where τ_i is the conductor of E_i for $i = 1, 2$.

(b) Show that $E_\tau = (E_1)_\tau \times (E_2)_\tau = (E_1)_{\tau_1} \times (E_2)_{\tau_2}$.

10. Suppose that G is a strongly indecomposable rtffr group of square-free rank. Suppose that t is an integer that is divisible by $h(G)$. Given indecomposable groups $H_1, \ldots, H_t \in \mathbf{P}_o(G)$ then $G^{(t)} \cong H_1 \oplus \cdots \oplus H_t$.

11. Suppose that $G \doteq G_1 \oplus \cdots \oplus G_t$ for some strongly indecomposable groups of square-free rank. Multiple appearances are allowed. Investigate the logical relations between the following statements.

   (a) $h(G) = 1$.

   (b) $h(G_i) = 1$ for each $i = 1, \ldots, t$.

   (c) Each G_i has the $\mathcal{L}$-refinement property.

   (d) Each G_i has the refinement property.

5.4 Questions for Future Research

Let G be a rtffr group, let E be a rtffr ring, and let $S = \text{center}(E)$ be the center of E. Let τ be the conductor of G. There is no loss of generality in assuming that G is strongly indecomposable.

1. Use the theory of rtffr groups to advance the understanding of the class group of an algebraic number field.

2. Use the theory of the class group of an algebraic number field to advance the understanding of the theory of rtffr groups.

3. Give a (finite abelian) group structure to $\Gamma(G)$, the set of isomorphism classes of groups locally isomorphic to G, and study the group.

4. Generalize the discussion given on class number $h(G)$ from strongly indecomposable rtffr groups G to just plain rtffr groups G.

5. There are results from the literature that show that $h(\mathcal{O})$, where $\mathcal{O}$ is the ring of algebraic integers in an algebraic number field, will grow in a certain way as the discriminant of $\mathcal{O}$ increases.

6. Make a study of the class group and class number of an algebraic number field and apply it to rtffr groups.

7. Make a study of the class number $h(G)$ of an rtffr group and apply it to the study of the class number of an algebraic number field.

8. See 5.1.9.   Determine $\ker \beta$.   Specifically, determine if $\ker \alpha$ is finitely generated. Determine if $\ker \beta$ is a direct summand of $\Gamma(E)$.

9. Use the Mayer-Vietoris Sequence and the Kunneth Sequence to study $\Gamma(E)$. See [69, Section 37].

10. Discuss how $\sqrt{\tau}$ is related to $h(G)$.

11. Nakayama's Lemma and the Chinese Remainder Theorem are used continually in this text. Explain why this is so by using some other kind of theory.

12. Give a complete accounting of the groups G that have the $\mathcal{L}$-property.

13. Give a complete accounting of the groups G that have the refinement property.

14. Let G be a strongly indecomposable rtffr group such that $E(G)$ is a Noetherian integral domain. Give group theoretic conditions for $h(G) = h(G')$ for quasi-isomorphic groups G and G'.

15. Give a more complete discussion of the integers $\nu_n(G)$, $e(G)$, and $\ell(G)$ given on page 118.

16. Improve the estimates in Theorem 5.1.22.

17. Do what is considered to be a great deal of fun in abelian groups. Relate the direct sum properties of G to integers associated with G. For example, G has the refinement property iff $h(G) = 1$. See also Theorem 5.1.26 and Theorem 5.1.28.

18. Characterize the groups G such that $E(G)$ is a locally semi-perfect ring.

19. Characterize semi-primary rtffr groups.

20. Functorially realize the group $\overline{G}$.

21. Extend the theory surrounding acd groups to semi-primary groups, or indicate where we cannot.

22. Let G be a strongly indecomposable rtffr group such that $E(G)$ is a commutative rtffr integral domain. Let $\overline{E}$ be the integral closure of $E(G)$. Investigate the converses of the statement "$h(\overline{E}) = 1$ if G has the refinement property."

23. Suppose that $\mathbb{Q}\mathrm{End}(G) = \mathbb{Q}[\sqrt{d}]$ for some negative square-free integer d. If $d \in \{-1, -2, -3, -7, -11, -19, -43, -67, -163\}$, then discuss the refinement property in G.

Chapter 6

Baer Splitting Property

The results in this section are from [38]. The short exact sequence

$$0 \to K \longrightarrow X \xrightarrow{\pi} H \to 0 \qquad (6.1)$$

of groups is said to be *split* if there is a map $j : H \to X$ such that $\pi j = 1_H$. Recall that

$$\mathbf{P}(G) = \{ \text{ groups } H \mid H \oplus H' \cong G^{(c)}$$
$$\text{for some cardinal } c \text{ and group } H'\}.$$

We will consider the splitting of the short exact sequence (6.1) under different hypotheses on X and the groups $H \in \mathbf{P}(G)$.

Closely related to the refinement property in rtffr groups is the *Baer splitting property*. The group G has the Baer splitting property if for each cardinal c, each surjection $\pi : G^{(c)} \to G$ is split. Free groups have the Baer splitting property and we will show that if $\operatorname{End}(G)$ is commutative then G has the Baer splitting property. Baer's Lemma 6.1.1 states that G has the Baer splitting property if $IG \neq G$ for each right ideal $I \subset \operatorname{End}(G)$. The purpose of this chapter is to study the Baer splitting property in detail.

6.1 Baer's Lemma

Given groups H and G we write

$$S_G(H) = \sum \{f(G) \mid f \in \operatorname{Hom}(G, H)\}.$$

We call $S_G(H)$ the *G-socle in H*, and we note that $S_G(H)$ is the largest
G-generated subgroup of H. If $H = S_G(H)$ then we say that H is *G-generated*, and we say that H is *finitely G-generated* if $H = f_1(G) + \cdots + f_n(G)$ for finitely many $f_1, \ldots, f_n \in \mathrm{Hom}(G, H)$.

The next result, known as *Baer's Lemma*, has assumed an important
role in the study of the transfer of properties between the rtffr group G
and its endomorphism ring $\mathrm{End}(G)$. One way of viewing this result is
that it shows us that a group G behaves like a projective module over a
ring if $IG \neq G$ for each right ideal $I \subset \mathrm{End}(G)$. The version we use is
found in [14].

LEMMA 6.1.1 [Baer's Lemma] *The following are equivalent for an
rtffr group G.*

1. *$IG \neq G$ for each right ideal $I \subset \mathrm{End}_S(G)$.*

2. *Each short exact sequence*

$$0 \to K \longrightarrow X \xrightarrow{\pi} G \to 0 \qquad (6.2)$$

in which $S_G(X) + K = X$ is split exact.

Proof: Assume part 1 and let $I = \pi\mathrm{Hom}(G, X)$. Then I is a right
ideal in the ring $\mathrm{End}(G)$. Let $y \in G$. There is an $x \in X$ such that
$\pi(x) = y$. Since $X = S_G(X) + K$ there are $f_1, \ldots, f_t \in \mathrm{Hom}(G, X)$,
$x_1, \ldots, x_t \in G$, and $z \in K$ such that

$$y = \pi(x) = \pi\left(\sum_{i=1}^t f_i(x_i) + z\right) = \sum_{i=1}^t \pi f_i(x_i) \in IG.$$

Thus $G = IG$. Inasmuch as $JG \neq G$ for proper right ideals $J \subset \mathrm{End}(G)$,
$I = \mathrm{End}(G)$. That is, there is a map $f \in \mathrm{Hom}(G, X)$ such that $\pi f = 1_G$,
so that (6.2) is split exact. This proves part 2.

Conversely assume part 1 is false. There is a proper right ideal $I \subset \mathrm{End}(G)$ such that $IG = G$. Write $I = \{f_1, f_2, \ldots\}$ and let $G_i \cong G$ for
each $i = 1, 2, \ldots$. There is a map

$$\pi : \oplus_{i=1}^\infty G_i \longrightarrow G$$

such that for each $x \in G_i$, $\pi(x) = f_i(x)$. Because G has finite rank, given
any map

$$g : G \longrightarrow \oplus_{i=1}^\infty G_i$$

there is a finite set $\{1, \ldots, n\}$ such that

$$g(G) \subset G_1 \oplus \cdots \oplus G_n$$

so that there are maps $g_i : G \to G_i$ such that $g = g_1 \oplus \cdots \oplus g_n$. Then

$$\pi g = f_1 g_1 + \cdots + f_n g_n \in I \neq \operatorname{End}(G).$$

Specifically $\pi g \neq 1_G$ for any g so that π is not split. Thus part 2 is false, which completes the proof.

COROLLARY 6.1.2 *The following are equivalent for an rtffr group G.*

1. *$IG \neq G$ for each proper right ideal $I \subset \operatorname{End}(G)$.*

2. *For each cardinal c, each short exact sequence*

$$0 \to K \longrightarrow G^{(c)} \xrightarrow{\pi} G \to 0 \qquad (6.3)$$

   is split exact.

The rtffr group G is *finitely faithful* if $IG \neq G$ for each proper ideal I of *finite index* in $\operatorname{End}(G)$. Of course faithful groups are finitely faithful. For rtffr groups the converse is true. For this reason we will use the term *faithful* and not *finitely faithful* throughout this chapter.

LEMMA 6.1.3 *Let G be an rtffr group. Then G is a faithful group iff G is a finitely faithful group.*

Proof: Suppose that $IG \neq G$ for each right ideal of finite index in $\operatorname{End}(G)$, and let $I \subset \operatorname{End}(G)$. Then I is contained in a maximal right ideal $M \subset \operatorname{End}(G)$ which by Lemma 4.2.3 has finite index in $\operatorname{End}(G)$. By hypothesis $IG \subset MG \neq G$ so that G is a faithful group. The converse is clear so the proof is complete.

There is at least one interesting previously encountered class of group that consists of faithful groups. To prove the next result apply Theorems 4.2.4 and 6.1.1.

THEOREM 6.1.4 [10, page 52, Theorem 5.9] *Let G be an rtffr group such that $\operatorname{End}(G)$ is commutative. Then G is a faithful group.*

A *faithfully E-flat group* is a faithful group G that is flat as a left $\text{End}(G)$-module.

EXAMPLE 6.1.5 Since each subring of $\mathbb{Q}$ is a *pid*, G is a faithfully E-flat group if $\text{End}(G) \subset \mathbb{Q}$. Given a prime $p \in \mathbb{Z}$, $I\mathbb{Z}(p^\infty) = \mathbb{Z}(p^\infty)$ for each nonzero ideal I in $\text{End}(\mathbb{Z}(p^\infty)) \cong \widehat{\mathbb{Z}}_p$. Thus $\mathbb{Z}(p^\infty)$ is not a faithful group even though $\text{End}(G)$ is a *pid*. Evidently, $\mathbb{Z}(p^\infty)$ is not E-flat.

The next example shows us that there are plenty of faithfully E-flat groups.

EXAMPLE 6.1.6 The group constructed according to Corner's Theorem 2.3.3 is faithfully flat.

Proof: Let E be an rtffr group and let G be the group constructed according to Corner's Theorem. There is a short exact sequence

$$0 \to E \longrightarrow G \longrightarrow \mathbb{Q}E \to 0$$

of left E-modules. Since E and $\mathbb{Q}E$ are flat left E-modules, there is an exact sequence

$$0 = \text{Tor}_E^1(\cdot, E) \longrightarrow \text{Tor}_E^1(\cdot, G) \longrightarrow \text{Tor}_E^1(\cdot, \mathbb{Q}E) = 0.$$

Then $\text{Tor}_E^1(\cdot, G) = 0$ and hence G is E-flat.

Let $I \subset \text{End}(G)$ be a right ideal. We may assume without loss of generality that I is a maximal right ideal. Lemma 4.2.3 states that E/I is a finite p-group for some prime $p \in \mathbb{Z}$ so $pE \subset I \subset E$. An application of $E/I \otimes \cdot$ to the above short exact sequence produces the exact sequence

$$\text{Tor}_E^1(E/I, \mathbb{Q}E) \to E/I \to G/IG \to \mathbb{Q}E/I\mathbb{Q}E.$$

Since $\mathbb{Q}E$ is a flat left E-module and since E/I is finite,

$$\text{Tor}_E^1(E/I, \mathbb{Q}E) = 0 = \mathbb{Q}E/I\mathbb{Q}E,$$

so that $G/IG \cong E/I \neq 0$. This completes the proof.

We will construct rtffr groups whose properties are from the set {faithful, not faithful, E-flat, not E-flat}.

EXAMPLE 6.1.7 There is an rtffr G that is E-flat but that is not faithful.

Proof: Let $\mathbf{k}$ be an algebraic number field such that $[\mathbf{k} : \mathbb{Q}] \geq 2$, and let $\mathcal{O}$ denote the ring of algebraic integers in $\mathbf{k}$. It is known that $\mathbb{Z}/p\mathbb{Z}$-dim$(\mathcal{O}/p\mathcal{O}) = [\mathbf{k} : \mathbb{Q}]$. (See [58].) Let $S = \mathbb{Z} + p\mathcal{O} \neq \mathcal{O}$, let

$$A = \begin{pmatrix} \mathbb{Q}S & \mathbb{Q}\text{Hom}(p\mathcal{O}, S) \\ \mathbb{Q}\text{Hom}(S, p\mathcal{O}) & \mathbb{Q}\mathcal{O} \end{pmatrix} = \begin{pmatrix} \mathbf{k} & \mathbf{k} \\ \mathbf{k} & \mathbf{k} \end{pmatrix},$$

and let $M = \begin{pmatrix} S \\ p\mathcal{O} \end{pmatrix}$. Then let

$$
\begin{aligned}
E &= \{q \in A \mid qM \subset M\} \\
&= \begin{pmatrix} S & \text{Hom}(p\mathcal{O}, S) \\ \text{Hom}(S, p\mathcal{O}) & \mathcal{O} \end{pmatrix} \\
&= \begin{pmatrix} S & p\mathcal{O} \\ p\mathcal{O} & \mathcal{O} \end{pmatrix}.
\end{aligned}
$$

Furthermore $I = \begin{pmatrix} S & p\mathcal{O} \\ p\mathcal{O} & p\mathcal{O} \end{pmatrix}$ is a right ideal in E such that

$$IM = \begin{pmatrix} S & p\mathcal{O} \\ p\mathcal{O} & p\mathcal{O} \end{pmatrix} \begin{pmatrix} S \\ p\mathcal{O} \end{pmatrix} = \begin{pmatrix} S \\ p\mathcal{O} \end{pmatrix} = M.$$

Theorem 2.3.4 states that there is an rtffr group G and a short exact sequence

$$0 \to M \longrightarrow G \longrightarrow A \oplus A \to 0 \tag{6.4}$$

such that $E \cong \text{End}(G)$. As in Example 6.1.6 we can prove that $IM = M$ implies that $IG = G$. Thus G is not faithful. Since M is the first column of E, M is a projective ($=$ flat) left E-module. Then as in Example 6.1.6, G is an E-flat group.

EXAMPLE 6.1.8 There is a faithful rtffr group that is not E-flat.

Proof: In Example 4.3.2 we constructed a group G such that $\text{End}(G)$ is an integral domain but not a Dedekind domain and $G \doteq \text{End}(G)$ is not a flat left $\text{End}(G)$. Since $\text{End}(G)$ is commutative G is faithful.

With these examples to guide us we will discuss faithful rtffr groups.

6.2   Splitting of Exact Sequences

The groups in this section need not have finite rank. We continue our study of the Baer splitting property by considering a more general concept. Let G be a torsion-free group.

1. Let $H \in \mathbf{P}(G)$. Then (G, H) has the *Baer splitting property* if the short exact sequence $0 \to K \longrightarrow X \xrightarrow{\pi} H \to 0$ is split exact whenever $S_G(X) + K = X$. The group G has the *Baer splitting property* iff for each cardinal c and group $H \in \mathbf{P}(G)$, each short exact sequence $0 \to K \longrightarrow G^{(c)} \xrightarrow{\pi} H \to 0$ is split exact.

2. Let $H \in \mathbf{P}(G)$. Then (G, H) has the *endlich Baer splitting property* if for each *positive integer c*, the short exact sequence $0 \to K \longrightarrow G^{(c)} \xrightarrow{\pi} H \to 0$ is split exact. The group G has the *endlich Baer splitting property* iff for each positive integer c, each short exact sequence $0 \to K \longrightarrow G^{(c)} \xrightarrow{\pi} G \to 0$ is split exact.

The term *Baer splitting property* was coined by U. Albrecht and H.P. Goeters [5] and the term *endlich Baer splitting property* first appears in [38]. Evidently G has the endlich Baer splitting property if G has the Baer splitting property. We will eventually prove that for an rtffr group G the endlich Baer splitting property implies the Baer splitting property.

The next two results show us that the set of groups H such that (G, H) has the Baer splitting property is closed under direct sums.

LEMMA 6.2.1 *Let G, H, H' be groups. If (G, H) and (G, H') have the Baer splitting property then $(G, H \oplus H')$ has the Baer splitting property.*

Proof: Suppose that (G, H) and (G, H') have the Baer splitting property and consider the short exact sequence

$$0 \to K \longrightarrow X \xrightarrow{\pi} H \oplus H' \to 0 \tag{6.5}$$

where $S_G(X) = X$. We will show that (6.5) is split.

Let $\rho : H \oplus H' \longrightarrow H$ be the canonical projection with $\ker \rho = H'$. Since (G, H) has the Baer splitting property there is an injection $g : H \to X$ such that $1_H = (\rho\pi)g = \rho(\pi g)$. Then

$$H \oplus H' = \pi g(H) \oplus H' \quad \text{and} \quad X = g(H) \oplus X'$$

where $X' = \ker \rho\pi$ and where $g(H) \cong H$. Let $g'' : \pi g(H) \cong g(H) \subset X$ and notice that

$$\pi g''(\pi g(H)) = \pi g(H).$$

We can assume without loss of generality that $\pi g''(x) = x$ for each $x \in \pi g(H)$.

The reader will show as an exercise that there is a short exact sequence

$$0 \to K \longrightarrow X' \xrightarrow{\pi'} H' \to 0$$

in which $S_G(X') = X'$ and π' is the restriction of π to X'. Since (G, H') has the Baer splitting property there is an injection $g' : H' \longrightarrow X'$ such that $\pi'g' = 1_{H'}$.

It is then an easy task to show that

$$\pi(g'' \oplus g')(\pi g(x) \oplus x') = \pi g''(\pi g(x)) + \pi g'(x') = \pi g(x) \oplus x'$$

for each $x \in H$ and $x' \in H'$. Thus $\pi(g'' \oplus g') = 1_{H \oplus H'}$ so that $(G, H \oplus H')$ has the Baer splitting property. This completes the proof.

THEOREM 6.2.2 *Let G, H be rtffr groups.*

1. *If (G, H) has the Baer splitting property then $(G, H^{(n)})$ has the Baer splitting property for each integer $n > 0$.*

2. *If (G, H) has the endlich Baer splitting property then $(G, H^{(n)})$ has the endlich Baer splitting property for each integer $n > 0$.*

Proof: Apply induction to $n > 0$ using Lemma 6.2.1 as the induction step.

The above lemma is the induction step in a transfinite induction.

THEOREM 6.2.3 [T.G. Faticoni] *(See [38].) Let G, H be rtffr groups. Then (G, H) has the Baer splitting property iff $(G, H^{(d)})$ has the Baer splitting property for each cardinal d.*

Proof: Let $H \in \mathbf{P}_o(G)$, let $H_i \cong H$ for each $i \in \mathcal{I}$, suppose that (G, H) has the Baer splitting property, let I be an index set, and let

$$0 \to K \longrightarrow X \xrightarrow{\pi} \bigoplus_{i \in I} H_i \to 0 \tag{6.6}$$

be a short exact sequence in which $S_G(X) + K = X$. Given a subset $J \subset I$ let

$$\rho_J : \bigoplus_{i \in I} H_i \longrightarrow \bigoplus_{j \in J} H_j$$

be the canonical projection. Let $\mathcal{B}$ be the set of pairs (J, g_J) such that $J \subset I$, $g_J : \oplus_{j \in J} H_j \to X$, and

$$\rho_J \pi g_J = 1_{\oplus_{j \in J} H_j}.$$

Given $i \in I$, the Baer splitting property produces a map $g_i : H_i \to X$ such that $\rho_i \pi g_i = 1_{H_i}$ so $\mathcal{B} \neq \emptyset$.

Partially order $\mathcal{B}$ by the usual inclusion of sets and restriction of functions. An elementary argument shows that each chain in $\mathcal{B}$ contains an upper bound in $\mathcal{B}$. Then by Zorn's Lemma $\mathcal{B}$ contains a maximal element (I_m, g_m). Assume for the sake of contradiction that $I_m \neq I$. Let $i \in I \setminus I_m$. The proof of Lemma 6.2.1 implies that there is a $g_i : H_i \to X$ such that

$$\rho_{I_m \cup \{i\}} \pi (g_m \oplus g_i) = 1_{\oplus_{\ell \in I_m \cup \{i\}} H_\ell}.$$

Thus $(I_m \cup \{i\}, g_m \oplus g_i) \in \mathcal{B}$. Since (I_m, g_m) is maximal in $\mathcal{B}$ it must be that $I_m = I$ and

$$\rho_I = 1_{\oplus_{i \in I} H_k} = \pi g_m.$$

Thus $(G, H^{(d)})$ has the Baer splitting property for each cardinal d.

COROLLARY 6.2.4 *Let G be a group.*

1. *G has the endlich Baer splitting property iff (G, H) has the endlich Baer splitting property for each $H \in \mathbf{P}_o(G)$.*

2. *G has the Baer splitting property iff (G, G) has the Baer splitting property.*

Proof: Parts 1 and 2 follow from Theorems 6.2.2 and 6.2.3.

6.3 G-Compressed Projectives

Baer's Lemma 6.1.1 shows that we should be interested in simple right $\text{End}(G)$-modules M such that $T_G(M) \neq 0$. From this point of view, introduced by U. Albrecht [4], it is natural to consider the set of right $\text{End}(G)$-modules M such that $T_G(M) = 0$.

Let $M \neq 0$ be a right $\text{End}(G)$-module. We say that M is G-compressed if $T_G(M/N) \neq 0$ for each proper $\text{End}(G)$-submodule $N \subset M$.

The right $\text{End}(G)$-module K is *finitely M-generated* if there is an integer n and an isomorphism $K \cong M^{(n)}/N$ for some subgroup $N \subset M^{(n)}$. The module M is *finitely G-compressed* $\text{T}_G(M/K) \neq 0$ for each proper finitely generated submodule K of M.

Let $E = \text{End}(G)$. Given an $H \in \mathbf{P}_o(G)$ the Arnold-Lady-Murley Theorem 2.4.1 shows that $\text{H}_G(H)$ is a finitely generated projective right $\text{End}(G)$-module. The Arnold-Lady-Murley Theorem 2.4.1 shows us that the *dual module of* $\text{H}_G(H)$ is

$$\text{Hom}_E(\text{H}_G(H), E) \cong \text{Hom}_E(\text{H}_G(H), \text{H}_G(G)) \cong \text{Hom}(H, G).$$

Given a projective $E = \text{End}(G)$-module P *the trace of* P in E is the ideal

$$\Delta_P = \sum \{f(P) \mid f \in \text{Hom}_E(P, E)\}.$$

If $H \in \mathbf{P}(G)$ then *the trace of* H is the trace ideal of $\text{H}_G(H)$

$$\Delta_H = \Delta_{\text{H}_G(H)}.$$

It is known that if P is a projective right E-module then

1. Δ_P is the smallest ideal in E such that $P\Delta_P = P$, and

2. $\Delta_P^2 = \Delta_P$.

Notice that condition 6 in the following result is similar to the condition that $IG \neq G$ for each right ideal $I \subset \text{End}(G)$.

THEOREM 6.3.1 *[38, T.G. Faticoni] The following are equivalent for* $H \in \mathbf{P}_o(G)$.

1. $\text{H}_G(H)^{(n)}$ *is G-compressed for each integer $n > 0$.*

2. $\text{H}_G(H)$ *is G-compressed.*

3. $\Delta_H \subset I$ *for each maximal right ideal $I \subset \text{End}(G)$ such that $IG = G$.*

4. $I\text{Hom}(H, G) = \text{Hom}(H, G)$ *for each maximal right ideal $I \subset \text{End}(G)$ such that $IG = G$.*

5. $T_G(M) \neq 0$ for each finitely $H_G(H)$-generated right $\text{End}(G)$-modules $M \neq 0$.

6. $NG \neq H$ for each proper $\text{End}(G)$-submodule $N \subset \text{Hom}_R(G, H)$.

7. (G, H) has the Baer splitting property.

8. (G, H) has the endlich Baer splitting property.

Proof: Throughout this proof let $E = \text{End}(G)$, fix $H \in \mathbf{P}_o(G)$, and let $P = H_G(H)$. Then P is a finitely generated projective right E-module. We will prove

$$1 \Rightarrow 2 \Rightarrow 3 \Leftrightarrow 4, \quad 3 \Rightarrow 1 \Leftrightarrow 5$$

$$2 \Leftrightarrow 6 \Rightarrow 7 \Rightarrow 8 \Rightarrow 6$$

$1 \Rightarrow 2$ is clear.

$2 \Rightarrow 3$ We prove the contrapositive. Assume there is a maximal right ideal $I \subset E$ such that $IG = G$ but $\Delta_P \not\subset I$. Let $\pi : E \to E/I$ be the natural projection map. There is a map $f : P \to E$ such that $f(P) \not\subset I$, and because I is a maximal right ideal of E, $f(P) + I = E$. Hence $\pi f : P \to E/I$ is a surjection. Furthermore, because $IG = G$,

$$T_G(P/\ker \pi f) \cong T_G(E/I) \cong G/IG = 0$$

so that P is not G-compressed.

$3 \Leftrightarrow 4$ We have observed that $P^* = \text{Hom}(H, G)$ and it is known that $\Delta_P = \Delta_{P^*}$ in general.

Assume part 3 is true. Given $IG = G$ then $\Delta_H \subset I$. Then because $\Delta_H = \Delta_{H_G(H)} = \Delta_{H_G(H)^*}$, $IH_G(H)^* = H_G(H)^*$. This proves part 4.

Conversely assume part 4 is true. Let $IG = G$. Then

$$IH_G(H)^* = I\text{Hom}(H, G) = \text{Hom}(H, G) = H_G(H)^*$$

implies that

$$\Delta_H = \Delta_{H_G(H)} = \Delta_{H_G(H)^*} \subset I.$$

This is part 3.

$3 \Rightarrow 1$ We prove the contrapositive. Suppose there is an integer $n > 0$ and an E-submodule $N \subset P^n$ such that $T_G(P^{(n)}/N) = 0$. Because P is finitely generated projective there is a maximal right E-submodule $N \subset M \subset P^n$, and because P^n/M is simple there is a maximal right ideal $I \subset E$ such that $P^n/M \cong E/I$. Then

$$0 = T_G(P^n/M) = T_G(E/I) = G/IG$$

so that $P^{(n)}$ is not *G*-compressed.

The proof of 5 $\Leftrightarrow$ 1 as an exercise.

2 $\Leftrightarrow$ 6 Let $N \subset \mathrm{H}_G(H)$ be a right E-submodule and apply the right exact functor $\mathrm{T}_G(\cdot)$ to the short exact sequence

$$0 \to N \xrightarrow{\imath} \mathrm{H}_G(H) \xrightarrow{\pi} \mathrm{H}_G(H)/N \to 0 \tag{6.7}$$

to form the diagram

$$
\begin{array}{ccccccc}
\mathrm{T}_G(N) & \xrightarrow{\mathrm{T}_G(\imath)} & \mathrm{T}_G\mathrm{H}_G(H) & \longrightarrow & \mathrm{T}_G(\mathrm{H}_G(H)/N) & \longrightarrow & 0 \\
& & \Big\downarrow{\scriptstyle \Theta_H} & & & & \\
& & H & & & &
\end{array}
$$

of groups. Notice that image $\Theta_H \mathrm{T}_G(\imath) = NG$. Because $H \in \mathbf{P}_o(G)$ the Arnold-Lady-Murley Theorem 2.4.1 implies that Θ_H is an isomorphism. Then $NG = H$ iff $\mathrm{T}_G(\imath)$ is a surjection iff $\mathrm{T}_G(\mathrm{H}_G(H)/N) = 0$. Then 2 $\Leftrightarrow$ 6 follows immediately.

6 $\Rightarrow$ 7 follows in the same way that we proved Baer's Lemma 6.1.1, but in proving 6 $\Rightarrow$ 7 use an $\mathrm{End}(G)$-submodule $N \subset \mathrm{H}_G(G)$ where we used a right ideal $I \subset \mathrm{End}(G)$.

7 $\Rightarrow$ 8 is clear.

8 $\Rightarrow$ 6 Suppose that $NG = H$ for some proper right $\mathrm{End}(G)$-submodule $N \subset \mathrm{H}_G(H)$. Since $\mathrm{H}_G(H)$ is finitely generated, there is a maximal right $\mathrm{End}(G)$-submodule $K \subset \mathrm{H}_G(H)$ that contains N. By Lemma 4.2.3 the simple module $\mathrm{H}_G(H)/K$ is a finite p-group, so that $p\mathrm{H}_G(H) \subset K \subset \mathrm{H}_G(H)$. Since H has finite rank $K/p\mathrm{H}_G(H) \subset \mathrm{H}_G(H)/p\mathrm{H}_G(H)$ is finite and thus K is finitely generated by the generators for $p\mathrm{H}_G(H)$ and representatives of the cosets for $K/p\mathrm{H}_G(H)$.

Write $K = \sum_{i=1}^{n} \pi_i \mathrm{End}(G)$ for some $\pi_1, \ldots, \pi_n \in K$. There is a short exact sequence

$$0 \to H' \longrightarrow \bigoplus_{i=1}^{n} G_i \xrightarrow{\pi} H \to 0 \tag{6.8}$$

where $G_k \cong G$ for each $i \in I$ and where π is the unique map such that $\pi(x) = \pi_i(x)$ for each $i = 1, \ldots, n$ and each $x \in G_i$. Notice that π is a surjection since $H = KG = \sum_{i=1}^{n} \pi_i(G_i)$. Furthermore, any map $\jmath : H \to \oplus_{i=1}^{n} G_i$ can be written as $\jmath = \sum_{i=1}^{n} \jmath_i$ for some maps $\jmath_i : H \to G_i$. Then for any map $g : G \to H$ we have

$$\pi \jmath g = \left(\sum_{i=1}^{n} \pi_i \right) \left(\sum_{i=1}^{n} \jmath_i \right) g = \sum_{i=1}^{n} \pi_i \jmath_i g.$$

Since $\pi_i \in K$ and since $\jmath_i : H \to G_i$ we have $\pi_\jmath g \in K$ for each $g :$ $G \to H$. Since $K \neq \mathrm{H}_G(H)$ the reader will show that $\pi_\jmath \neq 1_H$ for any $\jmath : H \to \oplus_{i=1}^n G_i$. Thus (6.8) does not split which proves that part 6 is false. This completes the logical circuit.

The next result illustrates the connection between G-compressed modules, faithful groups, and Baer's lemma.

THEOREM 6.3.2 [38, T.G. Faticoni].   *The following are equivalent for an rtffr group G.*

1. *$IG \neq G$ for each proper right ideal $I \subset E$.*

2. *$IG \neq G$ for each finitely generated proper right ideal $I \subset E$.*

3. *$\mathrm{T}_G(M) \neq 0$ for each finitely presented right $\mathrm{End}(G)$-module $M \neq 0$.*

4. *$\mathrm{T}_G(M) \neq 0$ for each finitely generated right $\mathrm{End}(G)$-module $M \neq 0$.*

5. *G has the Baer splitting property.*

6. *G has the endlich Baer splitting property.*

Proof: $6 \Leftrightarrow 5 \Leftrightarrow 4$ is Theorem 6.3.1(8), (7), (5).
$4 \Rightarrow 3$ is clear.
$3 \Rightarrow 2$ follows as in Theorem 6.3.1(5) $\Rightarrow$ (4).
$2 \Rightarrow 1$ Given part 2 and a right ideal $I \subset \mathrm{End}(G)$ there is a maximal right ideal $I \subset M \subset \mathrm{End}(G)$. By Lemma 4.2.3, M is finitely presented so that $IG \subset MG \neq G$. This proves part 1.
$1 \Rightarrow 5$ is Baer's Lemma 6.1.1. This completes the logical sequence.

In the presence of a flatness hypothesis the above result can be improved a little.

THEOREM 6.3.3 [3, U. Albrecht].   *Let G be an E-flat rtffr group. Then G has the Baer splitting property iff $\mathrm{T}_G(M) \neq 0$ for each right R-module $M \neq 0$.*

Proof: Say G has the Baer splitting property, and let $M \neq 0$ be a right $\mathrm{End}(G)$-module. Let $0 \neq M_o \subset M$ be any finitely generated $\mathrm{End}(G)$-submodule. Since G is E-flat, $\mathrm{T}_G(M_o) \subset \mathrm{T}_G(M)$, and by Theorem 6.3.2, $\mathrm{T}_G(M_o) \neq 0$. Then $\mathrm{T}_G(M) \neq 0$. The converse is Theorem 6.3.2 so the proof is complete.

6.4 Some Examples

The examples in this section illustrate the abundance of the splitting properties, and they show that the Baer splitting property and the endlich Baer splitting property are in general different notions.

EXAMPLE 6.4.1   1. Let G be rtffr group with commutative endomorphism ring. By Theorem 6.3.2 and Theorem 6.1.4, G has the Baer splitting property, so by Theorem 6.2.3, $G^{(c)}$ has the Baer splitting property for each cardinal c. However, let I_o be the ideal of those $f \in \text{End}(G^{(c)})$ such that $f(G)$ has finite rank. Then I_o is a proper ideal of $\text{End}(G^{(c)})$ such that $I_o G^{(c)} = G^{(c)}$. Thus the finite rank hypothesis in Theorem 6.3.2 is necessary.

   2. $G = \mathbb{Q}$ has the Baer splitting property, and $G = \mathbb{Z}(p^\infty)$ does not have the endlich Baer splitting property.

   3. D.M. Arnold and E.L. Lady [14] give an example of a completely decomposable abelian group G of rank 2 such that $IG = G$ for some proper pure ideal $I \subset \text{End}(G)$.

   4. The group G constructed in Example 6.1.7 is one for which $\{$right ideals $I \subset \text{End}(G) \,|\, IG = G\}$ is a finite set.

   In [4], U. Albrecht proves that the self-small E-flat group G is faithful iff $\text{T}_G(K) \neq 0$ for each right $\text{End}(G)$-module $K \neq 0$. The following example shows us that even if G is a faithful Corner group it may happen that $\text{T}_G(K) = 0$ for some nonzero $\text{End}(G)$-module K.

EXAMPLE 6.4.2 [36, T.G. Faticoni] There is a Corner group G such that

   1. $\text{T}_G(N) \neq 0$ for each finite right $\text{End}(G)$-module $N \neq 0$.

   2. There is a right $\text{End}(G)$-module $K \neq 0$ such that $\text{T}_G(K) = 0$.

Specifically G is faithful but $\text{T}_G(K) = 0$ for some right E-module $K \neq 0$.

   Proof: Let $p \in \mathbb{Z}$ be a prime and let E be the localization of $\mathbb{Z}[x]$ at the maximal ideal (x, p). That is,

$$E = \mathbb{Z}[x]_{(x,p)}.$$

Let M be the left E-module

$$\frac{E}{(x)} \oplus \frac{E}{(x^2)} \oplus \frac{E}{(x^3)} \oplus \cdots$$

where (x^k) is the ideal in E generated by x^k. One shows that M is a reduced torsion-free left E-module. By Nakayama's Lemma 1.2.3, $JM \neq M$ where $J = J(E)$. By Theorem 2.3.3 there is a short exact sequence

$$0 \longrightarrow M \longrightarrow G \longrightarrow \mathbb{Q}C \longrightarrow 0 \qquad (6.9)$$

constructed in Theorem 2.3.3 such that $E \cong \operatorname{End}(G)$. Since $p \in J(E) = J$ is the unique maximal ideal in E one proves in the now familiar way that $G \neq JG$ because $M \neq JM$.

Next let K be the left E-module

$$K = (E/pE)b_o \oplus (E/pE)b_1 \oplus (E/pE)b_2 \oplus \cdots$$

where $b_j x = b_{j-1}$ for each $j \geq 1$ and $b_o x = 0$. An application of $K \otimes_E \cdot$ to (6.9) produces the exact sequence

$$K \otimes_E M \longrightarrow K \otimes_E G \longrightarrow K \otimes_E \mathbb{Q}C \to 0.$$

Since $\mathbb{Q}C$ is divisible and since K is a torsion group $K \otimes_E \mathbb{Q}C = 0$. The reader will show that $Kx = K$. Since each element of M is annihilated by some power of x the usual argument shows us that $K \otimes_E M = 0$. Therefore $K \otimes_E G = 0$. Specifically, G is faithful but $\mathrm{T}_G(K) = 0$ for some right $\operatorname{End}(G)$-module $K \neq 0$. This completes the proof.

In the above example G is not self-small. The reader can produce a map $f : G \to G^{(\aleph_o)}$ whose image is not in any finite direct sum of copies of G.

A few remarks are in order before we end this section. U. Albrecht [3, Theorem 2.1, Corollary 2.2] extends Baer's Lemma by showing that the self-small right R-module G has the finite Baer splitting property iff $IG \neq G$ for each proper right ideal $I \subset E$ iff $\mathrm{T}_G(M) \neq 0$ for each nonzero finitely generated right $\operatorname{End}(G)$-module $M \neq 0$. Furthermore, he shows that if G is a self-small E-flat module, then G has the finite Baer splitting property iff $\mathrm{T}_G(M) \neq 0$ for each right $\operatorname{End}(G)$-module $M \neq 0$, [3, Corollary 2.3]. The faithful condition is central to the characterization of the rtffr groups G such that $\operatorname{Ext}^1(G, G)$ is a torsion-free abelian group, [11, 43]. D.M. Arnold and J. Hausen [12] use the endlich Baer splitting property to study modules G that have the summand intersection property, [41] and [12] use the endlich Baer splitting property to characterize those modules G such that $\operatorname{End}(G)$ is right semi-hereditary, and [3], [14], and [41] use the Baer splitting property to characterize those R-modules G such that $\operatorname{End}(G)$ is right Noetherian, right hereditary. K. Fuller [47] calls G *completely faithful* and U. Albrecht [3] calls G *fully faithful* if $\mathrm{T}_G(M) \neq 0$ for each nonzero right $\operatorname{End}(G)$-module M.

6.5 Exercises

G, H, K are rtffr groups, $p \in \mathbb{Z}$ is a prime, E is an rtffr ring, M, N, are rtffr E-modules (left or right, depending on the setting).

1. Prove any of the results in this section left as exercises for the reader.

2. Let $H \in \mathbf{P}(G)$ and suppose that (G, H) has the endlich Baer splitting property. If H is finitely G-generated then $H \in \mathbf{P}_o(G)$.

3. Refer to the proof of Theorem 6.2.3. Show that each chain in $\mathcal{C}$ contains an upper bound in $\mathcal{C}$.

4. Let M be a right $\mathrm{End}(G)$-module. Then $M^{(n)}$ is G-compressed for each integer $n > 0$ iff $T_G(N) \neq 0$ for each finitely M-generated right $\mathrm{End}(G)$-module $N \neq 0$.

5. Let P be a projective right E-module, let P^* be the dual of P, and let I be a left ideal of E. Prove that $IP = P$ iff $\Delta_P \subset I$ and that $\Delta_P = \Delta_{P^*}$.

6. Let $H \in \mathbf{P}_o(G)$. Show that $\mathrm{Hom}(G, H)^* = \mathrm{Hom}(H, G)$.

7. Prove that the following are equivalent.

   (a) $IG \neq G$ for each proper right ideal $I \subset E$.

   (b) $T_G(M) \neq 0$ for each nonzero finitely generated $M \in \mathbf{Mod}_{\mathrm{End}_R(G)}$.

8. If $IG \neq G$ for each proper finitely generated right ideal $I \subset \mathrm{End}(G)$ then G has the endlich Baer splitting property.

9. If $H \in \mathbf{P}_o(G)$ and if $N \subset \mathrm{H}_G(H)$ is such that $NG = H$ then $T_G(\mathrm{H}_G(H)/N) = 0$.

10. Complete the proof of Theorem 6.2.3 by showing that $\pi_J \neq 1_H$ for any $J : H \to G^{(c)}$.

11. If E is a reduced torsion-free ring of finite rank and if I is a maximal right ideal of E then E/I is finite.

12. Review the proof of Lemma 6.2.1. Show that the map $g'' : \pi g(H) \to H'' \subset X$ such that $g''(\pi g(x)) = g(x)$ is a well defined injection such that $\pi g'' = 1_{\pi g(H)}$.

13. Prove that the sequence $0 \to K \to X' \to H' \to 0$ given in Lemma 6.2.1 is exact.

14. If (G, H) has the endlich Baer splitting property then $(G, H^{(n)})$ has the endlich Baer splitting property for each integer $n > 0$.

6.6 Questions for Future Research

Let G be an rtffr group, let E be an rtffr ring, and let $S = \text{center}(E)$ be the center of E. Let τ be the conductor of G. There is no loss of generality in assuming that G is strongly indecomposable.

Given a ring E let $\Omega(E)$ denote the set of rtffr groups G such that $E \cong \text{End}(G)$. Some of the properties of E are seen in $\Omega(E)$.

1. Describe the rings E such that each $G \in \Omega(E)$ is a faithful group. Commutative and hereditary rings E satisfy this property. See [40] for more rings of this type.

2. In [43], it is shown that E is hereditary iff each group $G \in \Omega(E)$ is E-flat. Classify the rings E such that each group $G \in \Omega(E)$ is a faithfully E-flat module.

3. The group G has the Baer splitting property if for each cardinal c each surjection $\pi : G^{(c)} \longrightarrow G$ is split. Dualize the Baer splitting property in the obvious way. The group G has the *dual-Baer splitting property* if for each cardinal c, each injection $G \longrightarrow G^c$ is split. See [5]. Characterize those rtffr groups G that satisfy the dual-Baer splitting property. Develop a general theory as in Chapter 6.

4. Describe the rtffr rings E such that each $G \in \Omega(E)$ has the dual-Baer splitting property. See [40].

5. Characterize the G-compressed $\text{End}(E)$-modules. See Theorem 6.3.1.

6. Characterize the (rtffr) groups G such that $T_G(M) \neq 0$ for each nonzero right $\text{End}(G)$-module M.

7. Describe the rings E such that for each $G \in \Omega(E)$, $T_G(M) \neq 0$ for each nonzero right E-module M.

Chapter 7

$\mathcal{J}$-Groups, $\mathcal{L}$-Groups, and $\mathcal{S}$-Groups

The rtffr group G is called a $\mathcal{J}$-*group* if G is isomorphic to each subgroup of finite index in G, G is called an $\mathcal{L}$-*group* if G is locally isomorphic to each subgroup of finite index in G, and G is an $\mathcal{S}$-*group* if each subgroup of finite index in G is G-generated. Obviously

$$\mathcal{J}\text{-groups} \Rightarrow \mathcal{L}\text{-groups} \Rightarrow \mathcal{S}\text{-groups}.$$

In the first half of the chapter we will discuss how the torsion subgroup of $\mathrm{Ext}^1_{\mathbb{Z}}(G, G)$ is related to (finitely) faithful $\mathcal{S}$-groups. In the second half of the chapter we will scrutinize the converse relationships $\mathcal{L}$-groups $\Rightarrow$ $\mathcal{J}$-groups, and finitely faithful $\mathcal{S}$-groups $\Rightarrow$ $\mathcal{J}$-groups.

7.1 Background on Ext

Let G and H be groups. Call G an $\mathcal{S}$-*group* if each subgroup of finite index in G is G-generated. $\mathcal{S}$-groups are introduced by D.M. Arnold in [11]. Any group of the form $\mathbb{Z} \oplus H$ is an $\mathcal{S}$-group. If R is a pure subring of $\hat{\mathbb{Z}}_p$ for some prime p then R is an $\mathcal{S}$-group.

Recall that G is *finitely faithful* if $IG \neq G$ for each maximal right ideal of finite index in $\mathrm{End}(G)$. We investigate finitely faithful $\mathcal{S}$-groups in an effort to find partial solutions to the question *under what conditions on G are the groups quasi-isomorphic to G actually isomorphic to G?*

We will abbreviate notation as follows.

$$\mathrm{Ext}^1_{\mathbb{Z}}(G, H) = \mathrm{Ext}(G, H)$$

Then $\text{Ext}(G, H)$ consists of the short exact sequences

$$0 \to H \longrightarrow X \xrightarrow{\ \pi\ } G \to 0 \tag{7.1}$$

modulo an equivalence relation. See [46] for details. The split short exact sequences (7.1) correspond to 0 in $\text{Ext}(G, H)$. The short exact sequence (7.1) is *quasi-split* if there is a map $j : G \to X$ and an integer $n \neq 0$ such that $\pi j = n 1_G$. Let

$$\boxed{\ \ \text{tExt}(G, H) = \ \text{the torsion subgroup of } \text{Ext}(G, H).\ \ }$$

C.L. Walker's Theorem [46, Section 102] states that the quasi-split short exact sequences correspond to the torsion part of $\text{Ext}(G, H)$.

THEOREM 7.1.1 [85, C.L. Walker] *The group* $\text{tExt}(G, H)$ *is the set of equivalence classes of quasi-split short exact sequences (7.1). In particular given an equivalence class* $[\mathcal{E}] \in \text{tExt}(G, H)$ *of a short exact sequence* $\mathcal{E}$ *in (7.1) then* $n[\mathcal{E}] = 0$ *iff there is a map* $j : G \to X$ *such that* $\pi j = n 1_G$.

Consequently $\text{Ext}(G, H)$ is torsion-free iff each quasi-split short exact sequence (7.1) is split exact.

Given a prime $p \in \mathbb{Z}$ and a torsion-free group G, G/pG is a $\mathbb{Z}/p\mathbb{Z}$-vector space. *The p-rank of* G *is*

$$\boxed{\ \ r_p(G) = \mathbb{Z}/p\mathbb{Z}\text{-}\dim(G/pG).\ \ }$$

Since $\mathbb{Z}/p\mathbb{Z}$-linearly independent elements in G/pG lift to $\mathbb{Z}$-linearly independent elements in G, $r_p(G) \leq \text{rank}(G) < \infty$. For rtffr groups G and H, $\text{Ext}(G, H)$ is a divisible group so there are integers $\{r_p | p \in \mathbb{Z}$ is a prime $\}$ such that as groups

$$\text{Ext}(G, H) \cong \mathbb{Q}^{(\aleph_1)} \oplus \bigoplus_{\text{primes } p \in \mathbb{Z}} \mathbb{Z}(p^\infty)^{(r_p)}. \tag{7.2}$$

The numbers r_p are called the *p-rank of* $\text{Ext}(G, H)$. R.B. Warfield, Jr. [87] determines the values of the p-ranks r_p of $\text{Ext}(G, H)$.

THEOREM 7.1.2 [R.B. Warfield, Jr.] *(See [87].) Let* G *and* H *be torsion-free groups of finite p-rank for all primes* $p \in \mathbb{Z}$. *The p-rank* r_p *of* $\text{Ext}(G, H)$ *is given by*

$$r_p = r_p(G)r_p(H) - r_p(\text{Hom}(G, H))$$

for all primes $p \in \mathbb{Z}$. Consequently $\mathrm{Ext}(G, H)$ is p-torsion-free iff

$$r_p(G)r_p(H) = r_p(\mathrm{Hom}(G, H)).$$

7.2 Finite Projective Properties

In this section we characterize the torsion-freeness of $\mathrm{Ext}(G, H)$ via a projective property. Let G be a group and consider the short exact sequence

$$0 \to K \longrightarrow H \overset{\pi}{\longrightarrow} H/K \to 0. \tag{7.3}$$

1. Given a prime $p \in \mathbb{Z}$, G is *p-finitely H-projective* if G is projective relative to (7.3) whenever H/K is a finite p-group.

2. G is *finitely H-projective* if G is projective relative to (7.3) whenever H/K is a finite group.

3. G is *finitely projective* if G is finitely G-projective.

If $p \in \mathbb{Z}$ is prime, if $pG = G$, and if $pH \neq H$ then G is trivially p-finitely H-projective. The reader can show that if $\mathrm{Hom}(G, H) = 0$ then G is *not* p-finitely H-projective for any prime $p \in \mathbb{Z}$ such that $pG \neq G$ and $pH \neq H$.

EXAMPLE 7.2.1 Let $G \subset H \subset \mathbb{Q}$. Then G is p-finitely H-projective.
Proof: We may assume without loss of generality that $pH \neq H$ and that G is p-dense in H. Given a map $\bar{f} : G \to H/p^k H$ then $\bar{f}(p^k G) = 0$. Write $H/p^k H = \langle x + p^k H \rangle$. Since G is p-dense in H, $H/p^k H = (G + p^k H)/p^k H$ so we can assume that $x \in G$. Then $\bar{f}(x) = nx + p^k H$ and so $\bar{f}$ lifts to a map $f : G \to H$ over the inclusion map $G \subset H$.

We demonstrate a relationship between G-projectiveness and quasi-isomorphism. We will use without fanfare the fact that the torsion-free group G is (p)-finitely H-projective iff $\mathrm{H}_G(\cdot)$ is exact on the short exact sequence (7.3) whenever H/K is a finite p-group.

The reader will prove that if H/H' is a finite group then

$$\mathrm{Ext}(G, H) \cong \mathrm{Ext}(G, H'). \tag{7.4}$$

The $\mathbb{Z}$-adic completion of G is

$$\widehat{G} = \varprojlim_n G/nG$$

and the *p-adic completion of* G is

$$\widehat{G}_p = \varinjlim_{k>0} G/p^k G.$$

The group G is pure and dense in $\widehat{G}$ and G is p-pure and p-dense in $\widehat{G}_p$. That is, $G \cap n\widehat{G} = nG$ and $G + n\widehat{G} = \widehat{G}$ for each integer $n \neq 0$.

We will use the notation

$$\mathrm{H}_G(\cdot) = \mathrm{Hom}(G, \cdot).$$

THEOREM 7.2.2 [T.G. Faticoni and H.P. Goeters] *(See [44].) Let G and H be torsion-free groups and let $p \in \mathbb{Z}$ be a prime. The following are equivalent.*

1. *G is (p-)finitely H-projective.*

2. *$\mathrm{Ext}(G, H)$ is (p-)torsion-free.*

3. *$\mathrm{Hom}(G, H)$ is a (p-)pure and (p-)dense subgroup of $\mathrm{Hom}(G, \widehat{H})$, (of $\mathrm{Hom}(G, \widehat{H}_p)$).*

Proof: $1 \Leftrightarrow 2$ An application of $\mathrm{H}_G(\cdot)$ to the short exact sequence

$$0 \to H \xrightarrow{p} H \xrightarrow{\pi} H/pH \to 0$$

yields the exact sequence

$$\mathrm{H}_G(H) \xrightarrow{\pi^*} \mathrm{H}_G(H/pH) \to \mathrm{Ext}(G, H) \xrightarrow{p} \mathrm{Ext}(G, H)$$

in which $\pi^* = \mathrm{H}_G(\pi)$. Then $\mathrm{Ext}(G, H)$ is p-torsion-free iff multiplication by p is an injection iff π^* is a surjection iff G is p-finitely H-projective.

$2 \Leftrightarrow 3$. Since G is torsion-free and $\widehat{H}$ is pure-injective $\mathrm{Ext}(G, \widehat{H}) = 0$. Thus an application of $\mathrm{H}_G(\cdot)$ to the short exact sequence

$$0 \to H \xrightarrow{\jmath} \widehat{H} \xrightarrow{\pi} \widehat{H}/H \to 0$$

yields the exact sequence

$$\mathrm{H}_G(H) \xrightarrow{\jmath^*} \mathrm{H}_G(\widehat{H}) \xrightarrow{\pi^*} \mathrm{H}_G(\widehat{H}/H) \to \mathrm{Ext}(G, H) \to \mathrm{Ext}(G, \widehat{H}) = 0.$$

Then $\mathrm{Ext}(G, H)$ is p-torsion-free iff image π^* is p-divisible iff $\ker \pi^* = $ image $\jmath^*$ is p-pure and p-dense in $\mathrm{H}_G(\widehat{H})$. This completes the logical cycle.

EXAMPLE 7.2.3 [87, R.B. Warfield, Jr.] Let $G \subset H \subset \mathbb{Q}$ be groups. We showed in Example 7.2.1 that G is finitely H-projective. Consequently $\mathrm{Ext}(G, H)$ is torsion-free.

7.3 Finitely Projective Groups

The group G is *finitely faithful* if $IG \neq G$ for each right ideal I of finite index in $\mathrm{End}(G)$. Finitely faithful $\mathcal{S}$-groups were introduced by D.M. Arnold [11]. We will continue to use the terminology *finitely faithful $\mathcal{S}$-group* instead of *faithful $\mathcal{S}$-group* since *finitely faithful $\mathcal{S}$-group* seems to capture the spirit of our applications.

Rank one groups are finitely faithful $\mathcal{S}$-groups. Of course free groups of finite rank are finitely faithful $\mathcal{S}$-groups, but $G = \mathbb{Z}^{(\aleph_0)}$ is an $\mathcal{S}$-group without being faithful. If G is a finitely faithful $\mathcal{S}$-group then $G^{(n)}$ is a finitely faithful $\mathcal{S}$-group. (This follows from part 1 of Theorem 7.3.1 below.)

We will approach the problem of classifying the finitely faithful $\mathcal{S}$-groups by considering the finitely projective groups. We begin by showing that the finitely faithful $\mathcal{S}$-groups have finite p-rank for all primes $p \in \mathbb{Z}$. Let

$$0 \to H \longrightarrow G \xrightarrow{\pi} G/H \to 0 \tag{7.5}$$

be a short exact sequence in which G/H is finite.

THEOREM 7.3.1 *Let G be a reduced torsion-free group and let $p \in \mathbb{Z}$ be a prime. The following are equivalent.*

1. *G/pG is finite and G is projective relative to each short exact sequence (7.5) in which $pG \subset H \subset G$.*

2. *(a) If I is a right ideal of $\mathrm{End}(G)$ that contains some power of p then $IG \neq G$.*

   (b) If H is a subgroup of G such that $pG \subset H \subset G$ then H is G-generated.

Proof: Let $E = \mathrm{End}(G)$.

$1 \Rightarrow 2$ Assume part 1, and let $pE \subset I \subset E$ be a right ideal such that $IG = G$. By part 1,

$$E/pE \cong \mathrm{H}_G(G/pG) \cong \mathrm{Hom}(G/pG, G/pG)$$

so that E/pE is a simple Artinian ring.

Let I be a right ideal of E such that $pE \subset I \subset E$ and $IG = G$. There is no loss of generality in assuming that I is maximal in E and thus that $pE \subset I$. Since E/pE is a simple Artinian ring there is an $e^2 = e \in E/pE$ such that $e(E/pE) = I/pE$. Then

$$G/pG = IG/pG = (I/pE)(G/pG) = e(E/pE)(G/pG) = e(G/pG).$$

Inasmuch as $x(G/pG) \neq 0$ for each nonzero $x \in E/pE$ and since $(1 - e)(G/pG) = 0$ we conclude that $e = 1$. Thus $I = E$, as required by part (a).

Let $pG \subset H \subset G$. For each $x \in H \backslash pG$ there is a map $f_x : G \to H/pG$ such that $x + pG \in f_x(G)$. By hypothesis in part 1, f_x lifts to a map $g_x : G \to H$ such that $x \in g_x(G) + pG$. Then H is G-generated, which proves part (b). This completes the proof of $1 \Rightarrow 2$.

$2 \Rightarrow 1$. We first show that G/pG is finite. Let $I \subset E$ be the ideal such that

$$I/pE = \{f \in E/pE \,|\, \mathbb{Z}/p\mathbb{Z}\text{-}\dim(f(G/pG)) \text{ is finite }\}.$$

Given a subgroup $pG \subset H \subset G$ such that $H/pG \cong \mathbb{Z}/p\mathbb{Z}$ there is by part (b) a surjection $f_H : G \to H/pG$. Notice that $f_H \in I/pE$ and also that

$$G/pG = (I/pE)(G/pG) = IG/pG.$$

By part (a), $I = E$. Thus 1_G has finite image in G/pG and hence G/pG is finite.

Apply $H_G(\cdot)$ to the short exact sequence (7.5). To show that $\pi^* = H_G(\pi)$ is a surjection we will prove the next lemma.

LEMMA 7.3.2 Let G be a torsion-free group that satisfies Theorem 7.3.1(2). Then $H_G(\mathbb{Z}/p\mathbb{Z}) = \mathrm{Hom}(G, \mathbb{Z}/p\mathbb{Z})$ is a simple right E-module.

Proof: For any nonzero map $f : G \to \mathbb{Z}/p\mathbb{Z}$, $\ker f$ is a maximal subgroup of G that contains pG. Then

$$pE \subset \mathrm{ann}_E(f) = H_G(\ker f) \subset E.$$

We claim that $\mathrm{ann}_E(f)$ is a maximal right ideal of E. Suppose that $\mathrm{ann}_E(f) \subset I \subset E$ and that $\mathrm{ann}_E(f) \neq I$. Then by part 2(b) we have

$$pG \subset \ker f = H_G(\ker f)G \neq IG \subset G.$$

By the maximality of $\ker f$ in G we have $IG = G$ and hence part 2(a) implies that $I = E$. As claimed for each $f \in \mathrm{H}_G(\mathbb{Z}/p\mathbb{Z})$, $\mathrm{ann}_E(f)$ is a maximal right ideal in E.

Now assume to the contrary that there are $f \neq 0 \neq g \in \mathrm{H}_G(\mathbb{Z}/p\mathbb{Z})$ such that $fE \neq gE$ are distinct simple right E-modules. Then

$$fE \oplus gE \subset \mathrm{H}_G(\mathbb{Z}/p\mathbb{Z}).$$

By the above paragraph $\mathrm{ann}_E(f + g)$ is a maximal right ideal in E, whence $\mathrm{ann}_E(f \oplus g)$ is a maximal right ideal of E such that

$$\mathrm{ann}_E(f \oplus g) = \mathrm{ann}_E(f) \cap \mathrm{ann}_E(g) \subset \mathrm{ann}_E(f).$$

Thus $\mathrm{ann}_E(f \oplus g) = \mathrm{ann}_E(f) = \mathrm{ann}_E(g)$. Since

$$\mathrm{ann}_E(f \oplus g) = \mathrm{H}_G(\ker(f \oplus g))$$

and since $(f \oplus g)G$ is finite, part 2(b) implies that $\ker(f \oplus g)$ is G-generated, so that

$$\ker(f \oplus g) = \mathrm{H}_G(\ker(f \oplus g))G = \mathrm{ann}_E(f \oplus g)G = \mathrm{ann}_E(f)G = \ker f$$

as abelian groups. Similarly $\ker(f \oplus g) = \ker g$ so $g = uf$ for some automorphism u of $\mathbb{Z}/p\mathbb{Z}$. Since u is then multiplication by some integer, $fE = gE$. This contradiction to our choice of $fE \neq gE$ shows us that $\mathrm{H}_G(\mathbb{Z}/p\mathbb{Z}) = fE$ is a simple right E-module. This completes the proof of the Lemma.

To continue the proof of Theorem 7.3.1: Choose a subgroup $pG \subset H \subset G$ such that $H/pG \cong \mathbb{Z}/p\mathbb{Z}$. Consider the short exact sequence (7.5). By part 2(b) and because H/pG is simple there is a homomorphism $f : G \to H$ such that $f(G) + pG = H$. Then

$$0 \neq \pi f \in \mathrm{H}_G(H/pG) \cong \mathrm{H}_G(\mathbb{Z}/p\mathbb{Z})$$

so by Lemma 7.3.2, $\pi f E = \mathrm{H}_G(\mathbb{Z}/p\mathbb{Z})$. Therefore

$$\text{image } \pi^* = \pi \mathrm{H}_G(G) \supset \mathrm{H}_G(H/pG)$$

for each such H. Inasmuch as G/pG is a direct sum of copies of $\mathbb{Z}/p\mathbb{Z}$, image $\pi^* = \mathrm{H}_G(G/pG)$. That is, π^* is a surjection which proves part 1 and completes the proof.

7.4   Finitely Faithful $\mathcal{S}$-Groups

Using Theorem 7.3.1 we can characterize the finitely projective groups as being finitely faithful $\mathcal{S}$-groups.

THEOREM 7.4.1 [D.M. Arnold] *(See [11].) The following are equivalent for a torsion-free group G for which $r_p(G)$ is finite for all primes $p \in \mathbb{Z}$.*

1. *G is finitely projective.*

2. *$\mathrm{Ext}(G,G)$ is torsion-free.*

3. *G is a finitely faithful $\mathcal{S}$-group.*

4. *$r_p(G)^2 = r_p(\mathrm{End}(G))$ for each prime $p \in \mathbb{Z}$.*

5. *$\mathrm{End}(G)/p\mathrm{End}(G) \cong \mathrm{Mat}_{r_p(G)}(\mathbb{Z}/p\mathbb{Z})$ as rings for each prime $p \in \mathbb{Z}$.*

6. *Each quasi-split short exact sequence $0 \to G \to X \to G \to 0$ is split exact.*

Proof: $2 \Leftrightarrow 3$ follow from Theorem 7.2.2 and $2 \Leftrightarrow 6$ follows from Theorem 7.1.1.

$1 \Rightarrow 5$ Suppose that G is finitely projective. An application of $\mathrm{H}_G(\cdot)$ to the short exact sequence

$$0 \to G \xrightarrow{p} G \xrightarrow{\pi} G/pG \to 0$$

yields the isomorphisms

$$\mathrm{End}(G)/p\mathrm{End}(G) = \mathrm{image}\ \pi^* = \mathrm{H}_G(G/pG) \cong \mathrm{Hom}(G/pG, G/pG).$$

Since G is rtffr

$$\mathrm{Hom}(G/pG, G/pG) \cong \mathrm{Mat}_{r_p(G)}(\mathbb{Z}_p/p\mathbb{Z})$$

which proves 5.

$5 \Rightarrow 4$ By part 5

$$\mathrm{End}(G)/p\mathrm{End}(G) \cong \mathrm{Mat}_{r_p(G)}(\mathbb{Z}_p/p\mathbb{Z}) \cong \mathrm{Hom}(G/pG, G/pG)$$

as rings. Then

$$r_p(\mathrm{End}(G)) = \mathbb{Z}/p\mathbb{Z}\text{-}\dim(\mathrm{Hom}(G/pG, G/pG)) = r_p(G)^2.$$

This proves part 4.

$4 \Rightarrow 2$ follows from Theorem 7.1.2.

$2 \Rightarrow 1$ follows from Theorem 7.2.2, which completes the logical cycle.

For instance, if G is an rtffr group then $r_p(G)$ is finite for all primes $p \in \mathbb{Z}$.

COROLLARY 7.4.2 *Let G be an rtffr group. If G is a finitely faithful $\mathcal{S}$-group then*

1. *$G^{(k)}$ is a finitely faithful $\mathcal{S}$-group for each integer $k > 0$.*

2. *If $G = H \oplus H'$ then H is a finitely faithful $\mathcal{S}$-group.*

3. *If $G \doteq H$ then H is a finitely faithful $\mathcal{S}$-group.*

Proof: 1. Since G is a finitely faithful $\mathcal{S}$-group, $\text{Ext}(G, G)$ is torsion-free (Theorem 7.4.1), which implies that $\text{Ext}(G^{(k)}, G^{(k)})$ is torsion-free for each $k > 0$. Theorem 7.4.1 implies that $G^{(k)}$ is a finitely faithful $\mathcal{S}$-group.

2. If $G = H \oplus H'$ then $\text{Ext}(H, H)$ is a direct summand of the torsion-free group $\text{Ext}(G, G)$. Now apply Theorem 7.4.1.

3. If $G \doteq H$ then by Theorem 7.4.1, $\text{Ext}(G, G) \cong \text{Ext}(H, H)$ are torsion-free groups. This completes the proof.

COROLLARY 7.4.3 *If G is a finitely faithful $\mathcal{S}$-group then $\text{End}(G)$ is a semi-prime Noetherian ring.*

Proof: For each prime $p \in \mathbb{Z}$, $\text{End}(G)/p\text{End}(G)$ is a simple ring, so $\mathcal{N}(\text{End}(G)) \subset p\text{End}(G)$ for all p. Thus $p\mathcal{N}(\text{End}(G)) = \mathcal{N}(\text{End}(G))$ is divisible. Since G is reduced $\mathcal{N}(\text{End}(G)) = 0$. This completes the proof.

EXAMPLE 7.4.4 This is an example of an $\mathcal{S}$-group that is not faithful.

Proof: Let $0 \neq X \subset \mathbb{Q}$ be a subgroup such that $\text{End}(X) = \mathbb{Z}$, $\text{Hom}(X, \mathbb{Z}) = 0$, and let $G = \mathbb{Z} \oplus X$. Clearly G is an $\mathcal{S}$-group. Observe that $\text{End}(G) = \begin{pmatrix} \mathbb{Z} & 0 \\ \mathbb{Z} & \mathbb{Z} \end{pmatrix}$ is not semi-prime. Observe that for each integer n, $I_n = \begin{pmatrix} \mathbb{Z} & 0 \\ \mathbb{Z} & n\mathbb{Z} \end{pmatrix}$ is a right ideal of $\text{End}(G)$ such that $I_n G = G$. Thus G is not faithful.

In the above example $\text{Ext}(G, G)$ is not torsion-free. The next example gives a nonsplit quasi-split short exact sequence.

EXAMPLE 7.4.5 Let $\mathbb{Z} \subset X \subset \mathbb{Q}$ be such that $\text{Hom}(X, \mathbb{Z})$. Then $G = \mathbb{Z} \oplus X$ is an $\mathcal{S}$-group that is not faithful and so $\text{Ext}(X, \mathbb{Z})$ has nonzero torsion subgroup. Thus there is a quasi-split, nonsplit extension $0 \to \mathbb{Z} \to H \to X \to 0$ of groups. The embedding sends $\mathbb{Z}$ to $\mathbb{Z}(p \oplus 1) \in \mathbb{Z} \oplus X$. Notice that $\mathbb{Z}(p \oplus 1)$ is a pure subgroup, a quasi-summand, but not a direct summand of $\mathbb{Z} \oplus X$ so that $0 \to \mathbb{Z} \to H \to X \to 0$ is not split exact. Compare this to the fact that if $G \cong H$ then $H \cong \mathbb{Z} \oplus X$.

An example of torsion-free Ext is found in $\widehat{\mathbb{Z}}_p$.

EXAMPLE 7.4.6 Let $G \subset H \subset \widehat{\mathbb{Z}}_p$ be pure subgroups of $\widehat{\mathbb{Z}}_p$. Inasmuch as each map $f : G \to H$ lifts to a unique map $f : \widehat{\mathbb{Z}}_p \to \widehat{\mathbb{Z}}_p$ we see that

$$\mathbb{Z} \subset \text{Hom}(G, H) \subset \widehat{\mathbb{Z}}_p = \text{Hom}(\widehat{G}, \widehat{H}).$$

Moreover since H is torsion-free and pure in $\widehat{\mathbb{Z}}_p$, $\text{Hom}(G, H)$ is pure and dense in $\text{Hom}(\widehat{G}, \widehat{H})$. Then by Theorem 7.2.2, $\text{Ext}(G, H)$ is torsion-free.

The examples of $\mathcal{S}$-groups so far either have nonzero nilradical or they have commutative endomorphism ring. For finitely faithful $\mathcal{S}$-groups our examples thus far satisfy $\text{End}(G)/p\text{End}(G) \cong \mathbb{Z}/p\mathbb{Z}$. Note the next example.

EXAMPLE 7.4.7 [11, D.M. Arnold] This is an example of a finitely faithful $\mathcal{S}$-group such that $r_p(G) = 2$ and $\text{End}(G)/p\text{End}(G) \cong \text{Mat}_2(\mathbb{Z}/p\mathbb{Z})$.

Let E be a maximal $\mathbb{Z}$-order in the Hamiltonian Quaternions over $\mathbb{Q}$. If $p \neq 2 \in \mathbb{Z}$ is a prime then $E/pE \cong \text{Mat}_{2\times 2}(\mathbb{Z}/p\mathbb{Z})$ and the completion $\widehat{E}_p$ satisfies $\widehat{E} \cong \text{Mat}_{2\times 2}(\widehat{\mathbb{Z}})$. (See [69].) Let M be a pure and dense rtffr E-submodule such that

$$\begin{pmatrix} \mathbb{Z} & 0 \\ \mathbb{Z} & 0 \end{pmatrix} \subset M \subset \begin{pmatrix} \widehat{\mathbb{Z}} & 0 \\ \widehat{\mathbb{Z}} & 0 \end{pmatrix}$$

and using Theorem 2.3.4 construct a short exact sequence of left E-modules such that

$$0 \to M \longrightarrow G \longrightarrow \mathbb{Q}E \oplus \mathbb{Q}E \to 0.$$

(The reader can show that $E = \mathcal{O}(M)$.) Since $r_p(E) = 4 = r_p(G)^2$, G is a finitely faithful $\mathcal{S}$-group with noncommutative endomorphism ring E.

7.5 Isomorphism versus Local Isomorphism

In this section we engage in a little alphabet soup. Let G be an rtffr group.

1. G is a $\mathcal{J}$-*group* if G is isomorphic to each subgroup of finite index in G. $\mathcal{J}$ is for the author of *Jónsson's Theorem* 2.1.10.

2. G is an $\mathcal{L}$-*group* if G is locally isomorphic to each subgroup of finite index in G. $\mathcal{L}$ is for E.L. Lady who first used local (= near) isomorphism on abelian groups. See [56].

3. G is an $\mathcal{S}$-*group* if each subgoup of finite index in G is G-generated. $\mathcal{S}$ is for G-*socle*, $S_G(H)$. See Section 6.1.

Evidently G is an $\mathcal{S}$-group if G is an $\mathcal{L}$-group if G is a $\mathcal{J}$-group. We will investigate the implications $\mathcal{S}$-group $\Rightarrow$ $\mathcal{L}$-group $\Rightarrow$ $\mathcal{J}$-group. Examples of $\mathcal{J}$-groups include the following.

1. Groups of the form $\mathbb{Z} \oplus H$, with $H \subset \mathbb{Q}$, are $\mathcal{J}$-groups. See Theorem 2.1.12.

2. Groups of the form $\mathbb{Z} \oplus K$, with $K \neq 0$, are $\mathcal{S}$-groups. The reader can show that $\mathbb{Z} \oplus K$ is a $\mathcal{J}$-group if K is a $\mathcal{J}$-group.

3. If $X \subset \mathbb{Q}$ then $G = X^{(n)}$ is a $\mathcal{J}$-group for each integer $n > 0$.

It is interesting to note that we did not give an example of an $\mathcal{L}$-group that is not a $\mathcal{J}$-group. Call G a *Murley group* if p-rank$(G) \leq 1$ for each prime $p \in \mathbb{Z}$. We will show that certain classes of $\mathcal{J}$-groups are Murley groups.

THEOREM 7.5.1 [H.P. Goeters] *The following are equivalent for the rtffr group G.*

1. G *is a Murley group.*

2. G *is a $\mathcal{J}$-group and* $\text{End}(G)$ *is commutative.*

3. G *is an $\mathcal{S}$-group and* $\text{End}(G)$ *is commutative.*

Proof: $1 \Rightarrow 2$ Let G be a Murley group and let H be a subgroup of finite index in G. There is a chain of subgroups of G such that

$$H \subset H_1 \subset \cdots \subset H_k \subset G$$

and such that $H_i/H_{i-1} \cong \mathbb{Z}/p\mathbb{Z}$. Because $G/pG \cong H_k/pH_k$ for quasi-isomorphic groups, the groups H_i are Murley groups. But then $pH_1 \subset H \subset H_1$ and p-rank$(H_1) \leq 1$ so by induction

$$H = pH_1 \cong H_1 \cong H_2 \cong \cdots \cong H_k \cong G.$$

Hence G is a $\mathcal{J}$-group.

Since p-rank$(G) \leq 1$ for each prime $p \in \mathbb{Z}$, the isomorphisms $G/pG \cong \mathbb{Z}/p\mathbb{Z}$ lift to a pure imbedding $\widehat{G} \subset \widehat{\mathbb{Z}}$. Since $\widehat{\mathbb{Z}}$ is pure injective each $f : G \to G$ lifts to a unique map $\widehat{f} : \widehat{\mathbb{Z}} \to \widehat{\mathbb{Z}}$. That is, End$(G)$ imbeds in the commutative ring $\widehat{\mathbb{Z}}$. This proves part 2.

$2 \Rightarrow 3$ is clear.

$3 \Rightarrow 1$ Suppose that G is an $\mathcal{S}$-group and that End(G) is commutative. Then End$(G)/p$End(G) is a commutative ring for each prime $p \in \mathbb{Z}$. By Theorem 4.2.4 and part 3, G is a finitely faithful $\mathcal{S}$-group so by Theorem 7.4.1

$$\text{End}(G)/p\text{End}(G) \cong \text{Mat}_{r_p(G) \times r_p(G)}(\mathbb{Z}/p\mathbb{Z})$$

as rings for all primes $p \in \mathbb{Z}$. Consequently $r_p(G) \leq 1$ for all primes p so that G is a Murley group. This completes the proof.

EXAMPLE 7.5.2 This is an example of a strongly indecomposable $\mathcal{J}$-group that is not a Murley group. This example is inspired by D.M. Arnold's example in [11].

Choose a prime $p \neq 2 \in \mathbb{Z}$ and let E be a classical maximal $\mathbb{Z}_p$-order in the Hamiltonian Quaternions over $\mathbb{Q}$. Then each right ideal $I \subset E$ is principal, $I = xE$ for some $x \in E$.

Since $E_p/pE_p \cong \text{Mat}_{2 \times 2}(\mathbb{Z}/p\mathbb{Z})$, the completion $\widehat{E}_p$ satisfies $\widehat{E}_p \cong \text{Mat}_{2 \times 2}(\widehat{\mathbb{Z}}_p)$. As in Example 7.4.7 construct a pure and dense rtffr E-submodule M such that

$$\begin{pmatrix} \mathbb{Z}_p & 0 \\ \mathbb{Z}_p & 0 \end{pmatrix} \subset M \subset \begin{pmatrix} \widehat{\mathbb{Z}}_p & 0 \\ \widehat{\mathbb{Z}}_p & 0 \end{pmatrix}$$

and using Theorem 2.3.4 construct a short exact sequence of left E-modules such that

$$0 \to M \longrightarrow G \longrightarrow \mathbb{Q}E \oplus \mathbb{Q}E \to 0.$$

(The reader can show that $E = \mathcal{O}(M)$.) Since $r_p(E) = 4 = r_p(G)^2$, G is a finitely faithful $\mathcal{S}$-group but not a Murley group.

In fact given $H \doteq G$, there is a right ideal $I \subset E$ such that $H = IG$. Since $I \cong E$, $H \cong G$, so that G is an indecomposable $\mathcal{J}$-group.

7.6 Analytic Number Theory

The next series of results will demonstrate a beautiful connection between $\mathcal{J}$-groups, $\mathcal{L}$-groups, and $\mathcal{S}$-groups. However, they require some deep results on Analytic Number Theory whose proofs would take us too far afield. Thus we will state some results and reference their proofs.

The following theorem gives a rather general condition under which a finitely faithful $\mathcal{S}$-group is a $\mathcal{J}$-group.

Let i, j, k be such that $-1 = i^2 = j^2 = k^2 = ijk$ and let $\mathbf{k}$ be an algebraic number field. The algebra $D = \mathbf{k}1 \oplus \mathbf{k}i \oplus \mathbf{k}j \oplus \mathbf{k}k$ is called a *totally definite quaternion algebra* if each embedding of $\mathbf{k}$ into the complex numbers is an embedding of $\mathbf{k}$ into the reals. Observe that $\mathbf{k} = \text{center}(D)$, D is a division $\mathbf{k}$-algebra, and D has dimension 4 over its center $\mathbf{k}$. If we let $\mathbf{k} = \mathbb{Q}$ then D is the $\mathbb{Q}$-algebra of Hamiltonian Quaternions. Thus the classical $\mathbb{Q}$-algebra of Hamiltonian Quaternions is a totally definite quaternion algebra.

Let G be an rtffr group and write

$$\mathbb{Q}E(G) = A_1 \times \cdots \times A_t$$

for some nonzero simple Artinian $\mathbb{Q}$-algebras $A_1, \ldots, A_t$. We say that the rtffr group G is an *Eichler group* if A_i is not a totally definite quaterion algebra for each simple factor A_i of $\mathbb{Q}E(G)$. Note that Eichler groups are rtffr groups.

A proof of the next result is found in [44, Proposition III.6].

THEOREM 7.6.1 *If the Eichler group G is a finitely faithful $\mathcal{S}$-group then G is a $\mathcal{J}$-group.*

The next result shows just how rare an exception the totally definite quaternion algebra is.

COROLLARY 7.6.2 *Write the rtffr group G as*

$$G \cong G_1^{(n_1)} \oplus \cdots \oplus G_t^{(n_t)} \tag{7.6}$$

for some strongly indecomposable groups $G_1, \ldots, G_t$ and integers t, n_1, $\ldots$, $n_t > 0$ such that $G_i \cong G_j \Rightarrow i = j$. Suppose that G is a finitely faithful $\mathcal{S}$-group. Then G is a $\mathcal{J}$-group in either of the following cases.

1. *$n_i \neq 1$ for each $i = 1, \ldots, t$.*

2. *$\text{rank}(G_i)$ is not divisible by 4 for each $i = 1, \ldots, t$.*

Proof: Since we are assuming that G is a finitely faithful $\mathcal{S}$-group, it suffices to show that for each i, $\mathbb{Q}E(G_i^{(n_i)})$ is not a totally definite quaternion algebra.

1. In case $n_i \neq 1$ then $\mathbb{Q}E(G_i^{(n_i)})$ is a nontrivial ring of matrices so it is not a domain. Thus $\mathbb{Q}E(G_i^{(n_i)})$ is not a totally definite quaternion algebra.

2. By part 1 we can assume without loss of generality that $n_i = 1$. Since $\text{rank}(G_i)$ is not divisible by 4, and since $\mathbb{Q}E(G_i)$ is a division ring, $\text{rank}(E(G_i)) \neq 4$. (See Lemma 4.2.1.) Then $\mathbb{Q}E(G_i)$ is not a totally definite quaternion algebra.

The conditions $n_i \neq 1$ and $\text{rank}(G_i)$ not divisible by 4 are satisfied in case

1. $G = H^{(2)}$ for some group H, or

2. For each $i = 1, \ldots, t$,
$$\mathbb{Q}E(G_i^{(n_i)}) = \text{Mat}_{n_i}(\mathbf{k}_i)$$
for some algebraic number field $\mathbf{k}_i$.

Before we continue with our study of $\mathcal{L}$-groups we will introduce a bit of power.

THEOREM 7.6.3 [R.B. Warfield, Jr.] *(See [88].) Let G and H be rtffr groups. Then G and H are locally isomorphic iff $G^{(n)} \cong H^{(n)}$ for some integer $n > 0$.*

We need a pair of lemmas that are themselves interesting. The semiprime ring E is *integrally closed* if given a ring $E \subset R \subset \mathbb{Q}E$ such that R/E is finite then $E = R$. Equivalently E is integrally closed iff E is a finite product of classical maximal orders.

The following is a cancellation result for rtffr groups such that $A(G)$ is integrally closed. D.M. Arnold [10] has proved a similar result.

THEOREM 7.6.4 [T.G. Faticoni] *Let G be an Eichler group such that $E(G)$ is an integrally closed ring. If H is an rtffr group then*
$$G \oplus G \cong G \oplus H \implies G \cong H.$$

Proof: By Jónsson's Theorem 2.1.10, $G \doteq H$, so that $E(G) \doteq A(H)$. By Theorem 2.4.4 we have
$$E(G) \oplus E(G) \cong E(G) \oplus A(H)$$

as right modules over the integrally closed ring $E(G)$. There are classical maximal orders $\overline{E}_i$ such that

$$E(G) = \overline{E}_1 \times \cdots \times \overline{E}_t.$$

Since $E(G) \cong A(H)$ iff $\overline{E}_i \cong A(H)\overline{E}_i$ for each $i = 1, \ldots, t$ we have reduced to the case where $E(G)$ is a classical maximal order.

So assume that $E(G)$ is a classical maximal order with center S. Given a right ideal $I \subset E(G)$ let $\mathrm{nr}(I)$ be the *reduced norm of I* in S. (See [69, page 214].) From $E(G) \oplus E(G) \cong E(G) \oplus A(H)$ and [69, page 311, Corollary 35.11(ii)] we see that

$$\mathrm{nr}(E(G)) = \mathrm{nr}(E(G)) \cdot \mathrm{nr}(E(G)) = \mathrm{nr}(E(G)) \cdot \mathrm{nr}(A(H)) = \mathrm{nr}(A(H)).$$

Inasmuch as G is Eichler, [69, page 311, Corollary 35.11(ii)] implies that $E(G) \cong A(H)$, so that by Theorem 2.4.4, $G \cong H$.

COROLLARY 7.6.5 *Let G be an $\mathcal{L}$-group. Then $E(G)$ is an integrally closed ring.*

Proof: Let $E = E(G)$. By Lemma 3.1.8 there is a group $\overline{G} \doteq G$ such that $\overline{E} = E(\overline{G})$ is an integrally closed ring. Because G is an $\mathcal{L}$-group G is locally isomorphic to $\overline{G}$ and then Theorem 7.6.3 states that $G^{(n)} \cong \overline{G}^{(n)}$ for some integer $n > 0$. Then

$$\mathrm{Mat}_n(\mathrm{End}(G)) \cong \mathrm{End}(G^{(n)}) \cong \mathrm{End}(\overline{G}^{(n)}) \cong \mathrm{Mat}_n(\mathrm{End}(\overline{G}))$$

so that

$$\mathrm{Mat}_n(E) \cong \mathrm{Mat}_n(\overline{E})$$

is an integrally closed ring. Since *maximal order* is a Morita invariant property, [69], *integrally closed* is a Morita invariant property, so that E is integrally closed. This completes the proof.

LEMMA 7.6.6 *If G is an rtffr $\mathcal{L}$-group and if G is locally isomorphic to $H \oplus K$ then H is an $\mathcal{L}$-group.*

Proof: Let G be locally isomorphic to $H \oplus K$ and suppose that $H' \cong H$. Then G, $H \oplus K$, and $H' \oplus K$ are quasi-isomorphic. Since G is assumed to be an $\mathcal{L}$-group $H \oplus K$ and $H' \oplus K$ are locally isomorphic groups. Then by Theorem 2.5.7, H is locally isomorphic to H', whence H is an $\mathcal{L}$-group. This completes the proof.

LEMMA 7.6.7 *Let G be an rtffr $\mathcal{L}$-group. There is a set $\{G_1, \ldots, G_t\}$ of strongly indecomposable groups and integers $t, n_1, \ldots, n_t > 0$ such that*

1. $G_i \doteq G_j \Rightarrow i = j$,

2. G is locally isomorphic to

$$G' = G_t^{(n_t)} \oplus \cdots \oplus G_t^{(n_t)},$$

3. *for each $i = 1, \ldots, t$, G_i is an $\mathcal{L}$-group.*

Proof: Suppose that G is an $\mathcal{L}$-group.

1 and 2.   By Jónsson's Theorem 2.1.10 there are strongly indecomposable groups $\{G_1, \ldots, G_t\}$ and integers $n_1, \ldots, n_t > 0$ such that $G_i \doteq G_j \Rightarrow i = j$ and such that G is quasi-isomorphic to G'. Since G is an $\mathcal{L}$-group G is locally isomorphic to G'.

Part 3 follows from Lemma 7.6.6, so the proof is complete.

COROLLARY 7.6.8 *Let G be an rtffr $\mathcal{L}$-group. Then*

$$G = G_1 \oplus \cdots \oplus G_r$$

for some strongly indecomposable $\mathcal{L}$-groups $G_1, \ldots, G_r$.

Proof : Apply Arnold's Theorem 3.5.1(2) to the above lemma.

LEMMA 7.6.9 *Let G be an indecomposable rtffr $\mathcal{L}$-group. Then $\mathrm{End}(G)$ is a classical maximal order in the division ring $\mathbb{Q}\mathrm{End}(G)$.*

Proof: Let $E = \mathrm{End}(G)$ and let $\mathcal{N} = \mathcal{N}(E)$. By Lemma 7.6.7 an indecomposable $\mathcal{L}$-group is strongly indecomposable so that $\mathbb{Q}\mathrm{End}(G)$ is a local ring. That is, each $x \in \mathbb{Q}E \setminus \mathbb{Q}\mathcal{N}$ is a unit in $\mathbb{Q}E$, Lemma 1.3.3.

Since G is reduced there is an integer $n \neq 0$ such that $\cap_{k>0} n^k G = 0$. Let $I = nE + \mathcal{N}$. Then G and IG are locally isomorphic so that by Theorem 2.5.1, $\mathrm{H}_G(G) = E$ and $\mathrm{H}_G(IG)$ are locally isomorphic right E-modules. Lemma 2.5.4 states that the localizations E_n and $\mathrm{H}_G(IG)_n$ are isomorphic, so $\mathrm{H}_G(IG)_n = x E_n$ for some $x \in \mathrm{H}_G(IG)_n$. Since $x \notin \mathcal{N}_n$ is a nonzero-divisor in E_n, $\mathcal{N}_n \subset \mathrm{H}_G(IG)_n = x E_n$ implies that

$$\mathcal{N}_n = x \mathcal{N}_n = x^k \mathcal{N}_n \text{ for each integer } k > 0.$$

Moroeover there is an integer $k > 0$ such that $\mathcal{N}^k = 0$. Then

$$(x E_n)^k \subset (n E_n + \mathcal{N}_n)^k \subset n^k E_n + n^{k-1} \mathcal{N}_n + \cdots + n \mathcal{N}_n^{k-1} \subset n E_n$$

hence

$$\mathcal{N}_n = x^k \mathcal{N}_n \subset (xE)^k \mathcal{N}_n \subset n\mathcal{N}_n \subset \mathcal{N}_n$$

whence $n\mathcal{N} = \mathcal{N}$. Subsequently $\mathcal{N} = \cap_k n^k G = 0$, so that $\mathcal{N} = 0$.

Then by Corollary 7.6.5, $E/\mathcal{N}(E) = E$ is integrally closed. Inasmuch as the semi-prime local ring $\mathbb{Q}E$ is a division ring E is a classical maximal order in a division ring. This completes the proof.

We can use some of the above results to prove that an rtffr $\mathcal{L}$-group is almost faithfully E-flat.

LEMMA 7.6.10 *Let G be an rtffr $\mathcal{L}$-group and let $E = \mathrm{End}(G)$. If E is semi-prime then G is a faithful E-flat group.*

Proof: Suppose that E is semi-prime and let I be a right ideal of E. Apply $I \otimes_E \cdot$ to the exact sequence

$$0 \to G \longrightarrow \mathbb{Q}G \longrightarrow \mathbb{Q}G/G \to 0$$

to produce the following exact sequence of groups.

$$\mathrm{Tor}_E^1(I, \mathbb{Q}G/G) \longrightarrow I \otimes_E G \longrightarrow I \otimes_E \mathbb{Q}G.$$

By Corollary 7.6.5, E is integrally closed, hence hereditary, whence I is a projective (= flat) right E-module. Consequently $\mathrm{Tor}_E^1(I, \cdot) = 0$ and so $I \otimes_E G \longrightarrow I \otimes_E \mathbb{Q}G$ is an injection. Since $\mathbb{Q}G$ is a direct summand of $\mathbb{Q}E^{(n)}$ as a left E-module and since $\mathbb{Q}E$ is a flat E-module $\mathbb{Q}G$ is a flat left E-module. Thus

$$I \otimes_E \mathbb{Q}G \to E \otimes_E \mathbb{Q}G \cong \mathbb{Q}G$$

is an injection and so $I \otimes_E G \to G$ is an injection. Then G is E-flat.

Next suppose that I is a maximal right ideal of finite index in E such that $IG = G$. Since I is a finitely generated projective right $\mathrm{End}_R(G)$-module the Arnold-Lady-Murley Theorem 2.4.1 implies that $I \cong E$ is principal, say $I = xE$ for some $x \in E$. Then

$$xG = xEG = IG = G.$$

Since G has finite rank x is an automorphism of G. That is $I = E$ so that G is faithful. This completes the proof.

7.7   Eichler $\mathcal{L}$-Groups Are $\mathcal{J}$-Groups

To this point there have been three classes of groups that we have considered. We showed that the finitely faithful $\mathcal{S}$-groups are $\mathcal{J}$-groups except in one case, and we know that $\mathcal{J}$-groups are $\mathcal{L}$-groups. The main Theorems 7.7.2 and 7.7.3 will show us that Eichler $\mathcal{L}$-groups are $\mathcal{J}$-groups.

LEMMA 7.7.1 *Let G be an rtffr $\mathcal{L}$-group. If $G = H \oplus K$ then H is an $\mathcal{L}$-group.*

Proof: Suppose that G is an $\mathcal{L}$-group and let $H' \doteq H$. Then

$$H' \oplus K \doteq H \oplus K = G$$

implies that $H' \oplus K$, $H \oplus K$, and G are locally isomorphic since G is an $\mathcal{L}$-group. But then Theorem 2.5.7 implies that H' and H are locally isomorphic. Thus H is an $\mathcal{L}$-group.

Note the use of Analytic Number Theory in the proof of the next two theorems.

THEOREM 7.7.2 [T.G. Faticoni] *The following are equivalent for an indecomposable Eichler group G.*

1. *G is a $\mathcal{J}$-group.*

2. *G is an $\mathcal{L}$-group.*

3. *G is a finitely faithful $\mathcal{S}$-group.*

Proof: We have commented on the implication $1 \Rightarrow 2$.

$2 \Rightarrow 3$ Of course G is an $\mathcal{S}$-group if G is a $\mathcal{L}$-group. By Lemma 7.6.9, $\text{End}(G)$ is a semi-prime ring so Lemma 7.6.10 implies that the indecomposable $\mathcal{L}$-group is faithful. This proves part 3.

$3 \Rightarrow 1$ If G is a finitely faithful $\mathcal{S}$-group then by Corollary 7.4.2, $G^{(2)}$ is a finitely faithful $\mathcal{S}$-group, so that $G^{(2)}$ is a $\mathcal{J}$-group by Corollary 7.6.2. By Lemma 7.7.1, G is then an $\mathcal{L}$-group.

To show that G is a $\mathcal{J}$-group suppose that $G \doteq H$. Then H is locally isomorphic to G so that $G \oplus H \cong G \oplus G$ (Lemma 2.5.6). Because G is indecomposable, $\text{End}(G)$ is a classical maximal order in the division $\mathbb{Q}$-algebra $\mathbb{Q}\text{End}(G)$, Lemma 7.6.9. Then by Theorem 7.6.4, $G \cong H$. Hence G is a $\mathcal{J}$-group and part 1 is proved. This completes the proof.

THEOREM 7.7.3 [T.G. Faticoni] *Eichler $\mathcal{L}$-groups are $\mathcal{J}$-groups.*

Proof: Let G be an Eichler $\mathcal{L}$-group and let $H \doteq G$. Then H is locally isomorphic to G. By Corollary 7.6.8

$$G = G_1 \oplus \cdots \oplus G_r$$

for some strongly indecomposable $\mathcal{L}$-groups $G_1, \ldots, G_r$. The reader will prove as an exercise that each G_i is an Eichler group. By Arnold's Theorem 3.5.1(2)

$$H = H_1 \oplus \cdots \oplus H_r$$

where $H_1, \ldots, H_r$ are indecomposable groups such that G_i and H_i are locally isomorphic. In particular, $G_i \cong H_i$. By Theorem 7.7.2, the indecomposable Eichler $\mathcal{L}$-groups G_i are $\mathcal{J}$-groups. Hence $G_i \cong H_i$ for each $i = 1, \ldots, r$, whence $G \cong H$, and therefore G is a $\mathcal{J}$-group.

COROLLARY 7.7.4 *The Eichler group G is a $\mathcal{J}$-group iff it is an $\mathcal{L}$-group.*

There is one other interesting case where our groups overlap.

THEOREM 7.7.5 *The following are equivalent for an Eichler group G.*

1. *G is a finitely faithful $\mathcal{S}$-group.*

2. *End(G) is semi-prime and G is a $\mathcal{J}$-group.*

Proof: Assume part 1. Let G be a finitely faithful $\mathcal{S}$-group. By Corollary 7.4.2, $G^{(2)}$ is a finitely faithful $\mathcal{S}$-group, so by Theorem 7.6.1, $G^{(2)}$ is a $\mathcal{J}$-group. By Lemma 7.7.1, G is then an $\mathcal{L}$-group, so by Theorem 7.7.3, G is a $\mathcal{J}$-group. Lemma 7.4.3 states that End(G) is a semi-prime ring which proves part 2.

Conversely, assume that G is a $\mathcal{J}$-group with semi-prime endomorphism ring. By Lemma 7.6.10, G is faithful. Since G is clearly an $\mathcal{S}$-group we have proved part 1.

EXAMPLE 7.7.6 An example due to Eichler ([69, midpage 305]) produces a totally definite quaternion algebra D with center $\mathbf{k}$ an algebraic number field, a maximal $\mathcal{O}$-order E in D, where $\mathcal{O}$ is the ring of algebraic integers in $\mathbf{k}$, and a nonprincipal right ideal $E \subset L \subset D$ such that $\mathrm{nr}(L) = \alpha \mathcal{O}$ for some $\alpha \in \mathbf{k}$. Then L and E have the same reduced norm, but $L \not\cong E$. Compare this to the last lines of the proof of Theorem 7.6.4. This explains the necessity of the Eichler hypothesis in Theorem 7.6.4.

7.8   Exercises

G, H, K are rtffr groups, $p \in \mathbb{Z}$ is a prime, E is an rtffr ring, M, N, are rtffr E-modules (left or right, depending on the setting).

1. G is a $\mathcal{J}$-group if G is a pure subgroup of $\mathbb{Z}$.

2. Let $0 \to K \to X \xrightarrow{\pi} G \to 0$ be *quasi-split*. Then $X \cong G \oplus K$.

3. Let G be a faithful rtffr group. Show that if $I \subset \mathrm{End}(G)$ is a maximal right ideal then $IG \neq G$ iff $I = \mathrm{Hom}(G, IG)$.

4. Suppose that H/H' is finite. Then G is finitely H-projective iff G is finitely H'-projective.

5. If H/H' is a finitely group then the inclusion $H' \subset H$ induces a canonical isomorphism $\mathrm{Ext}(G, H) \cong \mathrm{Ext}(G, H')$ of groups.

6. Let G be a torsion-free group such that $r_p(G)$ is countably infinite. If G is p-finitely G-projective then $\mathrm{End}(G)$ is uncountable.

7. Let G and H be torsion-free groups and let $p \in \mathbb{Z}$ be a prime. If G is $(p\text{-})$finitely H-projective then for each integer $k > 0$

   (a) $G^{(k)}$ is $(p\text{-})$finitely H-projective.

   (b) G is $(p\text{-})$finitely $H^{(k)}$-projective.

8. Let G and H be rtffr groups. If G is finitely H-projective then each short exact sequence $0 \to K \longrightarrow H \xrightarrow{\pi} G \to 0$ is split exact.

9. Let G be an rtffr group and suppose that G/H is finite. Then G is a finitely faithful $\mathcal{S}$-group iff H is a finitely faithful $\mathcal{S}$-group.

10. Let G be an rtffr group. Then G is a finitely faithful $\mathcal{S}$-group iff G is a finitely faithful $\mathcal{L}$-group.

11. Let G and H be rtffr groups. Assume that H' has finite index in H.

    (a) Show that G is p-finitely H-projective iff G is p-finitely H'-projective.

    (b) Show that G is finitely H-projective iff G is finitely H'-projective.

12. Let G and H be torsion-free groups and let $p \in \mathbb{Z}$ be a prime. If G is $(p\text{-})$finitely H-projective then for each integer $k > 0$

(a) $G^{(k)}$ is $(p\text{-})$finitely H-projective.

(b) G is $(p\text{-})$finitely $H^{(k)}$-projective.

13. [C. Murley] If G is a pure subgroup of $\widehat{\mathbb{Z}}$ then G is a $\mathcal{J}$-group.

14. Show that an indecomposable $\mathcal{L}$-group is strongly indecomposable.

15. Show that an indecomposable $\mathcal{L}$-group is a finitely faithful $\mathcal{S}$-group.

7.9 Questions for Future Research

Let G be an rtffr group, let E be an rtffr ring, and let $S = \text{center}(E)$ be the center of E. Let τ be the conductor of G. There is no loss of generality in assuming that G is strongly indecomposable.

1. Dualize p-finitely H-projective groups and study the resulting class of groups.

2. If G is p-finitely projective then $\text{Hom}(G, \mathbb{Z}/p\mathbb{Z})$ is a simple $\text{End}(G)$-module. Study the modules $\text{Hom}(G, \mathbb{Z}/p^k\mathbb{Z})$. Study the module $\text{Hom}(G, \mathbb{Q})$. Study other interesting classes of $\text{End}(G)$-modules in terms of $\text{Hom}(G, \cdot)$.

3. Study the category of G-plexes when G is a p-finitely projective group. See [32] for relevant definitions.

4. Characterize group theoretically rtffr $\mathcal{J}$-groups, $\mathcal{N}$-groups, and especially $\mathcal{S}$-groups.

5. Delete as much of the Analytic Number Theory used in our work on $\mathcal{L}$-groups as is possible.

6. Describe the rtffr groups G such that $G \oplus G \cong G \oplus H \implies G \cong H$.

7. Characterize the endomorphism rings of rtffr $\mathcal{J}$-groups.

Chapter 8

Gabriel Filters

The material in this chapter comes from [40] where the filter of divisibility

$$\mathcal{D}(G) = \{\text{right ideals } I \subset \text{End}(G) \mid IG = G\}$$

is studied. This filter is nicely structured. We are most interested in the subfilters of $\mathcal{D}(M)$ called *Gabriel filters* that correspond to a hereditary torsion class of right E-modules K such that $K \otimes_E G = 0$. In particular, we will show that $\mathcal{D}(G)$ is in many interesting cases bounded by an *idempotent ideal* Δ. That is, there is an ideal Δ of $\text{End}(G)$ such that $\Delta^2 = \Delta$ and such that $IG = G$ implies $\Delta \subset I$.

8.1 Filters of Divisibility

Let E be an rtffr ring and let G be an rtffr group. Our work on the Baer Splitting Property in section 6.1 motivates us to investigate the set

$$\mathcal{D}_E(G) = \mathcal{D}(G) = \{\text{right ideals } I \subset \text{End}(G) \mid IG = G\}.$$

We call $\mathcal{D}(G)$ *the filter of divisibility of* G. Then G is a *faithful* right E-module iff $\mathcal{D}(G) = \{\text{End}(G)\}$. We will consider faithful modules to be uninteresting in this section as they ignore any interesting structure there might be in $\mathcal{D}(G)$.

Let the first example act as an intuition builder.

EXAMPLE 8.1.1 Let $X \subset Y$ be groups such that $\mathrm{Hom}(Y, X) = 0$ and $S_X(Y) = Y$. Let $G = X \oplus Y$, and let

$$\Delta = \mathrm{Hom}(X, G)\mathrm{Hom}(G, X) \subset \mathrm{End}(G).$$

(For example, we could choose $X = \mathbb{Z}$ and

$$Y = \left\{ \frac{n}{m} \,\middle|\, n, m \in \mathbb{Z} \text{ and } m \text{ is square-free} \right\}.)$$

Because $\mathrm{Hom}(Y, X) = 0$, Δ is a direct summand of $\mathrm{End}(G)$ such that

$$
\begin{aligned}
\Delta^2 &= \mathrm{Hom}(X, G)(\mathrm{Hom}(G, X)\mathrm{Hom}(X, G))\mathrm{Hom}(G, X) \\
&= \mathrm{Hom}(X, G)\mathrm{Hom}(X, X)\mathrm{Hom}(G, X) \\
&= \mathrm{Hom}(X, G)\mathrm{Hom}(G, X) \\
&= \Delta.
\end{aligned}
$$

Furthermore, because $\mathrm{Hom}(Y, X) = 0$ and $Y = S_X(Y)$ we have

$$
\begin{aligned}
\Delta G &= \mathrm{Hom}(X, G)\mathrm{Hom}(G, X)(X \oplus Y) \\
&= \mathrm{Hom}(X, G)\mathrm{Hom}(X, X)X \\
&= \mathrm{Hom}(X, G)X \\
&= S_X(X) \oplus S_X(Y) \\
&= X \oplus Y \\
&= G.
\end{aligned}
$$

Thus $\mathcal{D}(G)$ contains the *infinite* set

$$\{\Delta + m\mathrm{End}(G) \,\big|\, m \in \mathbb{Z}\}.$$

The reader will show that Δ *is the smallest right ideal in* $\mathcal{D}(G)$.
    We will let

$$E = \mathrm{End}(G) = \{q \in \mathbb{Q}\mathrm{End}(G) \,\big|\, qG \subset G\}.$$

Theorem 2.3.4 can be used to construct a short exact sequence

$$0 \longrightarrow G \longrightarrow H \overset{\pi}{\longrightarrow} \mathbb{Q}E \oplus \mathbb{Q}E \longrightarrow 0 \tag{8.1}$$

in which $E = \mathrm{End}(H)$. Since $IG = G$ and $I\mathbb{Q}E = \mathbb{Q}E$ for each right ideal of the form $I = \Delta + mE$ for some $m \neq 0 \in \mathbb{Z}$, tensoring (8.1) with E/I shows that

$$\frac{H}{IH} = 0.$$

Hence $IH = H$ for each $m \in \mathbb{Z}$. Thus there are infinitely many right ideals $I \subset E$ such that E/I is finite and $IH = H$.

But since $J\mathbb{Q}E \neq \mathbb{Q}E$ for each proper right ideal J in $\mathbb{Q}E$, $JH \neq H$ for each pure proper right ideal in E. Specifically, $\Delta H \neq H$ even though $IH = H$ for each right ideal $\Delta \subset I = \Delta + mE$ of finite index in E.

8.1.1 Hereditary Torsion Classes

A nonempty set S of right $\text{End}(G)$-modules is a *hereditary torsion class* if

1. S is closed under the formation of direct sums. That is, if $M_k \in S, k \in I$ then $\oplus_{k \in I} M_k \in S$.

2. S is closed under the formation of short exact sequences. That is, if $0 \longrightarrow K \longrightarrow M \longrightarrow N \longrightarrow 0$ is a short exact sequence then $K, N \in S$ iff $M \in S$.

The prototypical examples of hereditary torsion classes are the set of groups, the set of torsion groups, and the set of torsion p-groups for some prime $p \in \mathbb{Z}$.

Our motivation for considering hereditary torsion classes is the set

$$\mathbf{Tor}_E(L) = \{\text{right } E\text{-modules } K \mid K \otimes_E L = 0\}$$

where L is a left E-module.

PROPOSITION 8.1.2 *Let E be a ring and let L be a flat left E-module. Then $\mathbf{Tor}_E(L)$ is a hereditary torsion class.*

Proof: Let $M_k \in \mathbf{Tor}(L), k \in I$. Because $\cdot \otimes_E L$ is an additive functor

$$(\oplus_{k \in I} M_k) \otimes_E L \cong \oplus_{k \in I}(M_k \otimes_E L) = 0.$$

Next consider the short exact sequence

$$0 \longrightarrow K \longrightarrow M \longrightarrow N \longrightarrow 0$$

of right E-modules. An application of the exact functor $\cdot \otimes_E L$ yields the exact sequence

$$0 \longrightarrow K \otimes_E L \overset{\alpha}{\longrightarrow} M \otimes_E L \longrightarrow N \otimes_E L \longrightarrow 0.$$

It follows that $K \otimes_E L = N \otimes_E L = 0$ iff $M \otimes_E L = 0$. Hence $\mathbf{Tor}(L)$ is a hereditary torsion class.

EXAMPLE 8.1.3 $\mathbf{Tor}_E(L)$ is a hereditary torsion class if L is a projective left E-module. If $p \in \mathbb{Z}$ is a prime then $\text{End}(\mathbb{Z}(p^\infty)) = \widehat{\mathbb{Z}}_p$, so that $\mathbf{Tor}_E(\mathbb{Z}(p^\infty))$ is the set of $\widehat{\mathbb{Z}}_p$-modules of the form

$$H \oplus D$$

where H is a torsion p-group and D is a divisible $\widehat{\mathbb{Z}}_p$-module. But $\mathbf{Tor}(\mathbb{Z}(p^\infty))$ is not a hereditary torsion class since the quotient field $\mathbb{Q}_p$ of $\widehat{\mathbb{Z}}_p$ is in $\mathbf{Tor}(\mathbb{Z}(p^\infty))$ and $\widehat{\mathbb{Z}}_p \subset \mathbb{Q}_p$, but $\widehat{\mathbb{Z}}_p \notin \mathbf{Tor}(\mathbb{Z}(p^\infty))$.

An important example of hereditary torsion classes comes from idempotent ideals.

PROPOSITION 8.1.4 *Let E be a ring and let $\Delta \subset E$ be an ideal such that $\Delta^2 = \Delta$ and $\cdot \otimes_E \Delta$ is an exact functor. The set*

$$\{\text{right } E\text{-modules } K \mid K\Delta = 0\}$$

is a hereditary torsion class.

We leave the proof as an exercise for the reader.

There is a relationship between a hereditary torsion class and the cyclic submodules it contains. Let $\mathcal{X}$ be a set of right E-modules. The *hereditary torsion class generated by* $\mathcal{X}$, denoted by $\mathbf{Tor}_E(\mathcal{X})$, is the intersection of all of the hereditary torsion classes that contain $\mathcal{X}$. Given a class $\mathcal{T}$ of right E-modules let $\mathbf{Cyc}_E(\mathcal{T})$ denote *the cyclic modules contained in* $\mathcal{T}$. Given a hereditary torsion class $\mathcal{T}$ the reader can show that $C \in \mathbf{Cyc}_E(\mathcal{T})$ iff C is a cyclic submodule of some $M \in \mathcal{T}$.

LEMMA 8.1.5 $\mathbf{Tor}_E(\mathbf{Cyc}_E(\mathcal{T})) = \mathcal{T}$ *if $\mathcal{T}$ is a hereditary torsion class.*

Proof: Clearly $\mathbf{Tor}_E(\mathbf{Cyc}_E(\mathcal{T})) \subset \mathcal{T}$. Let $M \in \mathcal{T}$. There is a submodule $K \subset M$ such that

$$\frac{\oplus_{x \in M} x\text{End}(G)}{K} \cong M.$$

Since the hereditary torsion class $\mathbf{Tor}_E(\mathbf{Cyc}_E(\mathcal{T}))$ is closed under arbitrary direct sums and factor modules $M \in \mathbf{Tor}_E(\mathbf{Cyc}_E(\mathcal{T}))$. Thus $\mathbf{Tor}_E(\mathbf{Cyc}_E(\mathcal{T})) = \mathcal{T}$, and the proof is complete.

Some examples will serve to illustrate these ideas.

A module M is called *semi-Artinian* if each nonzero factor M/N of M has nonzero socle. The simple modules appearing in the socle of M/N are called *composition factors of M*. The reader is encouraged to prove these examples.

EXAMPLE 8.1.6 Let E be a ring and let $\mathcal{S}$ be a set of simple right E-modules. The hereditary torsion class $\mathbf{Tor}_E(\mathcal{S})$ is the set of semi-Artinian right E-modules M whose composition factors are are in $\mathcal{S}$.

EXAMPLE 8.1.7 Let G be a left E-module. Then

$$\mathbf{Tor}_E\{E/I \,|\, IG = G\} \quad = \quad \{M \,|\, \text{ for each } x \in M,\, xE \subset E/I$$
$$\text{for some } I \subset E \text{ such that } IG = G\}.$$

EXAMPLE 8.1.8 The group G in Example 6.4.2 satisfies $\mathrm{T}_G(M) \neq 0$ for each nonzero finitely generated (cyclic) right $\mathrm{End}(G)$-module M but there is a right $\mathrm{End}(G)$-module $K \neq 0$ such that $\mathrm{T}_G(K) = 0$. That is, $\{E/I \,|\, IG = G\} = \emptyset$ while $\{K \,|\, K \otimes_E G = 0\} \neq \emptyset$.

8.1.2 Gabriel Filters of Right Ideals

In this section we indirectly study sets of modules of the form $\mathbf{Cyc}_E(\mathcal{T})$ for some hereditary torsion class $\mathcal{T}$. Given a right ideal $I \subset E$ and $x \in E$ let

$$(I : x) = \{a \in E \,|\, xa \in I\}.$$

The reader can show that $(I : x) = \mathrm{ann}_E(x + I)$ where $x + I \in E/I$.
  A *Gabriel filter* on E is a set γ of right ideals of E such that

1. Given $I, J \in \gamma$ and $I \subset I'$ then $I \cap J,\, I' \in \gamma$.

2. Given $I \in \gamma$ and $x \in E$ then $(I : x) \in \gamma$.

3. Given a right ideal $J \subset E$ and $I \in \gamma$ such that $(J : x) \in \gamma$ for each $x \in I$ then $J \in \gamma$.

  The relationship between Gabriel filters and hereditary torsion classes follows.

LEMMA 8.1.9 *Let E be a ring.*

1. *Let γ be a Gabriel filter on E. Then the hereditary torsion class $\mathbf{Tor}_E\left\{\dfrac{E}{I} \,\middle|\, I \in \gamma\right\}$ is the set of right E-modules K such that each cyclic submodule of K has the form E/I for some $I \in \gamma$.*

2. *Let $\mathcal{T}$ be a hereditary torsion class of right E-modules. Then $\left\{\text{right ideals } I \subset E \,\middle|\, \dfrac{E}{I} \in \mathcal{T}\right\}$ is a Gabriel filter on E.*

Proof: 1. Let $\mathbf{Tor}_E\{E/I \mid I \in \gamma\} = \mathbf{Tor}_E(\gamma)$. Since $\mathbf{Tor}_E(\gamma)$ is closed under direct sums and quotients, the set of right E-modules K such that each cyclic submodule of K has the form E/I for some $I \in \gamma$ is contained in $\mathbf{Tor}_E(\gamma)$.

On the other hand, if $K \in \mathbf{Tor}_E(\gamma)$ then $xE \in \{E/I \mid I \in \gamma\}$ for each $x \in K$. Thus $\mathbf{Tor}_E(\gamma)$ is contained in the set of right E-modules K such that each cyclic submodule of K has the form E/I for some $I \in \gamma$, and hence the two sets coincide.

2. Let $\mathcal{T}$ be a hereditary torsion class, and let

$$\mathbf{Gab}_E(\mathcal{T}) \;=\; \left\{ \text{right ideals } I \subset E \;\Big|\; \frac{E}{I} \in \mathcal{T} \right\}.$$

One readily shows that if $I, J \in \mathbf{Gab}_E(\mathcal{T})$ and if $I \subset I'$ then $E/(I \cap J) \subset E/I \oplus E/J$, and E/I maps onto E/I'. Thus $I \cap J, I' \in \mathbf{Gab}_E(\mathcal{T})$.

Next let $I \in \mathbf{Gab}_E(\mathcal{T})$, let $J \subset E$ be a right ideal, and suppose that $(J : x) \in \mathbf{Gab}_E(\mathcal{T})$ for each $x \in I$. We have a short exact sequence

$$0 \longrightarrow \frac{I+J}{J} \longrightarrow \frac{E}{J} \longrightarrow \frac{E}{I+J} \longrightarrow 0.$$

Since $I \subset I + J$ and since the hereditary torsion class $\mathcal{T}$ is closed under quotients, $E/(I + J) \in \mathcal{T}$. Let $x \in I$. Since

$$\frac{I+J}{J} \supset \frac{xE+J}{J} = \frac{E}{(J:x)} \in \mathcal{T}$$

and since $\mathcal{T}$ is closed under direct sums and quotients, $(I + J)/J \in \mathcal{T}$. Finally, since $\mathcal{T}$ is closed under extensions we see that $E/J \in \mathcal{T}$. Thus $J \in \mathbf{Gab}_E(\mathcal{T})$, from which we see that $\mathbf{Gab}_E(\mathcal{T})$ is a Gabriel filter on E. This completes the proof.

PROPOSITION 8.1.10 *Let E be a ring, let $\mathbf{Gab}(E)$ be the set of Gabriel filters of E, and let $\mathbf{Tor}(E)$ be the set of hereditary torsion classes on E. There are inverse bijections*

$$\mathbf{GAB} \;:\; \mathbf{Tor}(E) \longrightarrow \mathbf{Gab}(E)$$
$$\mathbf{TOR} \;:\; \mathbf{Gab}(E) \longrightarrow \mathbf{Tor}(E)$$

given by $\mathbf{GAB}(\mathcal{T}) = \mathbf{Gab}_E(\mathcal{T})$ *and* $\mathbf{TOR}(\gamma) = \mathbf{Tor}_E(\gamma)$.

Proof: The reader can verify that $\mathbf{GAB}(\mathbf{TOR}(\gamma)) = \gamma$ for each Gabriel filter γ and that $\mathbf{TOR}(\mathbf{GAB}(\mathcal{T})) = \mathcal{T}$ for each hereditary torsion class $\mathcal{T}$.

Therefore we can study hereditary torsion classes of right E-modules by studying Gabriel filters on E.

The reader can verify the examples as exercises.

EXAMPLE 8.1.11 Let $\mathcal{T}$ be the class of right E-modules such that for each $x \in M \in \mathcal{T}$, xE is a bounded group (i.e., $(xE)m = 0$ for some integer $m \neq 0$). Then $\mathbf{Tor}(\mathcal{T})$ is the hereditary torsion class of modules whose cyclic submodules are bounded groups. The associated Gabriel filter $\mathbf{Gab}(\mathcal{T})$ is the set of right ideals $I \subset E$ such that E/I is a torsion group.

The following sets are important to our discussions in this Chapter. Let E be a ring and let M be a *left* E-module.

$$\mathcal{D}_E(M) = \{\text{right ideals } I \subset E \,|\, IM = M\}$$

$$\max_E(M) = \{\text{maximal right ideals } I \subset E \,|\, IM = M\}.$$

These sets are not necessarily Gabriel filters so we will look for Gabriel filters associated with them. Let

$$\delta_E(M) = \{\text{right ideals } I \subset E \,|\, (I : x)M = M \ \forall x \in E\}$$
$$\mu_E(M) = \text{the least Gabriel filter on } E \text{ that contains } \max_E(M).$$

Given $I \in \delta_E(M)$ then $I = (I : 1)$ so that $\delta_E(M) \subset \mathcal{D}_E(M)$. Since the ring E is understood we will drop it from our notation.

The next result will allow us to say more about the right ideals in $\mu(M)$ as well as being of independent interest.

LEMMA 8.1.12 Let E be an rtffr ring. Then $I \in \mu(M)$ iff E/I is a finite *right E-module and each composition factor of E/I is isomorphic to some simple module in $\{E/J \,|\, J \in \max_E(M)\}$.

Proof: Let $\mu'(M)$ be the set of right ideals $I \subset E$ such that $IM = M$, E/I is finite, and each composition factor of E/I is isomorphic to E/J for some $J \in \max_E(M)$. By Lemma 4.2.3, each $I \in \max_E(M)$ has finite index in E, so that $\max_E(M) \subset \mu'(M)$. The reader will show that $\mu'(M)$ is a Gabriel filter on E. By definition as the least such filter, $\mu(M) \subset \mu'(M)$.

Conversely, let $I \in \mu'(M)$. Then $IM = M$, E/I is finite and each composition factor of E/I is of the form E/J for some $J \in \max_E(M) \subset$

$\mu(M)$. We observe that E/I is a factor module of the direct sum of cyclic modules

$$\bigoplus_{x \in E} \frac{xE + I}{I}.$$

Because E is an rtffr ring, torsion cyclic E-modules are bounded, hence finite groups. Thus for each $x \in E$, $(xE + I)/I \cong E/(I : x)$ is finite. Let K be a simple submodule of $E/(I : x)$. Then $K \cong J/(I : x)$ for some $(I : x) \subset J \in \mu(M)$. By a simple induction of the composition length of $E/(I : x)$, each cyclic submodule of $J/(I : x)$ is of the form E/L for some $L \in \mu(M)$. Since $\mu(M)$ is a Gabriel filter $(I : x) \in \mu(E)$ for each $x \in E$, and so $I \in \mu(M)$. It follows that $\mu'(M) = \mu(M)$.

An ideal Δ in some ring is *idempotent* if $\Delta^2 = \Delta$. Idempotent ideals will be important in the sequel.

LEMMA 8.1.13 *Suppose that* $\Delta^2 = \Delta$ *is an ideal in* E. *Then*

$$
\begin{aligned}
\mathcal{D}(\Delta) &= \{\text{right ideals } I \subset E \,\big|\, I\Delta = \Delta\} \\
\delta(\Delta) &= \text{the largest Gabriel filter contained in } \mathcal{D}(\Delta) \\
\max(\Delta) &= \{\text{maximal right ideals } I \,\big|\, \Delta \subset I \subset E\}.
\end{aligned}
$$

In particular Δ *is the unique minimal element in* $\mathcal{D}(\Delta)$.

Proof: This follows from the definitions using $M = \Delta$.

It follows that if $\Delta^2 = \Delta$ then $\delta(\Delta)$ is generated by the set of cyclic right E-modules K such that $K \otimes_E \Delta = 0$.

There is an interesting characterization of the hereditary torsion class corresponding to $\delta(M)$.

LEMMA 8.1.14 *Let* M *be a left* E-module. *Then* $K \in \mathbf{Tor}(\delta(M))$ *iff* $L \otimes_E M = 0$ *for each cyclic submodule* $L \subset K$.

Proof: Suppose that $x \in L \in \mathbf{Tor}(\delta(M))$. Since $\delta(M)$ is a Gabriel filter $xE \cong E/I$ for some $I \in \delta(M)$. Then $E/I \otimes_E M = 0$ and since each submodule of L is a sum of cyclics $K \otimes_E M = 0$ for each $K \subset L$. The converse is clear so the proof is complete.

LEMMA 8.1.15 *Let* E *be an rtffr ring and let* M *be a left* E-module.

1. *If* $\gamma \subset \mathcal{D}(M)$ *is a Gabriel filter on* E *then* $\gamma \subset \delta(M)$.

2. *If* $I \in \mathcal{D}(M)$ *is an ideal then* $I \in \delta(M)$.

Proof: 1. Say $I \in \gamma \subset \mathcal{D}(M)$. Then for each $x \in E$, $(I : x) \in \gamma \subset \mathcal{D}(M)$ so that $(I : x)M = M$. Thus $I \in \delta(M)$ and hence $\gamma \subset \delta(M)$.

2. Let $I \in \mathcal{D}(M)$ be an ideal and let $x \in E$. Then $I \subset (I : x)$ so that

$$M = IM \subset (I : x)M \subset M$$

which implies that $(I : x) \in \mathcal{D}(M)$ for each $x \in E$. Hence $I \in \delta(M)$.

The next result shows that even though in general $\mu(M) \neq \delta(M) \neq \mathcal{D}(M)$ they still share the same maximal right ideals.

LEMMA 8.1.16 *Let M be a left E-module.*

1. $\max_E(M) = \max(\delta(M)) = \max(\mu(M))$.

2. $\mu(M) = \{$right ideals $I \subset E \mid IG = G$ and E/I is semi-Artinian$\}$.

Proof: 1. Since $\delta(M) \subset \mathcal{D}(M)$ and since $\max_E(M) \subset \mu(M)$ we have

$$\max(\delta(M)) \subset \max_E(M) \subset \max(\mu(M)).$$

Let $J \in \max(\mu(M))$ and let $x \in E$. Inasmuch as E/J is simple, $E/J \cong E/(J : x)$, so that

$$xE \otimes_E M = E/(J : x) \otimes_E M \cong E/J \otimes_E M \cong M/JM = 0.$$

Then Lemma 8.1.14 states that $J \in \delta(M)$. This proves part 1.

2. is left as an exercise.

8.2   Idempotent Ideals

As above E denotes an rtffr ring. Further restrict M to be an rtffr left E-module. We will show that the Gabriel filters $\delta(M)$ and $\mu(M)$ on E are determined by an *idempotent ideal* $\Delta \subset \widehat{E}$.

8.2.1   Traces of Covers

The following concepts are found in [6].

Let M be a left E-module and let $K \subset M$ be a submodule. We say that K *is superfluous in M* if $K + N = M \implies N = M$ for each submodule $N \subset M$.

For instance, Nakayama's Lemma 1.2.3 implies that $J(E)M$ is superfluous in each *finitely generated* left E-module M. However each proper submodule of $\mathbb{Z}(p^\infty)$ is superfluous in $\mathbb{Z}(p^\infty)$. In a ring E, the Jacobson radical $\mathcal{J}(E)$ is the largest superfluous E-submodule of E. No proper subgroup of $\mathbb{Z}$ is superfluous in $\mathbb{Z}$, but for each prime $p \in \mathbb{Z}$, $p\mathbb{Z}_p$ is superfluous in $\mathbb{Z}_p$. If $\mathcal{N}$ is a nilpotent ideal then $\mathcal{N}M$ is superfluous in M for *each* left E-module M.

The reader can prove the following facts as exercises.

1. If $N \subset K \subset L \subset M$ and if K is superfluous in L then K is superfluous in M and K/N is superfluous in M/N.

2. Suppose that P is a finitely generated projective left E-module. An E-submodule $K \subset P$ is superfluous iff $K \subset J(E)P$.

The left E-module M *possesses a projective cover* if there is a projective left E-module P and a surjection $\pi : P \rightarrow M$ such that $\ker \pi$ is superfluous in P. Our work in this section will show that *projective covers* are abundant when considering modules over rtffr rings.

The trace of the projective cover $\pi : P \rightarrow M$ is

$$\Delta_M = \text{ the trace ideal of } P \text{ in } E.$$

Evidently Δ_M is an idempotent ideal in $\mathcal{D}(M)$. Since $\Delta_M P = P$ we have $\Delta_M M = M$.

A Gabriel filter γ is *bounded* if for each $I \in \gamma$ there is a two sided ideal $J \in \gamma$ such that $J \subset I$. We say that γ is *bounded by (the idempotent ideal)* Δ if $\gamma = \{$ right ideals $I \subset E \,|\, \Delta \subset I\}$. In particular $\Delta \in \gamma$. For example, each Gabriel filter on a commutative ring is bounded. If Δ is an idempotent ideal in E then by Lemma 8.1.13, Δ is the unique minimal element in $\delta(\Delta)$. The filter of nonzero ideals on $\mathbb{Z}$ is bounded but is not bounded by a unique ideal.

LEMMA 8.2.1 *Let* $\pi : P \rightarrow M$ *be a projective cover of the left E-module M. Then* $\mathcal{D}(\Delta_M) = \delta(M)$ *iff* $\delta(M)$ *is bounded.*

Proof: If $\mathcal{D}(\Delta_M) = \delta(M)$ then by the above comment $\delta(M)$ is bounded by Δ_M.

Conversely suppose that $\delta(M)$ is bounded. By Lemma 8.1.13, $\delta(M)$ contains every ideal in $\mathcal{D}(M)$, and by the above comments, $\Delta_M M = M$. Then $\Delta_M \in \delta(M)$. It remains to show that $\delta(M) \subset \mathcal{D}(\Delta_M)$. Let $I \in \delta(M)$. There is an ideal $J \subset I$ such that $J \in \delta(M)$. Then $JM = M$ so that $JP + \ker \pi = P$. Because J is an ideal of E and because $\ker \pi$ is superfluous in P, $JP = P$. As usual $\Delta_M \subset J \subset I$ so that $I \in \mathcal{D}(\Delta_M)$. Hence $\mathcal{D}(\Delta_M) = \delta(M)$, which completes the proof.

EXAMPLE 8.2.2   1. Given a ring E then the canonical map $E \to E/J(E)$ is a projective cover.

2. Every module over a right Artinian ring possesses a projective cover, [6].

3. The ring E is called *semi-perfect* if each finitely generated left E-module possesses a projective cover. E is semi-perfect iff $E/J(E)$ is semi-simple Artinian and idempotents in $E/J(E)$ lift to idempotents of E. (See [6].) Local rings are semi-perfect.

EXAMPLE 8.2.3 The following is an example of a finitely generated module M over a ring E such that E is not semi-perfect, M is not projective, but M possesses a finitely generated projective cover. Let A be the ring of matrices $\begin{pmatrix} \mathbb{Z} & 0 \\ \mathbb{Z} & \mathbb{Z} \end{pmatrix}$, let $P = \begin{pmatrix} 0 & 0 \\ \mathbb{Z} & \mathbb{Z} \end{pmatrix} \oplus \begin{pmatrix} 0 & 0 \\ \mathbb{Z} & \mathbb{Z} \end{pmatrix}$, and let $K = \left\{ \begin{pmatrix} 0 & 0 \\ x & 0 \end{pmatrix} \oplus \begin{pmatrix} 0 & 0 \\ x & 0 \end{pmatrix} \,\middle|\, x \in \mathbb{Z} \right\}$. The reader can show that $J(A) = \begin{pmatrix} 0 & 0 \\ \mathbb{Z} & 0 \end{pmatrix}$ so that $K \subset J(A)P$ is superfluous in P. Then P/K is a doubly generated left E-module and the natural map $\pi : P \to P/K$ is a projective cover.

An important source of projective covers is the following result.

LEMMA 8.2.4 [40, T.G. Faticoni]. *For each index $i \in \mathcal{I}$ let E_i be a ring, let M_i be a finitely generated left E_i-module, let $\pi_i : P_i \to M_i$ be a projective cover of M_i over E_i, and let Δ_i be the trace ideal of P_i in E_i. If there is an integer n such that each M_i is generated by at most n elements then*

1. $\prod_{i \in \mathcal{I}} M_i$ *is generated by at most n elements over $\prod_{i \in \mathcal{I}} E_i$,*

2. $\prod_{i \in \mathcal{I}} P_i$ *is a finitely generated projective left $\prod_{i \in \mathcal{I}} E_i$-module whose trace ideal satisfies $\Delta = \prod_{i \in I} \Delta_i$, and*

3. $\prod_{i \in \mathcal{I}} \pi_i : \prod_{i \in \mathcal{I}} P_i \to \prod_{i \in \mathcal{I}} M_i$ *is a projective cover over $\prod_{i \in \mathcal{I}} E_i$.*

Proof: 1. By hypothesis, for each $i > 0$ there is a set of generators $\{x_{i1}, \ldots, x_{in}\} \subset M_i$. For each $j = 1, \ldots, n$ define a sequence

$$x_j = (x_{1j}, x_{2j}, \ldots) \in \prod_{i \in \mathcal{I}} M_i.$$

We claim that $\{x_1, \ldots, x_n\}$ generates $\prod_i M_i$.

Given $y \in \prod_i M_i$ write $y = (y_i)_i$. By our choice of x_{ij} for each i there are $r_j = (r_{ij})_i$, $j = 1, \ldots, n$, such that

$$y_i = \sum_{j=1}^{n} r_{ij} x_{ij}.$$

Then

$$y = (y_i)_i = \left(\sum_{j=1}^{n} r_{ij} x_{ij} \right)_i = \sum_{j=1}^{n} (r_{ij} x_{ij})_i = \sum_{j=1}^{n} (r_{ij})_i (x_{ij})_i = \sum_{j=1}^{n} r_j x_j.$$

As claimed $\{x_1, \ldots, x_n\}$ generates $M = \prod_i M_i$ over $E = \prod_i E_i$.

2. For each $i \in \mathcal{I}$, P_i is generated by at most n elements over E_i, so that by part 1, $P = \prod_i P_i$ is a finitely generated projective left E-module. Inasmuch as

$$\operatorname{Hom}_E(P, E) = \prod_{i \in \mathcal{I}} \operatorname{Hom}_{E_i}(P_i, E_i)$$

it follows that the trace ideal for P in E is the product $\Delta = \prod_{i \in \mathcal{I}} \Delta_i$.

3. Suppose that for each $i \in \mathcal{I}$, K_i is a superfluous E_i-submodule of P_i and let $K = \prod_i K_i$. Since K_i is superfluous in P_i, $K_i \subset \mathcal{J}(E_i) P_i$. Since $\mathcal{J}(\prod_i E_i) = \prod_i \mathcal{J}(E_i)$

$$K = \prod_i K_i \subset \mathcal{J} \left(\prod_i E_i \right) \left(\prod_i P_i \right) = \mathcal{J}(E) P.$$

Since P is finitely generated, Nakayama's Lemma 1.2.3 shows us that K is superfluous in P.

The reader can show that because $M = \prod_i M_i$ is finitely generated

$$\pi = \prod_i \pi_i : P \longrightarrow M$$

is a surjection. Inasmuch as $\ker(\pi) = \prod_i (\ker \pi_i)$, π is a projective cover of M over E. This completes the proof.

The reader might ask where in the realm of rtffr groups will we have a chance to see an infinite product of rings. The next few results will answer this question. Given a prime p let

$$p^\omega G = \bigcap_{n=1}^{\infty} p^k G.$$

Recall that idempotents lift modulo nilpotent ideals. (See [6].)

LEMMA 8.2.5 *Let E be an rtffr ring. Then for each prime p the p-adic completion $\widehat{E}_p$ of $E/p^\omega E$ is a semi-perfect ring.*

Proof: Let E_p be the localization of E at p. Because E has finite rank $E_p/p^k E_p \cong E/p^k E$ is a finite (hence Artinian) ring for each integer $k > 0$, so that idempotents lift module the Jacobson radical of $\widehat{E}_p/p^k \widehat{E}_p$. Furthermore $p \in \mathcal{J}(\widehat{E}_p)$.

Let $\bar{e}^2 = \bar{e} \in \widehat{E}_p/\mathcal{J}(\widehat{E}_p)$. Because

$$\mathcal{J}(\widehat{E}_p)/p^k \widehat{E}_p = \mathcal{J}(\widehat{E}_p/p^k \widehat{E}_p)$$

and because

$$p^{k-1}\widehat{E}_p/p^k \widehat{E}_p$$

is a nilpotent ideal of $\widehat{E}_p/p^k \widehat{E}_p$ for each integer $k > 0$ we can construct a sequence $(e_k)_k$ in $\widehat{E}_p$ such that

$$e_1 + \mathcal{J}(\widehat{E}_p) = \bar{e}, \qquad e_{k-1} - e_k \in p^{k-1}\widehat{E}_p, \qquad e_k^2 - e_k \in p^k \mathcal{J}(\widehat{E}_p).$$

Since $(e_k)_k$ clearly converges in the p-adic topology on $\widehat{E}_p$ its limit exists. That is, $\lim_k e_k = e \in \widehat{E}_p$. Moreover $e^2 - e \equiv e_k^2 - e_k \in p^k \widehat{E}_p$ for each k. Since $p^\omega \widehat{E}_p = 0$, $e^2 = e$. That is, e is an idempotent lifting of $\bar{e}$. Hence $\widehat{E}_p$ is semi-perfect.

LEMMA 8.2.6 *Let E be an rtffr ring, and let M be an rtffr right E-module. Then $\widehat{M}_p$ is generated as a $\widehat{\mathbb{Z}}_p$-module by at most rank(M) elements for each prime $p \in \mathbb{Z}$. Furthermore $\widehat{M}$ is generated as a $\widehat{\mathbb{Z}}$-module by at most rank(M) elements.*

Proof: Let $p \in \mathbb{Z}$ be a prime. Because $\mathbb{Z}/p\mathbb{Z}$-linearly independent subsets of M/pM lift to $\mathbb{Z}$-independent subsets of M, M/pM has $\mathbb{Z}/p\mathbb{Z}$-dimension at most rank(M). Thus $\widehat{M}_p/p\widehat{M}_p \cong M/pM$ has dimension at most rank(M), so that $\widehat{\mathbb{Z}}_p^{\mathrm{rank}(M)}$ maps onto $\widehat{M}_p/p\widehat{M}_p$. It follows that there is a free $\widehat{\mathbb{Z}}_p$-submodule K of $\widehat{M}_p$ generated by at most rank(M) elements over $\widehat{\mathbb{Z}}_p$, and such that $K + p\widehat{M}_p = \widehat{M}_p$. Thus K is dense in $\widehat{M}_p$. Since K is finitely generated it is p-adically complete so that $K = \widehat{M}_p$. Thus $\widehat{M}_p$ is generated by at most rank(M) elements. By Lemma 8.2.4, $\widehat{M} = \prod_p \widehat{M}_p$ is then generated by at most rank(M) elements as a $\widehat{\mathbb{Z}} = \prod_p \widehat{\mathbb{Z}}_p$-module. This proves the Lemma.

THEOREM 8.2.7 [40, T.G. Faticoni] *Let E be an rtffr ring and let M be an rtffr right E-module. Then $\widehat{M}$ possesses a finitely generated projective cover over $\widehat{E}$.*

Proof: Because E and M are rtffr groups, $\widehat{E} = \prod_p \widehat{E}_p$ and $\widehat{M} = \prod_p \widehat{M}_p$ where p ranges over the primes in $\mathbb{Z}$. Fix $p \in \mathbb{Z}$. By Lemma 8.2.6, $\widehat{M}_p$ is generated as a $\widehat{\mathbb{Z}}_p$-module by rank(M) elements. Consequently $\widehat{M}_p$ possesses a projective cover over the semi-perfect ring $\widehat{E}_p$ (Lemma 8.2.5), so that $\widehat{M} = \prod_p \widehat{M}_p$ possesses a finitely generated projective cover over the ring $\widehat{E}$ (Lemma 8.2.4). This completes the proof.

COROLLARY 8.2.8 *Let G be a rtffr group. Then $\widehat{G}$ possesses a finitely generated projective cover over $\widehat{\mathrm{End}(G)}$. Moreover $\widehat{G}$ is generated by at most* rank(G) *elements as a left $\widehat{\mathrm{End}(G)}$-module.*

8.2.2   Bounded Gabriel Filters

In this section we examine the existence and the usefulness of idempotent ideals Δ such that $\Delta M = M$. Specifically we show that $\mu(M)$ is finite iff it is bounded by an idempotent ideal.

Let us agree that

$$\delta(\widehat{M}) \;=\; \{\text{right ideals } I \subset \widehat{E} \,\big|\, (I:x)\widehat{M} = \widehat{M} \forall x \in \widehat{E}\}.$$

Then $\delta(\widehat{M})$ denotes the *largest Gabriel filter in* $\mathcal{D}_{\widehat{E}}(\widehat{M})$, the set of right ideals $I \subset \widehat{E}$ such that $I\widehat{M} = \widehat{M}$. (See Lemma 8.1.15.) Furthermore let

$$\mu(\widehat{M}) = \{\text{right ideals } I \subset \widehat{E} \,\big|\, \widehat{E}/I \text{ is finite and } I\widehat{M} = \widehat{M}\}.$$

Observe that $\mu(\widehat{M})$ is a Gabriel filter on $\widehat{E}$. We will show that Gabriel filters over p-adically complete torsion-free rings are especially nice.

THEOREM 8.2.9 *Let E be an rtffr ring and let M be an rtffr right E-module.*

1. $\delta(\widehat{M})$ *is a bounded Gabriel filter on $\widehat{E}$.*

2. $\mu(\widehat{M})$ *is a bounded Gabriel filter on $\widehat{E}$.*

Proof: 1. Let $I \in \delta(\widehat{M})$. Since $\widehat{E}$ is a finitely generated $\widehat{\mathbb{Z}}$-module (Lemma 8.2.6) we can write $\widehat{E}/I = \widehat{\mathbb{Z}}x_1 + \cdots + \widehat{\mathbb{Z}}x_n$ for some integer n and elements $x_i \in \widehat{E}/I$. An elementary argument shows us that

$$(I : x_1) \cap \cdots \cap (I : x_n) \subset \text{ann}_{\widehat{E}}(\widehat{E}/I) \subset I.$$

Since $\delta(\widehat{M})$ is a Gabriel filter, $(I : x_1) \cap \cdots \cap (I : x_n) \in \delta(\widehat{M})$, from which it follows that the ideal $\text{ann}_{\widehat{E}}(\widehat{E}/I)$ is in $\delta(\widehat{M})$. Thus $\delta(\widehat{M})$ is bounded.

2. Proceed as in part 1. This completes the proof.

COROLLARY 8.2.10 *Let E be an rtffr ring and let M be an rtffr right E-module. Then $\mu(M)$ is a bounded Gabriel filter on E.*

Proof: Let $I \in \mu(M)$. By Lemma 8.1.12, E/I is torsion so there is an integer $m \neq 0$ such that $mE1 = E(m1) \subset I$. Hence E/I is bounded, and since E is an rtffr, E/I is finite. Write $\{x_1, \ldots, x_n\} = E/I$. Then

$$mE \subset (I : x_1) \cap \cdots \cap (I : x_n) \subset \text{ann}_E(E/I) \subset I$$

so that $\text{ann}_E(E/I) \subset I$ is an ideal in $\mu(M)$. This completes the proof.

THEOREM 8.2.11 *Let E be an rtffr ring and let M be an rtffr right E-module. There is a finitely generated idempotent ideal $\Delta_{\widehat{M}} \subset \widehat{E}$ such that*

1. $\delta(\widehat{M}) = \{$right ideals $I \subset \widehat{E} \,|\, \Delta_{\widehat{M}} \subset I\}$. In particular, $\Delta_{\widehat{M}} \in \delta(\widehat{M})$.

2. $\mu(\widehat{M}) = \{$right ideals $I \subset \widehat{E} \,|\, \widehat{E}/I$ is finite and $\Delta_{\widehat{M}} \subset I\}$.

Proof: By Theorem 8.2.7, $\widehat{M}$ possesses a projective cover $\pi : P \to \widehat{M}$ over $\widehat{E}$. Let $\Delta_{\widehat{M}}$ be the trace ideal of P in $\widehat{E}$. Then $\Delta_{\widehat{M}}$ is the idempotent ideal in $\widehat{E}$ such that $\Delta_{\widehat{M}}P = P$, and $IP = P$ iff $\Delta_{\widehat{M}} \subset I$.

1. Since $\widehat{M} = \pi(P) = \pi(\Delta_{\widehat{M}}P) = \Delta_{\widehat{M}}\widehat{M}$, $\Delta_{\widehat{M}} \in \delta(\widehat{M})$, and so $\{$right ideals $\Delta_{\widehat{M}} \subset I \subset \widehat{E}\} \subset \delta(\widehat{M})$.

Conversely let $I \in \delta(\widehat{M})$. By Theorem 8.2.9, $\delta(\widehat{M})$ is bounded, so there an ideal $J \subset I$ such that $J\widehat{M} = \widehat{M}$. Because $\pi(JP) = J\widehat{M} = \widehat{M}$, $JP + \ker \pi = P$. Inasmuch as $\ker \pi$ is superfluous in P, and because JP is a left $\widehat{E}$-submodule of P, $JP = P$. Hence $\Delta_{\widehat{M}} \subset J$ and thus $\delta(\widehat{M}) = \{$right ideals $I \subset \widehat{E} \,|\, \Delta_{\widehat{M}} \subset I\}$.

Prove part 2 in a manner similar to part 1. This completes the proof.

The rtffr group H constructed at the end of Example 8.1.1 should dispel any notions that $\delta(M) = \mu(M)$ is bounded by an idempotent ideal when M is an rtffr right E-module.

EXAMPLE 8.2.12 This example shows that $\mu(M)$ need not be finite. Let $\pi : P \longrightarrow M$ be the projective cover constructed in Example 8.2.3. Then $\Delta_M = \begin{pmatrix} 0 & 0 \\ \mathbb{Z} & \mathbb{Z} \end{pmatrix}$, and

$$\delta(M) = \left\{ \begin{pmatrix} n\mathbb{Z} & 0 \\ \mathbb{Z} & \mathbb{Z} \end{pmatrix} \mid n \in \mathbb{Z} \right\}$$

is bounded by the idempotent ideal Δ_M. Furthermore,

$$\mu(M) = \left\{ \begin{pmatrix} n\mathbb{Z} & 0 \\ \mathbb{Z} & \mathbb{Z} \end{pmatrix} \mid n \neq 0 \in \mathbb{Z} \right\}$$

is not bounded by an idempotent ideal.

8.2.3   Finite Filters of Divisibility

In this section we study modules M for which $\mu(M)$ and $\delta(M)$ are finite sets.

The following Lemma shows us that $\mu(M)$ and $\mu(\widehat{M})$ are identical posets.

LEMMA 8.2.13 *Let E be an rtffr ring and let M be an rtffr right E-module. There are inverse poset isomorphisms $\mu(M) \cong \mu(\widehat{M})$ defined by*

$$I \mapsto \widehat{I} \text{ and } I \mapsto I \cap E.$$

Proof: Let $\widehat{I} \in \mu(\widehat{M})$ and let $I = \widehat{I} \cap E$. Then $\widehat{E}/\widehat{I}$ and E/I are finite. There is an integer $m \neq 0$ such that $m \in I, \widehat{I}$. Because I is dense in $\widehat{I}$ and because $\widehat{I} \in \mu(\widehat{M})$

$$IM + m\widehat{M} = (I + m\widehat{I})(M + m\widehat{M}) = \widehat{I}\widehat{M} = \widehat{M} = M + m\widehat{M}.$$

Then by the Modular Law and because $m \in I$ we have

$$\begin{aligned} M &= (M + m\widehat{M}) \cap M \\ &= (IM + m\widehat{M}) \cap M \\ &= IM + (m\widehat{M} \cap M) \\ &= IM + mM \\ &= IM. \end{aligned}$$

Thus the assignment $\widehat{I} \mapsto \widehat{I} \cap E$ defines a function $\mu(\widehat{M}) \longrightarrow \mu(M)$.

Conversely let $I \in \mu(M)$ and choose a nonzero integer $m \in I$. Then

$$\widehat{IM} = (I + m\widehat{I})(M + m\widehat{M}) = IM + m\widehat{M} = M + m\widehat{M} = \widehat{M}$$

so that $\widehat{I} \in \mu(\widehat{M})$. Thus the assignment $I \mapsto \widehat{I}$ defines a function $\mu(M) \longrightarrow \mu(\widehat{M})$.

Finally it is well known that $\widehat{I} \cap E = I$ for each right ideal I of finite index in E, and that $\widehat{I' \cap E} = I'$ for each right ideal I' of finite index in $\widehat{E}$. Thus the above maps are mutual inverses, which completes the proof.

Using the above correspondence we can characterize $\mu(M)$ in terms of an idempotent ideal.

THEOREM 8.2.14 *Let E be an rtffr ring and let M be an rtffr right E-module. There is a finitely generated idempotent ideal $\Delta_{\widehat{M}} \subset \widehat{E}$ such that*

$$\mu(M) = \{\text{right ideals } I \subset E \mid E/I \text{ is finite and } \Delta_{\widehat{M}} \subset \widehat{I} \subset \widehat{E}\}.$$

Moreover $\Delta_{\widehat{M}}$ is generated by at most $\operatorname{rank}(E)$ elements.

Proof: The proof is an application of Lemma 8.2.13 and Theorem 8.2.11. We leave the details to the reader.

We consider the Gabriel filters that are bounded by an idempotent ideal to be the most important examples of Gabriel filters for reasons that will be obvious in the next Theorem. Since each rtffr module M over E is associated with a projective cover π over $\widehat{E}$, the projective cover brings with it the trace ideal $\Delta_{\widehat{M}}$ of the projective. The next result gives at least one case in which $\Delta_{\widehat{M}}$ leads us to an idempotent ideal in E. This is the case when $\mu(M)$ is finite.

THEOREM 8.2.15 [40, T.G. Faticoni] *Let E be an rtffr ring and let M be an rtffr right E-module. The following are equivalent for M.*

1. $\mu(M)$ *is finite.*

2. $\mu(M) = \mathcal{D}(\Delta)$ *for some idempotent ideal Δ in E.*

3. $\Delta_{\widehat{M}}$ *has finite index in $\widehat{E}$.*

If M satisfies one and hence all of the above conditions then $\Delta = \Delta_{\widehat{M}} \cap E$.

Proof: $3 \Rightarrow 2$ Suppose that $\Delta_{\widehat{M}}$ has finite index in $\widehat{E}$. Then by Theorem 8.2.11

$$\mu(\widehat{M}) = \{\text{right ideals } I \subset \widehat{E} \,|\, \widehat{E}/I \text{ is finite and } \Delta_{\widehat{M}} \subset I\}$$
$$= \{\text{right ideals } I \subset \widehat{E} \,|\, \Delta_{\widehat{M}} \subset I\}.$$

If we let $\Delta = \Delta_{\widehat{M}} \cap E$ then Δ has finite index in E. The reader can show that Δ is idempotent since $\Delta_{\widehat{M}}$ is idempotent. Lemma 8.2.13 then implies that

$$\Delta \in \mu(M) = \{I \cap E \,|\, I \text{ is a right ideal in } \widehat{E} \text{ and } \Delta \subset I\}$$
$$= \{I \cap E \,|\, I \text{ is a right ideal in } \widehat{E} \text{ and } \Delta \subset I \cap E\}$$
$$= \{\text{right ideals } J \subset E \,|\, \Delta \subset J\}.$$

This proves part 2.

$2 \Rightarrow 1$ follows immediately from Lemma 8.1.12.

$1 \Rightarrow 3$ Because $\mu(M) = \{I_1, \cdots, I_n\}$ is finite, and because $\mu(M)$ is bounded, $\mu(M)$ contains a unique minimal element

$$\Delta = I_1 \cap \cdots \cap I_n \in \mu(M).$$

Then Lemma 8.2.13 shows us that $\widehat{\Delta}$ is the unique minimal element of $\mu(\widehat{M})$. We observe that $\widehat{\Delta}$ has finite index in $\widehat{E}$.

To see that $\Delta_{\widehat{M}} = \widehat{\Delta}$ notice that $\mu(M) \cong \mu(\widehat{M})$ implies that $\mu(\widehat{M})$ is finite, hence bounded by the idempotent ideal $\widehat{\Delta}$. We showed in Theorem 8.2.11 that $\Delta_{\widehat{M}}$ is the unique minimal element of $\mu(\widehat{M})$. This proves part 3 and completes the logical cycle.

Let E be the ring and let M be the module constructed possessing a projective cover $\pi : P \to M$ in Example 8.2.3. We showed that $\Delta_M = \begin{pmatrix} 0 & 0 \\ \mathbb{Z} & \mathbb{Z} \end{pmatrix}$. We then constructed an rtffr group $M \subset G \subset \widehat{M}$ such that each element of $\mu(G)$ has finite index in E. Specifically $\Delta_M \notin \mu(G)$. Furthermore, $\mu(G)$ is infinite and does not possess a unique minimal element.

EXAMPLE 8.2.16 The reader will show that $\mu(G)$ is finite if G is the group constructed in Example 6.1.7.

There is at least one class of rtffr ring in which each Gabriel filter is bounded by an idempotent ideal.

THEOREM 8.2.17 *Let E be an rtffr ring and let M be an rtffr left E-module such that $\mathrm{ann}_{\widehat{E}}(\widehat{M}) = 0$. (For example, let M be any module such that $E = \mathrm{End}(M)$.) If E is a semi-prime ring then $\mu(M)$ is finite.*

Proof: Suppose that E is semi-prime and that $\mathrm{ann}_E(M) = 0$. Then $\widehat{E}$ is semi-prime. Write

$$\Delta_{\widehat{M}} = \prod_p \Delta_p$$

where Δ_p is an idempotent ideal in the semi-prime ring $\widehat{E}_p$. Since $\mathrm{ann}_{\widehat{E}}(\widehat{M}) = 0$, $\mathrm{ann}_{\widehat{E}_p}(\Delta_p) = 0$. However, since $\widehat{E}_p$ is semi-prime there is an ideal I in $\widehat{E}$ such that $\Delta_p \oplus I \doteq \widehat{E}_p$. Since then $I\Delta_p \subset \Delta_p \cap I = 0$, $I = 0$ and so $\Delta_p \doteq \widehat{E}_p$. Now for all but at most finitely many primes p, $\widehat{E}_p$ is an integrally closed ring. In such a ring, proper ideals are invertible, not idempotent. See [69]. Say $\Delta_p J = \widehat{E}_p$. Then

$$\Delta_p = \Delta_p(\widehat{E}_p) = \Delta_p(\Delta_p J) = \Delta_p J = \widehat{E}_p.$$

That is $\Delta_p = \widehat{E}_p$ for all but at most finitely many primes p. Thus $\widehat{E}/\Delta_{\widehat{M}} = \prod_p \widehat{E}_p/\Delta_p$ is finite, and hence by the previous Theorem, $\mu(M)$ is finite. This completes the proof.

Let E be a rtffr ring and let M be a rtffr right E-module. We say that M is *faithful* if $IM \neq M$ for each right ideal I in E. Evidently M is faithful iff $\mathcal{D}(M) = \{E\}$. Since each right ideal in E is contained in a maximal right ideal and since each maximal right ideal I in E has finite index in E, we see that M is faithful iff $IM \neq M$ for each right ideal I of finite index in E, Lemma 4.2.3. It is clear that M is a faithful left E-module iff $\mu(M) = \{E\}$. As we showed in Chapter 6, the faithful property is connected to the splitting property of short exact sequences. The next result characterizes the faithful left E-modules.

THEOREM 8.2.18 *Let E be an rtffr ring and let M be an rtffr right E-module. The following are equivalent for M.*

1. $\mathcal{D}(M) = \{E\}$.

2. $\delta(M) = \{E\}$.

3. $\mu(M) = \{E\}$.

4. $\mu(\widehat{M}) = \{\widehat{E}\}$.

5. If I is a right ideal in E then there is an $x \in E$ such that $(I : x)M \neq M$.

6. $\Delta_{\widehat{M}} = \widehat{E}$.

7. Let K be a torsion (as a group) right E-module such that $K \otimes_E M = 0$. There exists a cyclic submodule $N \subset K$ such that $N \otimes_E M \neq 0$.

Proof: $1 \Rightarrow 2 \Rightarrow 3$ follows from the inclusions $\mu(M) \subset \delta(M) \subset \mathcal{D}(M)$.

$3 \Rightarrow 1$ Suppose that part 1 is false. There is a right ideal I such that $IM = M$. We may assume that I is a maximal right ideal of E and hence that $I \in \max(M) \subset \mu(M)$. This proves that part 3 is false.

$3 \Leftrightarrow 4$ follows from the isomorphism of posets $\mu(M) \cong \mu(\widehat{M})$ (Lemma 8.2.13).

$2 \Leftrightarrow 5$ follows from the definition of $\delta(M)$.

$3 \Leftrightarrow 6$ is Theorem 8.2.14.

$7 \Rightarrow 3$ Suppose that $\mu(M) \neq \{E\}$. Let $I \subset E$ be a maximal right ideal of E such that $IM = M$. Then $E/I = (x + I)E \cong E/(I : x)$ for each $x \in E$. It follows that $E/(I : x) = 0$ for each $x \in E$ and hence part 7 is false.

$3 \Rightarrow 7$ follows in a similar manner. This completes the logical cycle.

8.3   Gabriel Filters on Rtffr Rings

Let E denote a fixed rtffr ring and let

$$
\begin{aligned}
\Omega(E) &= \{ \text{rtffr groups } G \,|\, \mathrm{End}(G) \cong E \text{ as rings}\} \\
\Delta_{\widehat{G}} &= \text{the trace ideal of a projective cover} \\
&\quad\ \pi : P \to \widehat{G} \text{ over } \widehat{E}.
\end{aligned}
$$

(See Theorem 8.2.7.) In this section we will investigate the rtffr rings E such that each $G \in \Omega(E)$ is faithful. We characterize those idempotent ideals $\Delta \subset \widehat{E}$ such that $\Delta = \Delta_{\widehat{G}}$ for some $G \in \Omega(E)$.

8.3.1   Applications to Endomorphism Rings

Let G be an rtffr group. It is natural to investigate how the above Theorems can be used to explore the faithful property on G. In the next result we characterize those rtffr groups G for which $\delta(G)$ is a finite set.

THEOREM 8.3.1 *The following are equivalent for the rtffr group G.*

1. $\delta_E(G)$ is finite.

2. $\mu_E(G)$ is finite.

3. $\mu_{\widehat{E}}(\widehat{G})$ is finite.

4. $\Delta_{\widehat{G}}$ has finite index in $\widehat{\text{End}(G)}$.

5. There is an idempotent ideal Δ of finite index in $\text{End}(G)$ such that

$$\delta_E(G) = \{\text{right ideals } I \mid \Delta \subset I \subset \text{End}(G)\}.$$

In this case $\Delta = \Delta_{\widehat{G}} \cap \text{End}(G)$.

Proof: $5 \Rightarrow 1 \Rightarrow 2 \Rightarrow 3$ are clear, $3 \Rightarrow 4 \Rightarrow 5$ is Theorem 8.2.15. This completes the logical cycle.

The faithful rtffr groups are characterized in the following result.

THEOREM 8.3.2 *The following are equivalent for the rtffr group G.*

1. $\mathcal{D}(G) = \{\text{End}(G)\}$.

2. $\delta(G) = \{\text{End}(G)\}$.

3. $\mu(G) = \{\text{End}(G)\}$.

4. $\mu(\widehat{G}) = \{\widehat{\text{End}(G)}\}$.

5. If I is a right ideal in $\text{End}(G)$ then there is an $x \in \text{End}(G)$ such that $(I : x)G \neq G$.

6. $\Delta_{\widehat{G}} = \widehat{\text{End}(G)}$.

7. Let K be a torsion (as a group) right $\text{End}(G)$-module such that $K \otimes_{\text{End}(G)} G = 0$. There exists a cyclic submodule $N \subset K$ such that $N \otimes_{\text{End}(G)} G \neq 0$.

The Proof follows from Theorem 8.2.18.

The short exact sequence

$$0 \longrightarrow K \longrightarrow X \stackrel{\pi}{\longrightarrow} G \longrightarrow 0 \qquad (8.2)$$

is *quasi-split* if there is a map $\jmath : G \rightarrow X$ and an integer $n \neq 0$ such that $\pi\jmath = n1_G$. If (8.2) is quasi-split then $X \cong G \oplus K$. The property $\delta(G) = \mu(G)$ is characterized in terms of quasi-splitting.

The faithful property is intimately associated with the splitting of short exact sequences. Thus we include this connection.

THEOREM 8.3.3 *The following are equivalent for an rtffr group* G.

1. *If* $IG = G$ *for some right ideal* $I \subset \mathrm{End}(G)$ *then* $\mathrm{End}(G)/I$ *is finite. That is,* $\delta(G) = \mu(G)$.

2. *Each short exact sequence (8.2) in which* $X = \mathrm{S}_G(X) + K$ *is quasi-split.*

3. *Each short exact sequence (8.2) in which* $X \cong G^{(c)}$ *for some cardinal* c *is quasi-split.*

Proof: $1 \Rightarrow 2$ is proved in the same manner that we proved Baer's Lemma 6.1.1.

$2 \Rightarrow 3$ is clear.

$3 \Rightarrow 1$ Suppose that $IG = G$ for some proper right ideal I in $\mathrm{End}(G)$-submodule of $\mathrm{H}_G(H)$. Write $I = \sum_{k \in \mathcal{I}} \pi_k \mathrm{End}(G)$ for some index set $\mathcal{I}$, let $G_k = G$ for each $k \in \mathcal{I}$, and let $\pi : \oplus_{k \in \mathcal{I}} G_k \longrightarrow G$ be the unique map such that $\pi(x_k) = \pi_k(x_k)$ for each $x_k \in G_k$ and each $k \in \mathcal{I}$. Notice that π is a surjection since $G = IG = \sum_{k \in \mathcal{I}} \pi_k(G_k)$. Thus there is a short exact sequence (8.2) in which $X \cong \oplus_{k \in \mathcal{I}} G_k$. By part 5, (8.2) is quasi-split. There is a map $\jmath : G \rightarrow \oplus_{k \in \mathcal{I}} G_k$ and an integer $n \neq 0$ such that $\pi\jmath = n1_G$. Since the rtffr group G is self-small we can write $\jmath = \oplus_{k \in \mathcal{I}} \jmath_k$ for some maps $\jmath_k : G \rightarrow G_k$. Then

$$n1_G = \pi\jmath = \left(\sum_k \pi_k\right)\left(\sum_k \jmath_k\right) = \left(\sum_k \pi_k\jmath_k\right) \in I$$

by our choice of $\pi_k \in I$. Then $\mathrm{End}(G)/I$ is bounded, hence finite, and part 1 follows. This completes the logical cycle.

It is interesting to list the following result since it characterizes those rtffr groups G such that $IG = G$ for some proper nonzero pure right ideal $I \subset \mathrm{End}(G)$ in terms of a splitting result. We make the following elementary observation. Suppose that E is an rtffr ring and let $I \subset E$ be a right ideal such that E/I is a torsion group. There is an integer $n \neq 0$ such that $n1_G \in I$ so that $nE \subset I \subset E$. Since E has finite rank the bounded group E/I is finite.

COROLLARY 8.3.4 *The following are equivalent for the rtffr group* G.

1. *There is a proper nonzero pure right ideal* $I \subset \text{End}(G)$ *such that* $IG = G$.

2. *There is a short exact sequence*

$$0 \longrightarrow K \longrightarrow G^{(c)} \overset{\pi}{\longrightarrow} G \longrightarrow 0 \qquad (8.3)$$

that is not quasi-split.

Proof: This theorem is the contrapositive of Theorem 8.3.3.

8.3.2 Constructing Examples

Given a torsion-free ring R and a left R-module M let

$$\mathcal{O}_R(M) = \{r \in \mathbb{Q}R \mid rM \subset M\}.$$

LEMMA 8.3.5 *Let* E *be an rtffr ring and let* M *be an rtffr left* E-*module. Then*

$$\mathcal{O}_E(M) = E \text{ iff } \mathcal{O}_{\widehat{E}}(\widehat{M}) = \widehat{E}.$$

Proof: Suppose that $\mathcal{O}_{\widehat{E}}(\widehat{M}) = \widehat{E}$ and let $r \in \mathcal{O}_E(M)$. Then $rM \subset M$ and since M is a pure subgroup of the pure injective group $\widehat{M}$, $r\widehat{M} \subset \widehat{M}$. Thus

$$r \in \widehat{E} \cap \mathbb{Q}E = E$$

and hence $\mathcal{O}_E(M) \subset E$. The inclusion $E \subset \mathcal{O}_E(M)$ is obvious.

Conversely, assume that $\mathcal{O}_E(M) = E$. Evidently $\widehat{E} \subset \mathcal{O}_{\widehat{E}}(\widehat{M})$ and $\mathcal{O}_{\widehat{E}}(\widehat{M})/\widehat{E}$ is torsion. It is a short exercise to show that $\mathcal{O}_E(M)$ is a pure subgroup of $\mathcal{O}_{\widehat{E}}(\widehat{M})$. Then $E = \mathcal{O}_E(M)$ is pure and dense in $\widehat{E}$ and in $\mathcal{O}_{\widehat{E}}(\widehat{M})$, so that $\widehat{E} = \mathcal{O}_{\widehat{E}}(\widehat{M})$. This completes the proof.

The next result characterizes the idempotent ideals Δ in $\widehat{E}$ such that $\Delta = \Delta_G$ for some $G \in \Omega(E)$.

THEOREM 8.3.6 *Let* E *be an rtffr ring and let* Δ *be an idempotent ideal in* $\widehat{E}$. *The following are equivalent.*

1. $\Delta = \Delta_{\widehat{G}}$ for some $G \in \Omega(E)$.

2. Δ is finitely generated as a left $\widehat{E}$-module and $\widehat{E} = \mathcal{O}_{\widehat{E}}(\Delta)$.

Proof: $1 \Rightarrow 2$ Suppose that $\Delta = \Delta_{\widehat{G}}$ for some $G \in \Omega(E)$. By Theorem 8.2.14, Δ is finitely generated. Since Δ is an ideal in $\widehat{E}$, $\widehat{E} \subset \mathcal{O}_{\widehat{E}}(\Delta)$. On the other hand, by Lemma 8.3.5, $\Delta\widehat{G} = \widehat{G}$ implies that

$$\mathcal{O}_{\widehat{E}}(\Delta) \subset \mathcal{O}_{\widehat{E}}(\widehat{G}) = \widehat{E}.$$

Then $\mathcal{O}_{\widehat{E}}(\Delta) = \widehat{E}$ and part 2 follows.

$2 \Rightarrow 1$ Since Δ is a finitely generated torsion-free (=projective) $\widehat{\mathbb{Z}}$-module Δ is a $\mathbb{Z}$-adically complete group.

Let $\{x_1, \ldots, x_n\}$ be a set of generators of Δ as a left $\widehat{\mathbb{Z}}$-module. Let

$$M = (\mathbb{Q}Ex_1 + \cdots + \mathbb{Q}Ex_n) \cap \Delta.$$

Since $\mathbb{Z}x_i$ is dense in $\widehat{\mathbb{Z}}x_i$ for each $i = 1, \cdots, n$, M is pure and dense in $\Delta = \sum_i \widehat{E}x_i$. That is, $\widehat{M} = \Delta$.

By Theorem 8.2.7, $\widehat{M}$ possesses a projective cover $\pi : P \to \widehat{M}$ over $\widehat{E}$. Since $\Delta^2 = \Delta = \widehat{M}$

$$\pi(P) = \widehat{M} = \Delta = \Delta\Delta = \Delta\widehat{M} = \pi(\Delta P). \tag{8.4}$$

Hence $\Delta P + \ker \pi = P$. Since Δ is an ideal in $\widehat{E}$ and since $\ker \pi$ is superfluous in P, $\Delta P = P$. Then $\Delta_P \subset \Delta$ where Δ_P denotes the trace ideal of P. On the other hand, because Δ_P and Δ are idempotent ideals, $\Delta_P \subset \Delta$ implies that $\Delta_P\Delta = \Delta_P$. By (8.4)

$$\Delta = \pi(P) = \pi(\Delta_P P) = \Delta_P\pi(P) = \Delta_P\Delta = \Delta_P.$$

Thus $\Delta = \Delta_P$.

Finally by hypothesis $\mathcal{O}_{\widehat{E}}(\widehat{M}) = \mathcal{O}_{\widehat{E}}(\Delta) = \widehat{E}$ so that $\mathcal{O}_E(M) = E$ (Lemma 8.3.5). Theorem 2.3.3 constructs a short exact sequence

$$0 \longrightarrow M \longrightarrow G \longrightarrow \mathbb{Q}C \longrightarrow 0$$

of left E-modules such that $E \cong \operatorname{End}(G)$ naturally. It follows that $\widehat{M} = \widehat{G}$ so by the above paragraph $\Delta = \Delta_P = \Delta_{\widehat{G}}$. This completes the proof.

EXAMPLE 8.3.7 Let $E = \begin{pmatrix} \mathbb{Z} & 0 \\ \mathbb{Z} & \mathbb{Z} \end{pmatrix}$ and let $\Delta = \begin{pmatrix} \mathbb{Z} & 0 \\ \mathbb{Z} & 0 \end{pmatrix}$. Then Δ is a cyclic projective left E-module, $\Delta^2 = \Delta$, and $E = \mathcal{O}_E(\Delta)$. Consequently $\widehat{\Delta}$ is a finitely generated idempotent ideal of $\widehat{E}$ such that

$\widehat{E} = \mathcal{O}_{\widehat{E}}(\widehat{\Delta})$. By Theorem 8.3.6 there is an rtffr group G such that $\Delta \subset G \subset \widehat{\Delta} = \Delta_G$ and $E \cong \mathrm{End}(G)$. Furthermore $G/\Delta \cong \mathbb{Q}E \oplus \mathbb{Q}E$ so that $\Delta G \neq G$, while $\mu(G) = \{$right ideals $I \mid E/I$ is finite and $\Delta \subset I \subset E\}$ is infinite.

The following is an interesting example of an idempotent ideal of finite index in the ring.

EXAMPLE 8.3.8 Let $\mathcal{O}$ be a maximal order in the ring of Hamiltonian Quaternions $\mathbb{H}$. (See [69].) It is known that for primes $p \neq 2 \in \mathbb{Z}$, $\mathcal{O}/p\mathcal{O} \cong \mathrm{Mat}_{2\times2}(\mathbb{Z}/p\mathbb{Z})$. Let $p\mathcal{O} \subset \Delta \subset \mathcal{O}$ be the right ideal in $\mathcal{O}$ such that

$$\Delta/p\mathcal{O} = \begin{pmatrix} 0 & 0 \\ \mathbb{Z}/p\mathbb{Z} & \mathbb{Z}/p\mathbb{Z} \end{pmatrix}.$$

One shows that Δ is a finitely generated projective idempotent ideal in $\mathcal{O}_\mathbb{H}(\Delta)$ where $p\mathcal{O} \subset \mathcal{O}_\mathbb{H}(\Delta) \subset \mathcal{O}$ is such that

$$\mathcal{O}_\mathbb{H}(\Delta)/p\mathcal{O} = \begin{pmatrix} \mathbb{Z}/p\mathbb{Z} & 0 \\ \mathbb{Z}/p\mathbb{Z} & \mathbb{Z}/p\mathbb{Z} \end{pmatrix}.$$

Theorem 8.3.6 states that $\widehat{\Delta} = \Delta_{\widehat{G}}$ for some group G such that $\mathcal{O}_\mathbb{H}(\Delta)) \cong \mathrm{End}(G)$. Since $\mathcal{O}_\mathbb{H}(\Delta)/\Delta$ is finite and since $\widehat{G} = \widehat{\Delta}$, $\Delta G = G$, and hence $\delta(G) = \mu(G)$ is finite. Observe that G is strongly indecomposable.

EXAMPLE 8.3.9 Let $E = \mathbb{Z}$. Then $\widehat{E} = \prod_p \widehat{\mathbb{Z}}_p$. Let $\Delta = \oplus_p \widehat{\mathbb{Z}}_p$. Then one readily verifies that $\Delta^2 = \Delta$ and that $E = \mathcal{O}_E(\Delta)$. However $\Delta \neq \Delta_{\widehat{G}}$ for any rtffr group G such that $E \cong \mathrm{End}(G)$ since Δ is not a finitely generated ideal in $\widehat{E}$.

8.3.3   Faithful Rings

The ring E is a *faithful* ring if given a right ideal $I \subset E$ and a $G \in \Omega(E)$ such that $IG = G$ then $I = E$. D.M. Arnold and E.L. Lady [14] and [41] show that the rtffr ring is a faithful ring if E is commutative, right hereditary, or local. More examples of faithful rings are given at the end of this subsection.

The following theorem characterizes the faithful rtffr rings.

THEOREM 8.3.10 *Let E be an rtffr ring. The following are equivalent for E.*

1. E is a faithful ring.

2. If Δ is a finitely generated idempotent ideal of $\widehat{E}$ such that $\widehat{E} = \mathcal{O}_{\widehat{E}}(\Delta)$ then $\Delta = \widehat{E}$.

3. If P is a finitely generated projective left $\widehat{E}$-module such that $\widehat{E} = \mathcal{O}_{\widehat{E}}(P)$ then P generates $\widehat{E}$.

Proof: $1 \Rightarrow 2$ Assume part 2 is false. Then there is a proper finitely generated idempotent ideal $\Delta \subset \widehat{E}$ such that $\mathcal{O}_{\widehat{E}}(\Delta)$. By Theorem 8.3.6 there is a $G \in \Omega(E)$ such that $\Delta = \Delta_{\widehat{G}} \neq \widehat{E}$ and $\Delta\widehat{G} = \widehat{G}$. Then part 1 is false.

$2 \Rightarrow 3$ Assume part 2. Let P be a finitely generated projective left $\widehat{E}$-module such that $\widehat{E} = \mathcal{O}_{\widehat{E}}(P)$, and let Δ be the trace ideal of P in $\widehat{E}$. Then Δ is a finitely generated idempotent ideal of $\widehat{E}$ and by Lemma 8.3.5, $\widehat{E} = \mathcal{O}_{\widehat{E}}(\Delta)$. By part 2, $\Delta = \widehat{E}$, which proves part 3.

$3 \Rightarrow 1$ Assume part 3 and let $G \in \Omega(E)$. By Theorem 8.2.7 there is a finitely generated projective cover $\pi : P \to \widehat{G}$ over $\widehat{E}$. If we let Δ be the trace ideal of P then by Lemma 8.3.5

$$\widehat{E} \subset \mathcal{O}_{\widehat{E}}(P) = \mathcal{O}_{\widehat{E}}(\Delta) \subset \mathcal{O}_{\widehat{E}}(\widehat{G}) = \widehat{E}.$$

Then $\mathcal{O}_{\widehat{E}}(P) = \widehat{E}$, so that by part 3, P generates $\widehat{E}$. Hence $\Delta = \widehat{E}$ is the trace ideal of P in $\widehat{E}$. Inasmuch as $\Delta = \Delta_{\widehat{G}}$, Theorem 8.3.2 implies that G is a faithful group. This proves part 1 and completes the logical cycle.

The following are examples of faithful rings. The ring E is *local* if E possesses a unique maximal right ideal. In this case $\mathcal{J}(E)$ is the unique maximal right ideal of E and $E/\mathcal{J}(E)$ is a division ring.

EXAMPLE 8.3.11 A local rtffr ring is faithful.

Proof: Say $E/\mathcal{J}(E)$ is simple. It follows from Lemma 4.2.3 that $E/\mathcal{J}(E)$ is an elementary p-group for some prime $p \in \mathbb{Z}$. Then $pE \neq E$ and $qE = E$ for each prime $q \neq p \in \mathbb{Z}$. Thus $\widehat{E} = \widehat{E}_p$. One readily proves that $\widehat{\mathcal{J}(E)}_p = \mathcal{J}(\widehat{E}_p)$ so that $\widehat{E}_p$ is a local ring. Since a projective $\widehat{E}_p$-module is then free it generates $\widehat{E}_p$. Theorem 8.3.10 then implies that each $G \in \Omega(E)$ is faithful.

The ring E is *subcommutative* if each right ideal in E is an ideal in E. Commutative rings are subcommutative. If H is a maximal order in the ring of Hamiltonian Quaternions over $\mathbb{Q}$ then there is a unique prime ideal $2H \subset I \subset H$ such that H/I is a field. Then the localization H_2 and the completion $\widehat{H}_2$ are noncommutative subcommutative rings.

EXAMPLE 8.3.12 Each subcommutative rtffr domain is faithful.

Proof: Let E be an rtffr subcommutative domain, and let Δ be a finitely generated idempotent ideal of $\widehat{E}$ such that $\mathcal{O}_{\widehat{E}}(\Delta) = \widehat{E}$. By Theorem 8.3.10 it suffices to show that $\Delta = \widehat{E}$.

Since E is a subcommutative domain, each right ideal has finite index in E, and hence for each integer n, each right ideal of E/nE is an ideal. Thus $\widehat{E}$ is subcommutative. By Lemma 8.2.5, $\widehat{E}_p$ is a semi-perfect ring for each prime $p \in \mathbb{Z}$ so we can write

$$\widehat{E}_p = E_1 \oplus \cdots \oplus E_n$$

for some indecomposable (right) ideals $E_i \subset \widehat{E}_p$. This direct sum is one of indecomposable semi-perfect rings so each E_i is a local ring. Consequently

$$\Delta_p = \Delta_1 \oplus \cdots \oplus \Delta_n$$

where $\Delta_i \subset E_i$. But by Nakayama's Lemma the local Noetherian ring does not contain a proper nonzero ideal $I^2 = I$. Thus $\Delta_i = E_i$ for each $i = 1, \ldots, n$, hence $\Delta_p = \widehat{E}_p$, whence $\Delta = \widehat{E}$. Given our reductions and Theorem 8.3.10, E is a faithful ring.

We have shown in Theorem 8.2.17 that semi-prime rtffr rings are *bounded rings*. That is, each right ideal contains a nonzero ideal. We characterize the rtffr bounded rings by first characterizing those rtffr rings E such that $\delta(G)$ is finite for each $G \in \Omega(E)$.

THEOREM 8.3.13 [40, T.G. Faticoni] *The following are equivalent for the rtffr ring E.*

1. $\mu(G)$ *is finite for each* $G \in \Omega(E)$.

2. $\delta(G)$ *is finite for each* $G \in \Omega(E)$.

3. *If* $\Delta \subset \widehat{E}$ *is a finitely generated idempotent ideal such that* $\widehat{E} = \mathcal{O}_{\widehat{E}}(\Delta)$ *then* $\widehat{E}/\Delta$ *is finite.*

Proof: $1 \Leftrightarrow 2$ is Theorem 8.3.2.

$1 \Rightarrow 3$ Assume part 1, and let Δ be a finitely generated idempotent ideal in $\widehat{E}$ such that $\widehat{E} = \mathcal{O}_{\widehat{E}}(\Delta)$. By Theorem 8.3.6 there is a group $G \in \Omega(E)$ such that $\Delta_{\widehat{G}} = \Delta$. By Theorem 8.3.2 and part 1, $\Delta_{\widehat{G}}$ has finite index in $\widehat{E}$. This proves part 3.

$3 \Rightarrow 1$ Assume part 3, let $G \in \Omega(E)$, and consider $\mu(G)$. By Lemma 8.2.13 and Theorem 8.3.2, $\mu(G) \cong \mu(\widehat{G}) = \{$ right ideals $I \mid \widehat{E}/I$ is finite and $\Delta_{\widehat{G}} \subset I \subset \widehat{E}$}. It therefore suffices to show that $\Delta_{\widehat{G}}$ has finite index in $\widehat{E}$.

Since G is an rtffr group, $\widehat{G}$ is a finitely generated left module over $\widehat{E}$, and thus $\Delta_{\widehat{G}}$ is a finitely generated $\widehat{\mathbb{Z}}$-submodule of $\widehat{E}$. By Lemma 8.3.5, $\mathcal{O}_{\widehat{E}}(\Delta) = \widehat{E}$, so that by part 3, $\Delta_{\widehat{G}}$ has finite index in $\widehat{E}$. Given our reductions $\mu(G)$ is finite.

In Example 8.2.12 we constructed a group G such that $\delta(G) = \mu(G)$ is infinite.

8.4   Gabriel Filters on $\mathbb{Q}\mathrm{End}(G)$

In this section we characterize in terms of quasi-summands of G the Gabriel filter of right ideals I of $\mathbb{Q}\mathrm{End}(G)$ such that $I\mathbb{Q}G = \mathbb{Q}G$. Let Example 8.2.12 act as an intuition builder.

8.4.1   Central Quasi-summands

In the previous sections we discussed the right ideals I of finite index in $\mathrm{End}(G)$ such that $IG = G$. In this section we will investigate the pure right ideal $I \subset \mathrm{End}(G)$ such that G/IG is torsion, or equivalently such that $(\mathbb{Q}I)\mathbb{Q}G = \mathbb{Q}G$. This naturally leads us to consider

$$\mathcal{D}(\mathbb{Q}G) = \{\text{right ideals } I \text{ of } \mathbb{Q}\mathrm{End}(G) \,\big|\, I(\mathbb{Q}G) = \mathbb{Q}G\}.$$

$$\delta(\mathbb{Q}G) = \{\text{right ideals } I \text{ of } \mathbb{Q}\mathrm{End}(G) \,\big|\, (I : x)(\mathbb{Q}G) = \mathbb{Q}G$$
$$\text{for each } x \in \mathbb{Q}\mathrm{End}(G)\}.$$

Inasmuch as $\mathbb{Q}\mathrm{End}(G)$ is Artinian and since the Gabriel filter $\delta(\mathbb{Q}G)$ contains finite intersections there is an idempotent ideal $\Delta_{\mathbb{Q}G} \subset \mathbb{Q}\mathrm{End}(G)$ such that

$$\delta(\mathbb{Q}G) = \mathcal{D}(\Delta_{\mathbb{Q}G}) = \{\text{ right ideals } I \subset \mathbb{Q}\mathrm{End}(G) \,\big|\, \Delta_{\mathbb{Q}G} \subset I\}.$$

EXAMPLE 8.4.1 Let X and Y be rtffr groups such that $Y = \mathrm{Hom}(X, Y)(X)$, $0 = \mathrm{Hom}(Y, X)$, and

$$\mathrm{End}(X \oplus Y) = \begin{pmatrix} \mathbb{Z} & 0 \\ \mathbb{Z} & \mathbb{Z} \end{pmatrix}.$$

See, e.g., Example 8.1.1. Then $\Delta = \begin{pmatrix} 0 & 0 \\ \mathbb{Z} & \mathbb{Z} \end{pmatrix}$ is such that $\Delta G = G$ for any group $G \doteq X \oplus Y$. Thus $\mathbb{Q}\Delta\mathbb{Q}G = \mathbb{Q}G$. The reader can show that $\mathbb{Q}\Delta$ is the smallest element of $\delta(\mathbb{Q}G)$.

Given a quasi-direct sum decomposition $G \doteq H \oplus H' \subset \mathbb{Q}G$ there are idempotents $e_H, e_{H'} \in \mathbb{Q}\text{End}(G)$ such that

$$e_H G \doteq H, \quad e_{H'}G \doteq K, \quad \text{and} \quad e_H \oplus e_{H'} = 1_G.$$

Call H a *nilpotent quasi-summand of* G if each composition of maps $H \to H' \to H$ is naturally a nilpotent endomorphism of G. In this case we call $G \doteq H \oplus H'$ a *nilpotent quasi-decomposition of* G. The reader can show that if H is a nilpotent quasi-summand of G and if K is a nilpotent quasi-summand of H then K is a nilpotent quasi-summand of G. Evidently, if $G \doteq H \oplus H'$ is a nilpotent quasi-decomposition of G then $\{H, H'\}$ is a nilpotent set.

The first lemma characterizes nilpotent quasi-summands in terms of idempotents in $\mathbb{Q}\text{End}(G)/\mathcal{N}(\mathbb{Q}\text{End}(G))$.

LEMMA 8.4.2 *Let* $G \doteq H \oplus H'$ *and let* $e_H : G \longrightarrow H$ *be the canonical idempotent associated with* H. *Consider maps* $f : H \longrightarrow H'$ *as endomorphisms of* G *in the natural way. The following are equivalent for* H.

1. H *is a nilpotent quasi-summand of* G.

2. $\text{Hom}(H, H') \subset \mathcal{N}(\mathbb{Q}\text{End}(G))$.

3. e_H *is central modulo* $\mathcal{N}(\mathbb{Q}\text{End}(G))$.

Proof: $1 \Rightarrow 2$ Let $a : H \longrightarrow H'$ and $x : G \longrightarrow G$ be elements of $\mathbb{Q}\text{End}(G)$. Then $a = e_{H'}ae_H$ and since H is a nilpotent quasi-summand $e_H x e_{H'} a e_H$ is a nilpotent endomorphism of G. Hence

$$(xa)^{m+1} = (xa)(e_E x e_B a e_E)^m = 0$$

for large enough m. That is, a generates a nil left ideal of $\mathbb{Q}\text{End}(G)$.

$2 \Rightarrow 3$ Let $x \in \mathbb{Q}\text{End}(G)$. By part 2, $e_H x e_{H'}$, $e_{H'} x e_H \in \mathcal{N}(\mathbb{Q}\text{End}(G))$. Then

$$
\begin{aligned}
e_H x - x e_H &= (e_H x e_H + e_H x e_{H'}) - (e_H x e_H + e_{H'} x e_H) \\
&= e_H x e_{H'} - e_{H'} x e_H \\
&\in \mathcal{N}(\mathbb{Q}\text{End}(G))
\end{aligned}
$$

so that e_H is central modulo $\mathcal{N}(\mathbb{Q}\mathrm{End}(G))$.

$3 \Rightarrow 1$ Let $H \neq H'$ and let $H \xrightarrow{a} H' \xrightarrow{b} H$. Then $ba = e_H b(e_{H'} a e_H)$. By part 3,

$$0 = e_H(e_{H'}a) \equiv (e_{H'}a)e_H(\mathrm{mod}\ \mathcal{N}(\mathbb{Q}\mathrm{End}(G)))$$

so that $ba \equiv 0(\mathrm{mod}\ \mathcal{N}(\mathbb{Q}\mathrm{End}(G)))$. Then ba is nilpotent, and hence H is a nilpotent quasi-summand of G. This completes the logical cycle.

Given a quasi-summand H of G let

$$
\begin{aligned}
\Delta_{G,H} &= \mathbb{Q}\mathrm{Hom}(H,G)\mathbb{Q}\mathrm{Hom}(G,H) \\
&= \text{the ideal generated by compositions} \\
& \qquad G \longrightarrow H \longrightarrow G.
\end{aligned}
$$

Evidently $\Delta_{G,H}$ is an ideal in $\mathbb{Q}\mathrm{End}(G)$. Because H is a quasi-summand of G, $\mathbb{Q}\mathrm{Hom}(G,H)\mathrm{Hom}(H,G) = \mathbb{Q}\mathrm{End}(H)$ so that

$$
\begin{aligned}
\Delta_{G,H}^2 &= (\mathbb{Q}\mathrm{Hom}(H,G)\mathrm{Hom}(G,H))(\mathbb{Q}\mathrm{Hom}(H,G)\mathrm{Hom}(G,H)) \\
&= \mathbb{Q}\mathrm{Hom}(H,G)\mathbb{Q}\mathrm{End}(H)\mathbb{Q}\mathrm{Hom}(G,H) \\
&= \mathbb{Q}\mathrm{Hom}(H,G)\mathrm{Hom}(G,H) \\
&= \Delta_{G,H}.
\end{aligned}
$$

That is, $\Delta_{G,H}$ is an idempotent ideal of $\mathbb{Q}\mathrm{End}(G)$.

Let E be a ring. Recall that $\mathbf{GAB}(E)$ denotes the set of Gabriel filters of right ideals on E, and let $\mathbf{IDEM}(E)$ denote the set of idempotent ideals $\Delta \subset E$. Partially order $\mathbf{GAB}(\mathbb{Q}\mathrm{End}(G))$ and $\mathbf{IDEM}(\mathbb{Q}\mathrm{End}(G))$ by inclusion.

Let $\mathbf{Nil}(G)$ denote the set of quasi-isomorphism classes $[H]$ of nilpotent quasi-summands H of G, and define a partial order $\leq$ on $\mathbf{Nil}(G)$ by declaring that $[K] \leq [H]$ if K is quasi-isomorphic to a quasi-summand of H. Because G has finite rank $\leq$ is a partial order on $\mathbf{Nil}(G)$.

The next result is a pair of bijections that characterize the nilpotent quasi-summands of $\mathbb{Q}\mathrm{End}(G)$ in terms of idempotent ideals of $\mathbb{Q}\mathrm{End}(G)$ and Gabriel filters of right ideals in $\mathbb{Q}\mathrm{End}(G)$.

LEMMA 8.4.3 *Define maps*

$$
\begin{aligned}
\Delta &: \mathbf{GAB}(\mathbb{Q}\mathrm{End}(G)) \longrightarrow \mathbf{IDEM}(\mathbb{Q}\mathrm{End}(G)) \\
D &: \mathbf{IDEM}(\mathbb{Q}\mathrm{End}(G)) \longrightarrow \mathbf{Nil}(G)
\end{aligned}
$$

· by

$$
\begin{aligned}
\Delta(\gamma) &= \text{\emph{the unique minimal element of }} \gamma \\
D(\Delta) &= [eG], \text{\emph{where }} \Delta \text{\emph{ is generated by the idempotent}} \\
&\quad \text{\emph{e in }} \mathbb{Q}\mathrm{End}(G).
\end{aligned}
$$

Then Δ and D are isomorphisms of posets.

Proof: Given a Gabriel filter γ on the Artinian ring $\mathbb{Q}\mathrm{End}(G)$ the intersection of all right ideals in γ is an ideal $\Delta = \Delta(\gamma)$ that is evidently the unique minimal element in γ. This Δ is necessarily an idempotent ideal. Thus $\gamma = \mathcal{D}(\Delta)$ for some unique idempotent ideal Δ of $\mathbb{Q}\mathrm{End}(G)$. The function Δ is the one that sends $\gamma \mapsto \Delta(\gamma)$. It is an easy exercise to show that the function $\Delta(\cdot)$ is a bijection.

Next given $\Delta^2 = \Delta \subset \mathbb{Q}\mathrm{End}(G)$ there is a unique $e^2 = e \in \mathbb{Q}\mathrm{End}(G)$ such that e is central modulo $\mathcal{N}(\mathbb{Q}\mathrm{End}(G))$ and

$$
\Delta = \mathbb{Q}\mathrm{End}(G)e\mathbb{Q}\mathrm{End}(G)
$$

([75, Corollary VIII.6.4]). Then by Lemma 8.4.2, eG is a nilpotent quasi-summand of G. The function D is the one that sends $\Delta \mapsto [eG]$. The reader can show that $D : \mathbf{IDEM}(\mathbb{Q}\mathrm{End}(G)) \longrightarrow \mathbf{Nil}(G)$ is a bijection.

EXAMPLE 8.4.4 In the ring $E = \begin{pmatrix} \mathbb{Q} & 0 \\ \mathbb{Q} & \mathbb{Q} \end{pmatrix}$ the reader can show that

$$
\mathbf{IDEM}(E) =
$$
$$
\left\{ 0, \begin{pmatrix} \mathbb{Q} & 0 \\ \mathbb{Q} & 0 \end{pmatrix}, \begin{pmatrix} 0 & 0 \\ 0 & \mathbb{Q} \end{pmatrix}, \begin{pmatrix} 0 & 0 \\ \mathbb{Q} & \mathbb{Q} \end{pmatrix}, \begin{pmatrix} \mathbb{Q} & 0 \\ 0 & 0 \end{pmatrix}, E \right\}.
$$

The Gabriel filters on E are those that contain one of the elements in $\mathbf{IDEM}(E)$.

Given $\Delta \in \mathbf{IDEM}(E)$ let $e_\Delta^2 = e_\Delta = e \in E$ be the unique idempotent such that $\Delta = EeE$. Then eG is the nilpotent quasi-summand of G that corresponds to Δ.

Let H be a nilpotent quasi-summand of G. We say that H is a $\mathbb{Q}\delta$-quasi-summand of G if $\Delta_{G,H}\mathbb{Q}G = \mathbb{Q}G$, or equivalently if $\Delta_{G,H} \in \delta(\mathbb{Q}G)$. If H is a $\mathbb{Q}\delta$-quasi-summand of G and if no proper quasi-summand of H is a $\mathbb{Q}\delta$-quasi-summand then we say that H is a minimal $\mathbb{Q}\delta$-quasi-summand. The lemma follows immediately from Lemma 8.4.3.

PROPOSITION 8.4.5 *Let G be an rtffr group and let H be a nilpo-tent quasi-summand of G. Then H is a minimal $\mathbb{Q}\delta$-quasi-summand of G iff $\Delta_{G,H}$ is the unique minimal element $\Delta_{\mathbb{Q}G}$ of $\delta(\mathbb{Q}G)$.*

LEMMA 8.4.6 *Let G be an rtffr group.*

1. *A minimal $\mathbb{Q}\delta$-quasi-summand of G is unique up to quasi-isomorphism.*

2. *If H' is a (minimal) $\mathbb{Q}\delta$-quasi-summand of H and if H is a (minimal) $\mathbb{Q}\delta$-quasi-summand of G then H' is a minimal $\mathbb{Q}\delta$-quasi-summand of G.*

Proof: 1. Suppose that H and H' are minimal $\mathbb{Q}\delta$-quasi-summands of G then by Proposition 8.4.5, $\Delta_{G,H}$ and $\Delta_{G,H'}$ equal the unique minimal element $\Delta_{\mathbb{Q}G}$ of $\delta(\mathbb{Q}G)$. By Lemma 8.4.2, $H \doteq H'$.

We leave the proof of part 2 as an exercise for the reader.

The next result reduces the study of $\delta(G)$ to the groups G such that $\delta(G) = \mu(G)$. That is, to those groups G such that each $I \in \delta(G)$ has finite index in $\text{End}(G)$.

LEMMA 8.4.7 *Let G be an rtffr group. Then $G \doteq H \oplus H'$ where H is a minimal $\mathbb{Q}\delta$-quasi-summand of G that is unique up to quasi-isomorphism and such that $IH \neq H$ for each proper right ideal $I \subset \mathbb{Q}\text{End}(H)$.*

Proof: By Lemma 8.4.6, the unique minimal element $\Delta_{\mathbb{Q}G} \in \delta(\mathbb{Q}G)$ corresponds to a minimal $\mathbb{Q}\delta$-quasi-summand H of G. Since $\Delta_{\mathbb{Q}G}$ is unique H is unique up to quasi-isomorphism. Furthermore, the ideal $\Delta_{\mathbb{Q}H} \subset \mathbb{Q}\text{End}(H)$ corresponds to a minimal $\mathbb{Q}\delta$-quasi-summand H' of H. Then H' is a $\mathbb{Q}\delta$-quasi-summand of G. The minimality of H shows us that $H' = H$, and hence that $\Delta_{\mathbb{Q}H} = \mathbb{Q}\text{End}(H)$. That is, $I\mathbb{Q}H \neq \mathbb{Q}H$ for each proper right ideal $I \subset \mathbb{Q}\text{End}(H)$. This completes the proof.

8.4.2 $\mathbb{Q}$-Faithful Groups

We say that G is a $\mathbb{Q}$-*faithful group* if $I\mathbb{Q}G \neq \mathbb{Q}G$ for each proper right ideal $I \subset \mathbb{Q}\text{End}(G)$, or equivalently if $\mathcal{D}(\mathbb{Q}G) = \{\mathbb{Q}\text{End}(G)\}$. Evidently G is a $\mathbb{Q}$-faithful group iff $IG = G \Rightarrow I$ has finite index in $\text{End}(G)$. We will use a variation on the Baer splitting property to characterize $\mathbb{Q}$-faithful groups. Theorem 8.3.3 implies that G is a $\mathbb{Q}$-faithful rtffr group iff given a short exact sequence

$$0 \longrightarrow K \longrightarrow X \overset{\pi}{\longrightarrow} G \longrightarrow 0 \tag{8.5}$$

in which $S_G(X) + K = X$, then (8.5) is *quasi-split*. That is, there is a map $\jmath : G \longrightarrow X$ and an integer $n \neq 0$ such that $\pi \jmath = n1_G$. Theorem 8.2.17 shows us that if E is semi-prime then each $G \in \Omega(E)$ is a $\mathbb{Q}$-faithful rtffr group. Another important class of $\mathbb{Q}$-faithful rtffr groups is the class of strongly indecomposable rtffr groups.

THEOREM 8.4.8 *If $G \neq 0$ is a strongly indecomposable rtffr group then $I\mathbb{Q}G \neq \mathbb{Q}G$ for each proper right ideal $I \subset \mathbb{Q}\mathrm{End}(G)$.*

Proof: Let G be a strongly indecomposable rtffr group and suppose that $I\mathbb{Q}G = \mathbb{Q}G$ for some proper right ideal $I \subset \mathbb{Q}\mathrm{End}(G)$. Because $\mathbb{Q}\mathrm{End}(G)$ is a local ring (see the comment preceding Jónsson Theorem 2.1.10), $I \subset \mathcal{J}(\mathbb{Q}\mathrm{End}(G))$ so that $\mathcal{J}(\mathbb{Q}\mathrm{End}(G))\mathbb{Q}G = \mathbb{Q}G$. Since $\mathbb{Q}G$ is finitely generated Nakayama's Lemma shows that $\mathbb{Q}G = 0$. This completes the proof.

COROLLARY 8.4.9 *Let G be a strongly indecomposable rtffr group. Then $\mu(G) = \delta(G)$.*

THEOREM 8.4.10 *The following are equivalent for an rtffr group G.*

1. *G is $\mathbb{Q}$-faithful.*

2. *If $G \doteq H \oplus H'$ and if $\mathbb{Q}\mathrm{Hom}(H, H')H = H' \neq 0$ then some nonzero strongly indecomposable quasi-summand of H is isomorphic to a quasi-summand of H'.*

3. *Each short exact sequence (8.5) in which $S_G(X) + K = X$ is quasi-split.*

Proof: $1 \Leftrightarrow 2$ G is a $\mathbb{Q}$-faithful group iff $I\mathbb{Q}G \neq \mathbb{Q}G$ for each proper right ideal $I \subset \mathbb{Q}\mathrm{End}(G)$ iff $\Delta_{\mathbb{Q}G} = \mathbb{Q}\mathrm{End}(G)$ iff G is a minimal $\mathbb{Q}\delta$-quasi-summand of G (Lemma 8.4.3), iff no proper quasi-summand of G is a $\mathbb{Q}\delta$-quasi-summand of G. The reader can show that this is equivalent to part 2.

$1 \Leftrightarrow 3$ Suppose that G is a $\mathbb{Q}$-faithful group, and let (8.5) be a short exact sequence such that $S_G(X) + K = X$. Then $IG = G \Rightarrow \mathrm{End}(G)/I$ is finite for each ideal $I \subset \mathrm{End}(G)$. Let I be the right ideal $I = \pi\mathrm{Hom}(G, X) \subset \mathrm{End}(G)$. Since $S_G(X) + K = X$, $G = \pi\mathrm{Hom}(G, X)G = IG$. Then $\mathrm{End}(G)/I$ is finite. That is, there is an integer $n \neq 0$ and a map $\jmath \in \mathrm{Hom}(G, X)$ such that $\pi\jmath = n1_G$. This proves part 3.

Conversely assume part 3. Suppose that $IG = G$ for some proper right ideal I of $\mathrm{End}(G)$. Let $\pi_1, \pi_2, \cdots$ denote the generators of I, let $G =$

G_k for each $k = 1, 2, \ldots$, and define $\pi : \oplus_{k=1}^{\infty} G_k \to G$ to be the unique map such that $\pi(x) = \pi_k(x)$ for each $x \in G_k$ and each $k = 1, 2, \ldots$. Then

$$
\begin{aligned}
G = IG &= \left(\sum_{k=1}^{\infty} \pi_k \mathrm{End}(G) \right) G \\
&= \sum_{k=1}^{\infty} \pi_k \mathrm{End}(G) G \\
&= \sum_{k=1}^{\infty} \pi_k(G_k) \\
&= \sum_{k=1}^{\infty} \pi(G_k) = \pi \left(\bigoplus_k G_k \right).
\end{aligned}
$$

By part 3 there is a map $\jmath : G \longrightarrow \oplus_k G_k$ and an integer $n \neq 0$ such that $\pi \jmath = n 1_G$. Since G has finite rank we will write $\jmath = \sum_{k=1}^{m} \jmath_k$ for some finite list of maps $\jmath_k : G \longrightarrow G_k$. Then $\jmath_k \in \mathrm{End}(G)$ for each $k = 1, \ldots, m$ and hence

$$
n 1_G = \pi \jmath = \sum_{k=1}^{m} \pi_k \jmath_k \in I.
$$

Therefore $\mathrm{End}(G)/I$ is finite, $\mathbb{Q}\mathrm{End}(G) = \mathbb{Q}I$, and thus part 1 is proved. This completes the logical cycle.

For the next result, for each subgroup H of G let H_* be the pure subgroup of G generated by H. We leave it as an exercise for the reader to show that if $I \subset \mathrm{End}(G)$ is a maximal right ideal then $IG \neq G$ iff $I = \mathrm{Hom}(G, IG)$. Let $\mathcal{N} = \mathcal{N}(\mathrm{End}(G))$.

THEOREM 8.4.11 *The following are equivalent for the rtffr group G.*

1. *G is a $\mathbb{Q}$-faithful group.*

2. *$I = \mathrm{Hom}(G, IG_*)$ for each maximal pure right ideal $I \subset \mathrm{End}(G)$.*

3. *$\mathcal{N}(\mathrm{End}(G)) = \mathrm{Hom}(G, \mathcal{N}G_*)$.*

4. *If $\pi : P \to \mathbb{Q}G$ is a projective cover over $\mathbb{Q}\mathrm{End}(G)$ then P generates $\mathbb{Q}\mathrm{End}(G)$.*

Proof: $1 \Rightarrow 2$ Assume that G is $\mathbb{Q}$-faithful and let I be a maximal pure right ideal in $\mathrm{End}(G)$. Evidently $I \subset \mathrm{Hom}(G, IG_*)$ and $\mathrm{Hom}(G, IG_*)$ is a pure right ideal in $\mathrm{End}(G)$. By our assumption on I, $I = \mathrm{Hom}(G, IG_*)$.

$2 \Rightarrow 3$ Assume that $I = \text{Hom}(G, IG_*)$ for each maximal pure right ideal in $\text{End}(G)$. Since $Q\mathcal{N} = \mathcal{J}(\text{QEnd}(G))$, $Q\mathcal{N}$ is the intersection of the maximal right ideals of $\text{QEnd}(G)$. Thus $\mathcal{N}$ is the intersection of the maximal pure right ideals of I. Then

$$\mathcal{N} \subset \text{Hom}(G, \mathcal{N}G_*)$$
$$\subset \cap\{\text{Hom}(G, IG_*) \,|\, I \subset \text{End}(G) \text{ is a maximal pure right ideal}\}$$
$$= \cap\{I \,|\, I \subset \text{End}(G) \text{ is a maximal pure right ideal}\} \text{ by part 2}$$
$$= \mathcal{N}.$$

Thus $\mathcal{N} = \text{Hom}(G, \mathcal{N}G_*)$, which proves part 3.

$3 \Rightarrow 1$ Suppose that G is not a $\mathbb{Q}$-faithful group. By the comments beginning section 8.4.1 on page 204 there is a pure ideal $\Delta \subset \text{End}(G)$ such that $\Delta\mathbb{Q}G = \mathbb{Q}G$. Since $\text{QEnd}(G)/\mathcal{N}$ is a semi-simple Artinian ring there is an ideal $\mathcal{N} \subset I \subset \text{QEnd}(G)$ such that $\Delta \cap I = \mathcal{N}$ and $\Delta + I = \text{QEnd}(G)$. It follows that

$$I\mathbb{Q}G = I(\Delta\mathbb{Q}G) \subset (I \cap \Delta)\mathbb{Q}G = \mathcal{N}\mathbb{Q}G$$

so that I is a nonnilpotent subset of $\text{Hom}(G, \mathcal{N}G_*)$. Hence $\mathcal{N} \neq \text{Hom}(G, \mathcal{N}G_*)$ which completes the contrapositive.

$1 \Leftrightarrow 4$ In the notation of part 4, P generates $\text{QEnd}(G)$ iff $\text{QEnd}(G)$ is its trace ideal $\Delta_{\mathbb{Q}G}$ iff $I\mathbb{Q}G \neq \mathbb{Q}G$ for each proper right ideal I in $\text{QEnd}(G)$ (see page 204), iff G is a $\mathbb{Q}$-faithful group. This completes the logical cycle.

8.4.3 $\mathbb{Q}$-Faithful E-Flat Groups

The results in this subsection show that when G is an E-flat $\mathbb{Q}$-faithful group then the results in the previous section can be strengthened.

The next result shows us that when G is an E-flat group then we lose very little information about $\delta(G)$ when we pass to $\delta(\mathbb{Q}G)$.

LEMMA 8.4.12 Let G be an rtffr E-flat group. If $\mathbb{Q}I \subset \text{QEnd}(G)$ and $(\mathbb{Q}I)\mathbb{Q}G = \mathbb{Q}G$ then $(\mathbb{Q}I \cap \text{End}(G))G = G$. In particular

$$(\Delta_{\mathbb{Q}G} \cap \text{End}(G))G = G.$$

Proof: Let $I \in \delta(\mathbb{Q}G)$. Since G is a flat left $\text{End}(G)$-module an application of $\cdot \otimes_{\text{End}(G)} G$ to the inclusion

$$\frac{\text{End}(G)}{(I \cap \text{End}(G))} \subset \frac{\text{QEnd}(G)}{I}$$

yields the inclusions

$$\frac{G}{(I \cap \operatorname{End}(G))G} \cong \frac{\operatorname{End}(G)}{(I \cap \operatorname{End}(G))} \otimes_{\operatorname{End}(G)} G$$

$$\subset \frac{\mathbb{Q}\operatorname{End}(G)}{I} \otimes_{\operatorname{End}(G)} G$$

$$\cong \frac{\mathbb{Q}G}{I\mathbb{Q}G} = 0.$$

Thus $G = (I \cap \operatorname{End}(G))G$ for each $I \in \delta(\mathbb{Q}G)$. This completes the proof.

EXAMPLE 8.4.13 Let H and H' be strongly indecomposable rtffr groups such that $\operatorname{Hom}(H', H) = 0$, and such that $S_H(H') = \operatorname{Hom}(H, H')H$ is a full subgroup of H', in which the factor group $H'/S_H(H')$ is infinite. Let $G \doteq H \oplus H'$. Then H is a nilpotent quasi-summand of G and $\Delta_{G,H} = \mathbb{Q}\operatorname{Hom}(H, G)\mathbb{Q}\operatorname{Hom}(G, H)$ is in $\delta(\mathbb{Q}G)$. Observe that since $\operatorname{Hom}(H, H')H \neq H'$ then G is not an E-flat group since otherwise

$$G = (\Delta_{G,H} \cap \operatorname{End}(G))G$$
$$= \operatorname{Hom}(H, G)\operatorname{Hom}(G, H)G = \operatorname{Hom}(H, G)H \neq G.$$

This is an example of a quasi-isomorphism class (G) of rtffr groups in which no $G' \in (G)$ is an E-flat group.

THEOREM 8.4.14 *The following are equivalent for the E-flat rtffr group.*

1. G is a $\mathbb{Q}$-faithful group.

2. $I = \operatorname{Hom}(G, IG)$ for each maximal pure right ideal $I \subset \operatorname{End}(G)$.

3. $\mathcal{N}(\operatorname{End}(G)) = \operatorname{Hom}(G, \mathcal{N}G)$.

Proof: Given a pure right ideal $I \subset \operatorname{End}(G)$ there is an inclusion

$$\frac{G}{IG} \cong \frac{\operatorname{End}(G)}{I} \otimes_{\operatorname{End}(G)} G \subset \frac{\mathbb{Q}\operatorname{End}(G)}{\mathbb{Q}I} \otimes_{\operatorname{End}(G)} G = \frac{\mathbb{Q}G}{\mathbb{Q}I\mathbb{Q}G}.$$

Thus $IG = IG_*$, whence the Theorem follows immediately from Theorem 8.4.11.

8.5   Exercises

G, H, K are rtffr groups, $p \in \mathbb{Z}$ is a prime, E is an rtffr ring, M, N, are rtffr E-modules (left or right, depending on the setting).

1. Let $H \in \mathbf{P}(G)$ and suppose that (G, H) has the endlich Baer splitting property. If H is finitely G-generated then $H \in \mathbf{P}_o(G)$.

2. Refer to the proof of Theorem 6.2.3. Show that each chain in $\mathcal{C}$ contains an upper bound in $\mathcal{C}$.

3. Let M be a right End(G)-module. Then $M^{(n)}$ is G-compressed for each integer $n > 0$ iff $\mathrm{T}_G(N) \neq 0$ for each finitely M-generated right End(G)-module $N \neq 0$.

4. Let P be a projective right E-module, let P^* be the dual of P, and let I be a left ideal of E. Prove that $IP = P$ iff $\Delta_P \subset I$ and that $\Delta_P = \Delta_{P^*}$.

5. Prove Proposition 8.1.4.

6. Let $H \in \mathbf{P}_o(G)$. Show that $\mathrm{Hom}(G, H)^* = \mathrm{Hom}(H, G)$.

7. If $IG \neq G$ for each proper finitely generated right ideal $I \subset \mathrm{End}(G)$ then G has the endlich Baer splitting property.

8. If $H \in \mathbf{P}_o(G)$ and if $N \subset \mathrm{H}_G(H)$ is such that $NG = H$ then $\mathrm{T}_G(\mathrm{H}_G(H)/N) = 0$.

9. Let G be a self-small right E-module. Then each $H \in \mathbf{P}_o(G)$ is G-small.

10. Prove Example 8.1.7.

11. Prove Proposition 8.1.10.

12. A torsion-free group of finite rank is self-small.

13. If E is a reduced torsion-free ring of finite rank and if I is a maximal right ideal of E then E/I is finite.

14. If (G, H) has the endlich Baer splitting property then $(G, H^{(n)})$ has the endlich Baer splitting property for each integer $n > 0$.

15. Prove Theorem 8.2.14.

16. Let G be a right E-module. Then $\mathbf{Cyc}(\mathrm{tor}(G)) = \{\mathrm{End}(G)/I \mid IG = G\}$.

17. Let $X = \mathbb{Z}$, $Y = \mathbb{Z}[\frac{1}{p}]$, $G = X \oplus Y$, and let $\Delta = \mathrm{Hom}(X, G)\mathrm{Hom}(G, X)$. Show that G is a cyclic projective right $\mathrm{End}(G)$-module, that $\Delta G = G$, and that Δ is a direct summand of $\mathrm{End}(G)$.

18. Let E be a ring, let $\mathbf{Gab}(E)$ be the set of Gabriel filters of E, and let $\mathbf{Tor}(E)$ be the set of hereditary torsion classes on E. There are inverse bijections $\mathbf{Gab} : \mathbf{Tor}(E) \longrightarrow \mathbf{Gab}(E)$ and $\mathbf{Tor} : \mathbf{Gab}(E) \longrightarrow \mathbf{Tor}(E)$.

19. Let Δ be an idempotent ideal in E. Then $\mathbf{Gab}(\Delta) = \{$right ideals $I \subset E \mid \Delta \subset I\}$ is a Gabriel filter on E. The hereditary torsion class corresponding to $\mathbf{Gab}(\Delta)$ is the set $\mathbf{Tor}_E(\Delta)$ of modules annihilated by Δ.

20. Show that $I \in \mu(M)$ iff E/I is a semi-Artinian right E-module and each composition factor of E/I is isomorphic to some simple module in $\{E/J \mid J \in \max_E(M)\}$.

21. Let M be a right E-module. Then $\mu(M) = \delta_E(M) \cap \mu(0)$.

22. Let M be a left E-module. If E/I is an Artinian module for each $I \in \delta_E(M)$ then $\delta_E(M) = \mu(M)$.

23. If $N \subset K \subset L \subset M$ and if K is superfluous in M then L is superfluous in M and K/N is superfluous in M/N.

24. Suppose that P is a finitely generated projective left E-module. An E-submodule $K \subset P$ is superfluous iff $K \subset J(E)P$.

25. If $\mathcal{N}$ is a nilpotent ideal then $\mathcal{N}M$ is superfluous in M for *each* left E-module M.

26. If $\Delta^2 = \Delta \subset E$ then the hereditary torsion class corresponding to $\mathcal{D}(\Delta)$ is the set of right E-modules K such that $K\Delta = 0$.

27. Show that $\mu(M)$ equals the set of right ideals $I \subset E$ such that E/I is a semi-Artinian E-module whose composition factors are isomorphic to E/J for some $J \in \max_E(M)$.

28. Show that if $\pi : P \to M$ is a projective cover then M is finitely generated iff P is finitely generated.

29. Show that $\mu(\widehat{M})$ is a finitely Gabriel filter on $\widehat{E}$.

30. Show that $\mu'(M) = \{$right ideals $I \subset E \mid E/I$ is a finite right E-module such that $IM = M$ and each composition factor of E/I is isomorphic to E/J for some $J \in \max_E(M)\}$ is a Gabriel filter on the reduced torsion-free ring E of finite rank.

31. Show that $\widehat{I} \cap E = I$ for each right ideal I of finite index in E, and that $\widehat{I' \cap E} = I'$ for each right ideal I' of finite index in $\widehat{E}$.

32. Let E be a reduced torsion-free ring of finite rank and let M be a reduced torsion-free left E-module of finite rank. There is an idempotent ideal $\Delta_M \subset \widehat{E}$ such that

$$\mu(M) = \{\text{right ideals } I \subset E \mid E/I \text{ is finite and } \Delta_M \subset \widehat{I} \subset \widehat{E}\}.$$

33. Show that $\mu(G)$ is finite for the group G constructed in Example 6.1.7.

34. Let E be a Corner ring (not necessarily of finite rank) and let P be a finitely generated projective left E-module. Let Δ_P be the trace of P in E. Then $\mathcal{O}_E(\Delta_M) = \mathcal{O}_E(P)$.

35. Show that $\mathcal{O}_E(M)$ is pure and dense in $\mathcal{O}_{\widehat{E}}(\widehat{M})$.

36. Let E be an rtffr ring and let $\widehat{E}x_1 + \cdots + \widehat{E}x_n = \widehat{M}$. Show that $\sum_{i=1}^n Ex_i$ is dense in $\sum_{i=1}^n \widehat{E}x_i$.

37. Let E be a local rtffr ring such that $pE \neq E$ for some prime $p \in \mathbb{Z}$. Prove that $\widehat{\mathcal{J}(E)}_p = \mathcal{J}(\widehat{E}_p)$.

38. Let E be an rtffr ring.

    (a) If E is a semi-prime ring then $\widehat{E}$ is a semi-prime ring.

    (b) If E is a subcommutative ring then $\widehat{E}$ is a subcommutative ring.

39. Let $0 \to K \to X \xrightarrow{\pi} G \to 0$ be *quasi-split* . Then $X \cong G \oplus K$.

40. Show that if H is a nilpotent quasi-summand of G and if H' is a nilpotent quasi-summand of H then H' is a nilpotent quasi-summand of G.

41. Let G be a faithful rtffr group. Show that if $I \subset \text{End}(G)$ is a maximal right ideal then $I = \text{Hom}(G, IG)$ iff $IG \neq G$.

42. Let G be an rtffr group and suppose that G/H is finite. Then G is a finitely faithful $\mathcal{S}$-group iff H is a finitely faithful $\mathcal{S}$-group.

43. Let G be an rtffr group. Then G is a finitely faithful $\mathcal{S}$-group iff G is a faithful $\mathcal{L}$-group.

8.6   Questions for Future Research

Let G be an rtffr group, let E be an rtffr ring, and let $S = \text{center}(E)$ be the center of E. Let τ be the conductor of G. There is no loss of generality in assuming that G is strongly indecomposable.

1. Give an example G in which $\{$right ideals $I \subset \text{End}(G) \,|\, IG = G\}$ is not a Gabriel filter. That is, find an rtffr group G such that $\mathcal{D}(G) \neq \delta(G)$.

2. Give an internal characterization of the class $\textbf{Tor}(G)$ of right $E = \text{End}(G)$-modules K such that $K \otimes_E G = 0$.

3. Describe all Gabriel filters on the rtffr ring E.

4. Describe all hereditary torsion classes $\mathcal{T}$ of right E-modules.

5. Let $E = \text{End}(G)$ and let $I \subset E$ be a right maximal right ideal. If $IG = G$ then $\text{Hom}_E(I, E) = E = \text{End}_E(I)$. Describe the set of right ideals $I \subset E$ such that $\text{Hom}_E(I, E) = I$.

6. Describe the idempotent ideals Δ in E.

7. Describe the superfluous subgroups of G.

8. Describe the rtffr groups G that possess a projective cover over $\text{End}(G)$.

9. Characterize the groups G such that $\text{End}(G)$ is semi-perfect.

10. Further investigate $\delta(G)$ and $\mu(G)$.

11. Characterize those G such that $\delta(G)$ is finite.

12. Suppose $\delta(G) = \mathcal{D}(\Delta)$ for some idempotent ideal Δ. Describe Δ in terms of G.

13. Further characterize the faithful rings E.

14. Describe the Gabriel filters on $\mathbb{Q}E$ and their associated hereditary torsion classes in terms of G.

15. Study (E-flat) $\mathbb{Q}$-faithful rtffr groups.

Chapter 9

Endomorphism Modules

In Chapter 4 we began a recurring theme, that certain *E-properties* coincide for rtffr groups. In this chapter we emphasize that coincidence. We will consider rtffr groups that are *E*-finitely generated, *E*-finitely presented, *E*-projective, *E*-generators, or *E*-flat. Some of these properties are classified up to quasi-isomorphism. We will also consider the homological dimensions of *G* as a left End(*G*)-module. The author's research interests will of course greatly influence the list of properties that are examined here.

9.1 Additive Structure of Rings

A left *E*-module *M* is said to be a *finitely generated rtffr E-module* if *M* is a finitely generated left or right *E*-module whose additive structure $(M, +)$ is an rtffr group. What follows is an extension of the Beaumont-Pierce-Wedderburn Theorem 3.1.6 from rings to finitely generated rtffr modules. Such a discussion is fundamental to our subsequent discussion of rtffr modules over rtffr rings.

LEMMA 9.1.1 *Let E be an rtffr ring and let* $T = E/\mathcal{N}(E)$. *Let M be an rtffr left E-module and suppose that there is an E-submodule* $K \subset M$ *such that* M/K *is a finitely generated rtffr T-module. Then*

1. $(M/K) \oplus M' \stackrel{.}{\cong} T^{(n)}$ *for some finitely generated rtffr left T-module* M' *and some integer n, and*

2. $M \stackrel{.}{\cong} (M/K) \oplus K$ *as groups.*

Proof: 1. Since M/K is a finitely generated rtffr left *T*-module $\mathbb{Q}M/\mathbb{Q}K$ is a finitely generated projective left $\mathbb{Q}T$-module. There is a finitely generated projective $\mathbb{Q}T$-module P and an integer m such that

$$(\mathbb{Q}M/\mathbb{Q}K) \oplus P \cong \mathbb{Q}T^{(m)}.$$

Since P is finitely generated there is a finitely generated T-submodule $M' \subset P$ such that $\mathbb{Q}M' = P$. Then

$$(M/K) \oplus M' \subset \mathbb{Q}T^{(m)}$$

and since both M/K and M' are finitely generated T-modules there is an integer $n \neq 0$ such that $n(M/K \oplus M') \subset T^{(m)}$. Furthermore since

$$\frac{T^{(m)}}{n(M/K \oplus M')}$$

is a finitely generated T-module that is also a torsion group there is an integer $n' \neq 0$ such that

$$n'T^{(m)} \subset n(M/K \oplus M') \subset T^{(m)}.$$

This proves part 1.

2. Since M/K is a finitely generated left T-module there are $x_1, \ldots, x_r \in M$ such that

$$M \doteq Tx_1 + \cdots + Tx_r + K.$$

Thus there is a T-module mapping

$$\pi : T^{(r)} \longrightarrow M \longrightarrow M/K$$

which lifts to a surjection

$$\pi : \mathbb{Q}T^{(r)} \longrightarrow \mathbb{Q}M/\mathbb{Q}K$$

of projective $\mathbb{Q}T$-modules over the semi-simple ring $\mathbb{Q}T$. There is a $\mathbb{Q}T$-module map

$$f : \mathbb{Q}M/\mathbb{Q}K \longrightarrow \mathbb{Q}T^{(r)}$$

such that

$$\pi f = 1_{\mathbb{Q}M/\mathbb{Q}K}.$$

Since M/K is a finitely generated T-module there is an integer $n \neq 0$ such that image $nf \subset T^{(r)}$ so that $\pi(nf) = n1_{M/K}$. It follows that $M \cong (M/K) \oplus K$. This completes the proof.

To classify the additive structure of rtffr semi-prime rings up to quasi-isomorphism it suffices to classify the additive structure of a classical maximal order E over its center S. The next result follows immediately from [69, Theorem 39.14] which implies that a classical maximal order E is a projective S-module so that $E \cong S^{(m-1)} \oplus J$ for some ideal $J \subset S$.

LEMMA 9.1.2 *Let E be a prime rtffr ring and let $S = \text{center}(E)$. There is an integer $m \neq 0$ such that $E \cong S^{(m)}$ as S-modules.*

Our classification of the additive structure of a semi-prime rtffr ring E is reduced to the case where $E = S$ is a Dedekind domain. The ring E is a *strongly indecomposable ring* if $(E, +)$ is strongly indecomposable. The next result is due to R. A. Beaumont and R.S. Pierce [18].

THEOREM 9.1.3 *Let E be a Dedekind domain. There is a strongly indecomposable Dedekind domain $R \subset \mathbb{Q}E$ such that $E \doteq R^{(m)}$ for some integer m. Moreover $R \cong \text{center}(\text{End}(E, +))$ as rings.*

Proof: Suppose that E is a Dedekind domain. Let

$$\mathcal{R} = \{\text{Dedekind domains } R \subset \mathbb{Q}E \mid RE \doteq E \text{ and} \\ RE \text{ is a finitely generated } R\text{-module }\}.$$

Since E is an rtffr Dedekind domain $E \in \mathcal{R} \neq \emptyset$ and since E has finite rank $\mathcal{R}$ contains a Dedekind domain R of minimal rank. Since a finitely generated torsion-free R-module is projective and since RE is a finitely generated torsion-free R-module $E \doteq RE \doteq R^{(m)}$ for some integer $m \neq 0$.

We claim that R is strongly indecomposable as a group. Observe that R is a cyclic left $\text{End}(R, +)$-module. Given an ideal $I \neq 0 \subset \text{End}(R, +)$ then $I\mathbb{Q}R \neq 0$ is an ideal in $\mathbb{Q}R$. Since $\mathbb{Q}R$ is a field, $I\mathbb{Q}R = \mathbb{Q}R$, so that $I^2 \neq 0$. Thus $\text{End}(R, +)$ is a prime ring. By Lemma 9.1.2, $\text{End}(R, +)$ is finitely generated over the integral domain $S = \text{center}(\text{End}(R, +))$. Then R is a finitely generated rtffr S-module and thus $R \doteq S^{(\ell)}$ for some integer $\ell > 0$. By the minimality of R, S has finite index in R, and since S is a pure subgroup of $\text{End}(R, +)$, $S = R$.

Finally write $R \doteq B^{(r)}$ for some strongly indecomposable group B and integer $r > 0$. Since $R = S$ commutes with the idempotent projections mapping R onto each copy of B, B is an ideal in R, hence R/B is finite, whence $R \doteq B$ is strongly indecomposable.

9.2   *E*-Properties

In this section we will visit each of the *E-properties* mentioned in the introduction of this chapter.

9.2.1 E-Rings

Given a ring E with identity there is a natural *embedding of rings*

$$\lambda : E \to \text{End}(E, +)$$

such that for $f \in E$, $\lambda(f)(x) = fx$ for each $x \in E$. That is, $\lambda(f)$ is left multiplication by f on E. We ask *When is λ an isomorphism?*

The ring E is called an *E-ring* if λ is an isomorphism. Examples of E-rings include $\mathbb{Z}$, $\mathbb{Q}$, $\mathbb{Z}/n\mathbb{Z}$ for integers n. Other examples are hard to come by in the rtffr case. But once you find them they can be used to illuminate difficult points in our research.

The existence of E-rings with a variety of ring structures is demonstrated in the following result. However, the constructed E-rings have countably infinite rank. See Theorem J.1.3 for a proof.

THEOREM 9.2.1 *[42, T.G. Faticoni] Each countable reduced torsion-free commutative ring is a pure subring of an E-ring.*

The countable hypothesis in the above theorem can be relaxed as M. Dugas, A. Mader, and C. Vinsonhaler [28] show us that *each cotorsion-free commutative ring is a pure subring of an E-ring.* Thus there are plenty of E-rings. For instance, if $x_1, x_2, x_3, \ldots$ are commuting indeterminants then $\mathbb{Z}[x_1, x_2, x_3, \ldots]/(x_1^2, x_2^3, x_3^4, \ldots)$ and $\mathbb{Z}[x_1, x_2, x_3, \ldots]$ are contained in E-rings E_1 and E_2 respectively. These constructions show that in general the ring theoretic structure of an E-ring can be quite complicated. However the rtffr E-rings have a relatively uncomplicated ring structure.

We will call the ring $\overline{E}$ a *Dedekind E-ring* if $\overline{E}$ is a Dedekind domain and an E-ring. See Appendix K.

THEOREM 9.2.2 *Let E be an rtffr integral domain. Then E is an E-ring iff E is a strongly indecomposable ring. In this case there is a unique Dedekind E-ring $\overline{E}$ such that $E \subset \overline{E}$ and $\overline{E}/E$ is finite.*

Proof: Suppose that E is not strongly indecomposable. Then there are nonzero subgroups A and B of E such that $E/(A \oplus B)$ is finite. There is a group map $e : E \to E$ such that $e^2 = ne$ for some integer $n \neq 0$ and $eE \doteq A$. Then $e, n - e \in \text{End}(E)$ are such that $e(n - e) = 0$. Since E is an integral domain, $\text{End}(E) \neq E$, and then E is not an E-ring. Consequently if the domain E is an rtffr E-ring then E is strongly indecomposable.

Conversely, suppose that E is strongly indecomposable rtffr integral domain. By Theorem 9.1.3, $E \cong S = \text{center}(\text{End}(E))$. Since E is an integral domain $\text{End}(E)$ is a prime ring so that $\mathbb{Q}\text{End}(E) \cong \text{Mat}_n(D)$ for some division ring D. Let $e^2 = e \neq 0 \in \mathbb{Q}\text{End}(E)$. Then $E \cong eE \oplus (1 - e)E$ and since E is strongly indecomposable $E \doteq eE$. That is, $e = 1_E$. Then $\mathbb{Q}E \subset \mathbb{Q}\text{End}(E) = D$ is a $\mathbb{Q}S$-vector space and $\mathbb{Q}E = \mathbb{Q}S$ is a D-vector space. Since these rings are finite dimensional $D = \mathbb{Q}S = \mathbb{Q}E$ and hence $E = S = \text{End}(E)$. Therefore E is an E-ring, and the proof is complete.

Our next result extends the Beaumont-Pierce Theorem 9.1.3 on the additive structure of rtffr prime rings.

THEOREM 9.2.3 *Let E be a prime rtffr ring. Then $S = \text{center}(\text{End}(E))$ is a strongly indecomposable E-ring and $E \doteq \overline{S}^{(n)}$ as S-modules where $\overline{S}$ is a strongly indecomposable Dedekind E-ring such that $S \doteq \overline{S}$.*

Proof: We proved in Theorems 9.1.2 and 9.1.3 that $E \doteq \overline{S}^{(m)}$ where S is a strongly indecomposable integral domain. By Theorem 9.2.2, S is an E-ring, and since $S \doteq \overline{S}$, $\overline{S}$ is a strongly indecomposable Dedekind E-ring.

COROLLARY 9.2.4 *Let E be a semi-prime rtffr ring. There are strongly indecomposable Dedekind E-rings $R_1, \cdots, R_t$ and integers $n_1, \ldots, n_t \geq 0$ such that $R_i \doteq R_j$ implies that $i = j$ and such that*

$$E \doteq R_1^{(n_1)} \oplus \cdots \oplus R_t^{(n_t)}$$

as groups.

The additive structure of rtffr E-rings is contained in the following theorem. A proof can be found in [10, Corollary 14.7] but try to prove it yourself without peeking.

THEOREM 9.2.5 [73, R. Bowshell, P. Schultz] *Let E be an rtffr ring. Then E is an E-ring iff there are strongly indecomposable Dedekind E-rings $\overline{E}_1, \ldots, \overline{E}_t$ such that*

1. *$\text{Hom}(\overline{E}_i, \overline{E}_j) = 0$ for each $1 \leq i \neq j \leq t$ and*

2. *$n(\overline{E}_1 \oplus \cdots \oplus \overline{E}_t) \subset E \subset \overline{E}_1 \oplus \cdots \oplus \overline{E}_t$ for some integer $n > 0$.*

We have reduced the classification of rtffr E-rings to the case where the rtffr ring E is a Dedekind E-ring. The key to understanding Dedekind E-rings is an understanding of the distribution of the maximal ideals in the ring of algebraic integers in an algebraic number field. See Theorem K.2.4 for a proof.

THEOREM 9.2.6 [T.G. Faticoni] *Suppose that the rtffr ring E is a Dedekind domain and let $\mathcal{O}$ denote the ring of algebraic integers in $\mathbb{Q}E$. Then E is an E-ring iff given a proper subfield $\mathbb{Q} \subset \mathbf{k} \subset \mathbb{Q}E$ there are maximal ideals $M, M' \subset \mathcal{O}$ such that $ME \neq E$, $M'E = E$, and $M \cap \mathbf{k} = M' \cap \mathbf{k}$.*

The field $\mathbf{k}$ is a *minimal field extension of* $\mathbb{Q}$ if $\mathbb{Q}$ and $\mathbf{k}$ are the only subfields of $\mathbf{k}$. For minimal field extensions $\mathbf{k}$ of $\mathbb{Q}$ the classification of E-rings in $\mathbf{k}$ is a bit simpler.

THEOREM 9.2.7 *Let E be a Dedekind domain and let $\mathcal{O}$ be the ring of algebraic integers in $\mathbb{Q}E$. Further suppose that $\mathbb{Q}E$ is a minimal field extension of $\mathbb{Q}$. Then E is an E-ring if there is a prime $p \in \mathbb{Z}$ and maximal ideals $M, M' \subset \mathcal{O}$ such that $ME \neq E$, $M'E = E$, and $p \in M \cap M'$.*

Let $p \in \mathbb{Z}$ be a prime. The group G is *p-local* if $pG \neq G$ and $qG = G$ for each prime $q \neq p \in \mathbb{Z}$. Theorem 9.2.6 and some Algebraic Number Theory will show that each algebraic number field $\mathbf{k}$ is the field of fractions of some Dedekind E-ring E. The question of which fields are of the form $\mathbb{Q}E$ for some p-local Dedekind E-ring E is answered masterfully by R.S. Pierce [65] and R.S. Pierce and C. Vinsonhaler [66]. The number field $\mathbf{k}$ is said to be *p-realizable* if $\mathbf{k} = \mathbb{Q}E$ for some p-local Dedekind E-ring E. See [65, Theorem 2.1] for a proof of the following theorem.

THEOREM 9.2.8 [R.S. Pierce] *Let Γ be the Galois group $\Gamma = \mathrm{Gal}(\mathbf{k}/\mathbb{Q})$ where $\mathbf{k}$ is a finite Galois extension of $\mathbb{Q}$. If Γ has no nontrivial, normal, cyclic subgroups, and if Γ is not isomorphic to the symmetric group S_4 or the alternating groups A_4, then $\mathbf{k}$ is p-realizable for all primes $p \in \mathbb{Z}$.*

9.2.2 E-Finitely Generated Groups

The group G is *E-cyclic* if $G = \mathrm{End}(G)x$ for some $x \in G$, and G is *E-finitely generated* if G is a finitely generated left $\mathrm{End}(G)$-module. Each ring is an E-cyclic group, and in general if G is a finitely generated left module over some ring then G is E-finitely generated. Evidently Noetherian rtffr E-rings and their nonzero ideals are E-finitely generated groups. Furthermore if M is a left E-module then $E \oplus M$ is an E-cyclic group. Fortunately the class of E-finitely generated rtffr groups is closed under quasi-isomorphism. Given an integer $m > 0$ we have constructed in Example 4.3.2 a group G that requires exactly $m \geq 2$ generators as a left $\text{End}(G)$-module. The group in question is quasi-equal to an E-cyclic group. Thus the number of generators for an E-finitely generated rtffr group can vary wildly in the quasi-isomorphism class of the group. We will present J.D. Reid's [67] classification of the E-finitely generated rtffr groups. Our presentation differs from that in [67].

LEMMA 9.2.9 *Let E be a semi-prime rtffr ring and let M and M' be torsion-free left E-modules with M' finitely generated. If $\pi : M \to M'$ is a surjection then $M \overset{\cdot}{\cong} \ker \pi \oplus M'$.*

Proof: Observe that π lifts to a surjection $\pi : \mathbb{Q}M \to \mathbb{Q}M'$ of left modules over the semi-simple ring $\mathbb{Q}E$. Since every $\mathbb{Q}E$-module is projective $\mathbb{Q}M \cong \ker \pi \oplus \mathbb{Q}M'$. Thus there is a $\mathbb{Q}E$-module map $\sigma : \mathbb{Q}M' \to \mathbb{Q}M$ such that $\pi\sigma = 1_{\mathbb{Q}M'}$. Since M' is finitely generated there is an integer $n \neq 0$ such that $n\sigma(M') \subset M$. Inasmuch as

$$e = \sigma\pi = (\sigma\pi)^2 \in \text{End}_{\mathbb{Q}T}(\mathbb{Q}M)$$

and since $ne \in \text{End}_T(M)$, we have $n1_M = n(1-e) \oplus ne$. Hence

$$(n1_M)M \subset [n(1-e)](M) \oplus (ne)(M) \subset \ker(\pi|_M) \oplus (n\sigma)(M') \subset M$$

which completes the proof.

LEMMA 9.2.10 *Let E be a semi-prime rtffr and let M be a finitely generated torsion-free left E-module. There are integers $t, n_1, \ldots, n_t > 0$ and Dedekind rtffr E-rings $R_1, \ldots, R_t$ such that*

$$M \overset{\cdot}{\cong} R_1^{(n_1)} \oplus \cdots \oplus R_t^{(n_t)}.$$

Proof: Since M is finitely generated there is an integer n and a surjection $\pi : E^{(n)} \to M$. By Lemma 9.2.9, $E^{(n)} \overset{\cdot}{\cong} M \oplus \ker \pi$, and by Theorem 9.2.4, there are integers $t, m_1, \ldots, m_t > 0$ and strongly indecomposable Dedekind E-rings $R_1, \ldots, R_t$ such that

$$E \overset{\cdot}{\cong} R_1^{(m_1)} \oplus \cdots \oplus R_t^{(m_t)}.$$

Then by Jónsson's Theorem 2.1.10, there are integers $t, n_1, \ldots, n_t$ such that

$$M \doteq R_1^{(n_1)} \oplus \cdots \oplus R_t^{(n_t)}$$

which completes the proof.

The next result is a classification of the group structure of a finitely generated left module over an rtffr ring. Recall that X_* is the purification of X in a larger group.

THEOREM 9.2.11 [T.G. Faticoni] *Let E be an rtffr ring and let M be an rtffr left E-module. If M is finitely generated then $M \doteq M_o \oplus N$ where*

1. *$M_o = R_1^{(n_1)} \oplus \cdots \oplus R_t^{(n_t)}$ for some integers $t, n_1, \ldots, n_t > 0$ and Dedekind E-rings $R_1, \ldots, R_t$,*

2. *$N = (\mathcal{N}(E)M_o)_* = $ the purification of $\mathcal{N}(E)M_o$ in M, and*

3. *$(\mathcal{N}(E)M_o)_*/\mathcal{N}(E)M_o$ is finite.*

Proof: Let E be an rtffr ring and let M be a finitely generated rtffr left E-module. By the Beaumont-Pierce-Wedderburn Theorem 3.1.6 there is a subring $T \subset E$ such that $E \doteq T \oplus \mathcal{N}(E)$ as groups and it is clear that

$$M_o = M/(\mathcal{N}(E)M)_*$$

is a finitely generated left T-module. Then by Lemma 9.2.9

$$M \doteq M_o \oplus (\mathcal{N}(E)M)_*$$

as T-modules. Furthermore by Lemma 9.2.10 there are integers $t, n_1, \ldots, n_t > 0$ and Dedekind E-rings $R_1, \ldots, R_t$ such that

$$M_o = R_1^{(n_1)} \oplus \cdots \oplus R_t^{(n_t)}.$$

Evidently the module $M/\mathcal{N}(E)M$ is finitely generated over the Noetherian ring $E/\mathcal{N}(E)$ so $(\mathcal{N}(E)M)_*/\mathcal{N}(E)M$ is a finitely generated torsion $E/\mathcal{N}(E)$-module. Since $(\mathcal{N}(E)M)_*/\mathcal{N}(E)M$ is a bounded quotient of an rtffr group, it is finite. This proves the theorem.

THEOREM 9.2.12 [J.D. Reid] *(See [67].) Let G be an rtffr group and let $E = \text{End}(G)$. The following are equivalent.*

1. *G is E-finitely generated.*

2. There are integers $t, n_1, \ldots, n_t > 0$, strongly indecomposable Dedekind *E*-rings $R_1, \ldots, R_t$, and a group N such that

$$G \doteq R_1^{(n_1)} \oplus \cdots \oplus R_t^{(n_t)} \oplus N$$

and where $N = \mathrm{S}_{R_1 \oplus \cdots \oplus R_t}(N)$.

Proof: Assume part 1. Let $E = \mathrm{End}(G)$ and suppose that G is an *E*-finitely generated rtffr group. Let $G_o = G/(\mathcal{N}(E)G)_*$. By Lemma 9.2.9

$$G \stackrel{\cdot}{\cong} G_o \oplus (\mathcal{N}(E)G_o)_*$$

where $(\mathcal{N}(E)G_o)_*/\mathcal{N}(E)G_o$ is finite. Let $N' = \mathcal{N}(E)G_o$. Inasmuch as $(\mathcal{N}(E)G_o)_*/N'$ is finite, and since $\mathcal{N}(E)G_o = \mathrm{S}_{G_o}(N')$, $G \stackrel{\cdot}{\cong} G_o \oplus N'$ for some group N' such that $\mathrm{S}_{G_o}(N') = N'$.

By Theorem 9.2.11 there are integers $t, n_1, \ldots, n_t > 0$ and strongly indecomposable Dedekind *E*-rings $R_1, \ldots, R_t$ such that

$$G_o \stackrel{\cdot}{\cong} R_1^{(n_1)} \oplus \cdots \oplus R_t^{(n_t)}.$$

Then $N' \stackrel{\cdot}{\cong} \mathrm{S}_{R_1 \oplus \cdots \oplus R_t}(N')$ so we let $N = \mathrm{S}_{R_1 \oplus \cdots \oplus R_t}(N')$. Hence $G = G_o \oplus N$ for some rtffr group N such that $N = \mathrm{S}_{R_1 \oplus \cdots \oplus R_t}(N)$. This proves part 2.

Conversely, assume part 2. Suppose that $G \doteq G_o \oplus N$ where G_o and N satisfy the conditions in part 2. If we let $R = R_1 \times \cdots \times R_t$ then G_o is a finitely generated R-module, and hence G_o is an *E*-finitely generated group. Let $x_1, \ldots, x_r \in G_o$ be generators for G_o over R. Inasmuch as R and $\mathrm{Hom}(G_o, N)$ embed naturally in $\mathbb{Q}\mathrm{End}(G)$ we have the following equations.

$$
\begin{aligned}
G &\doteq G_o \oplus \mathrm{S}_R(N) \\
&\doteq G_o \oplus \mathrm{Hom}(G_o, N)G_o \\
&= \mathrm{End}(G_o)x_1 + \cdots + \mathrm{End}(G_o)x_r + \mathrm{Hom}(G_o, N)G_o \\
&= \mathrm{End}(G_o)x_1 + \cdots + \mathrm{End}(G_o)x_r + \\
&\qquad \mathrm{Hom}(G_o, N)\mathrm{End}(G_o)x_1 + \cdots + \mathrm{Hom}(G_o, N)\mathrm{End}(G_o)x_r \\
&= \mathrm{End}(G_o)x_1 + \cdots + \mathrm{End}(G_o)x_r + \\
&\qquad \mathrm{Hom}(G_o, N)x_1 + \cdots + \mathrm{Hom}(G_o, N)x_r \\
&\doteq \mathrm{End}(G)x_1 + \cdots + \mathrm{End}(G)x_r.
\end{aligned}
$$

Thus G, and hence any group quasi-isomorphic to G is *E*-finitely generated. This concludes the proof.

Several of the *E-properties* imply the E-finitely generated property for rtffr groups. The group G is *E-projective* if G is a projective left $\text{End}(G)$-module, and G is an *E-generator* if G generates the category of left $\text{End}(G)$-modules. G is an *E-progenerator* if G is a finitely generated projective generator in the category of left $\text{End}(G)$-modules. Groups of the form $\mathbb{Z} \oplus N$ are E-projective while E-rings are E-progenerator groups.

COROLLARY 9.2.13 *Let G be a strongly indecomposable E-finitely generated rtffr group. Then $G \doteq E$ for some Dedekind E-ring E.*

Proof: If G is a strongly indecomposable E-finitely generated rtffr group then by Theorem 9.2.12, $G \doteq R_1 = E$ for some Dedekind E-ring E.

COROLLARY 9.2.14 *Let G be an rtffr group. Then G is E-finitely generated iff G is quasi-isomorphic to an E-cyclic group.*

Proof: Let $E = \text{End}(G)$ and suppose that G is an E-finitely generated rtffr group. By Theorem 9.2.12 there are positive integers $t, n_1, \ldots, n_t$, strongly indecomposable E-rings $R_1, \ldots, R_t$, and a group N such that

$$N = S_{G_o}(N)$$
$$G_o = R_1^{(n_1)} \oplus \cdots \oplus R_t^{(n_t)}, \text{ and}$$
$$G \doteq G_o \oplus N.$$

It suffices to show that $G_o \oplus N$ is an E-cyclic group. Let x_i be a generator of R_i as an R_i-module and let

$$x = x_1 \oplus \cdots \oplus x_t.$$

Since these are *group homomorphisms*, $\text{End}(G_o \oplus N)x$ contains each copy of R_i in G_o. Thus $\text{End}(G_o \oplus N)x \supset G_o$. Furthermore, by hypothesis

$$N = S_{G_o}(N) = \text{Hom}(G_o, N)G_o = \text{Hom}(G_o, N)x$$

so that $\text{End}(G_o \oplus N)x \supset N$. Therefore $G_o \oplus N = \text{End}(G_o \oplus N)x$.

COROLLARY 9.2.15 *An E-projective rtffr group is E-finitely generated.*

Proof: Let $E = \text{End}(G)$, suppose that G is E-projective, and write $G \oplus P \cong E^{(\mathcal{I})}$ for some index set $\mathcal{I}$. Since G has finite rank there is a finite subset $\{x_1, \ldots, x_n\} \subset G$ and a finite subset $\mathcal{I}_o \subset \mathcal{I}$ such that $\mathbb{Q}G = \sum_{i=1}^{n} \mathbb{Q}x_i$ and $\{x_1, \ldots, x_n\} \subset E^{(\mathcal{I}_o)}$. Then

$$G = E^{(\mathcal{I})} \cap \mathbb{Q}G \subset E^{(\mathcal{I}_o)}.$$

The reader can show that because G is a direct summand of $E^{(\mathcal{I})}$, G is a direct summand of $E^{(\mathcal{I}_o)}$, thus proving that G is finitely generated as a left E-module. This proves the lemma.

COROLLARY 9.2.16 *Let G be an E-generator rtffr group and let $S = \text{center}(\text{End}(G))$. Then G is E-finitely generated and G is a finitely generated projective S-module.*

Proof: Let $E = \text{End}(G)$, suppose that G is an E-generator, and let $S = \text{End}_E(G)$. Since G is a generator for E, G is a finitely generated projective right $S = \text{End}_E(G)$-module. But $S \subset \text{End}(G)$ so that G is E-finitely generated.

For strongly indecomposable rtffr groups we can prove a strong relationship between these E-*properties*. The rtffr group G is E-*finitely presented* if G is a finitely presented left E-module.

THEOREM 9.2.17 [T.G. Faticoni] *Let G be a strongly indecomposable rtffr group. The following are equivalent.*

1. *G is quasi-isomorphic to a Dedekind E-ring.*

2. *G is quasi-isomorphic to an E-cyclic group.*

3. *G is an E-finitely generated group.*

4. *G is an E-finitely presented group.*

5. *G is quasi-isomorphic to an E-projective group.*

6. *G is quasi-isomorphic to an E-generator group.*

7. *G is quasi-isomorphic to an E-progenerator group.*

Proof: $1 \Rightarrow 2 \Rightarrow 3$ is clear.
$3 \Rightarrow 4$ By Lemma 9.2.13, $G \doteq R$ for some strongly indecomposable Dedekind E-ring R, so that $\text{End}(G) \doteq R$ is a Noetherian integral domain. The E-finitely generated G is thus E-finitely presented.

$4 \Rightarrow 5$ and 6 By Lemma 9.2.13, $G \doteq R$ for some strongly indecomposable Dedekind E-ring R. This proves part 5 and 6.

5 or $6 \Rightarrow 1$ Suppose that G is an E-projective rtffr group or an E-generator rtffr group. By Lemmas 9.2.15 and 9.2.16, G is E-finitely generated, and since G is a strongly indecomposable group, Lemma 9.2.13 implies that $G \doteq R$ for some Dedekind E-ring R. This proves part 1.

Inasmuch as $1 \Rightarrow 7 \Rightarrow 5$ is clear the proof is complete.

An example will show that some of these implications do not possess a converse.

EXAMPLE 9.2.18 There is an E-cyclic rtffr group that is not E-projective.

Proof: Let $\mathbf{k}$ be a quadratic field extension of $\mathbb{Q}$ and let $\mathcal{O}$ denote the ring of algebraic integers in $\mathcal{O}$. There is a prime $p \in \mathbb{Z}$ such that $p\mathcal{O} = MM'$ for some maximal ideals $M \neq M'$ in $\mathcal{O}$. Let

$$S = \mathcal{O}_M \cap \bigcap \{\mathcal{O}_q \,|\, q \neq p \in \mathbb{Z} \text{ is a prime } \}.$$

Since $M\mathcal{O}_M \neq \mathcal{O}_M$, $MS \neq S$, and since $M'\mathcal{O}_x = \mathcal{O}_x$ for each $x \in \{I, \text{ primes } q \neq p\}$, $M'S = M'$. By Theorem 9.2.7, S is an E-ring and one can show that $\mathcal{O}/q\mathcal{O} \cong S/qS \cong (\mathbb{Z}/q\mathbb{Z})^{(2)}$ for each prime $q \neq p$. Furthermore $S/pS \cong \mathbb{Z}/p\mathbb{Z}$. Thus $\mathbb{Z}$ is a pure subring of S. By counting p-ranks of the rank one group $N = S/\mathbb{Z}$ we see that $qN \neq N$ for each prime $q \neq p$ while $pN = N$. Therefore $G = S \oplus N$ is E-finitely generated. Note that N is not an S-module even though it is S-generated. Thus G is E-cyclic but not E-projective.

9.2.3 E-Projective Groups

LEMMA 9.2.19 Let S be a commutative rtffr ring and let $I \doteq S$ be an ideal. If I is a projective S-module then $I \oplus I = S \oplus K$ for some S-module K.

Proof: By hypothesis S/I is a finite commutative ring so that $S/I = T_1 \times \cdots \times T_r$ for some local rings $T_1, \ldots, T_r$. Thus $I/I^2 = L_1 \oplus \cdots \oplus L_r$ where L_i is a T_i-module. Since the projective S-module I generates S, L_i generates T_i, and since T_i is a local ring, there is a surjection $L_i \to T_i$. Therefore there is a surjection $I \to I/I^2 \to S/I$ that lifts to a map $f : I \to S$ such that $f(I) + I = S$. It follows that $I \oplus I \cong S \oplus \ker(f \oplus 1_I)$ where $(f \oplus 1_I) : I \oplus I \to S$ is the obvious surjection. This proves the lemma.

Consider the problem of classifiying the E-projective rtffr groups. The investigation is complicated by the fact that if N is any rtffr group then $\mathbb{Z} \oplus N$ is E-cyclic and E-projective. Another example of an E-projective group is as follows.

Let S be an E-ring, and let N be an S-module. We say that N is an *S-linear* module if

$$\text{Hom}_S(S, N) = \text{Hom}(S, N).$$

Our S-linear modules are referred to as *E-modules* by A. Mader and C. Vinsonhaler [61]. Any group is a $\mathbb{Z}$-linear module, a $\mathbb{Q}$-vector space is a $\mathbb{Q}$-linear module, and one can prove that if I is an ideal of finite index in an E-ring S then I is an S-linear module.

EXAMPLE 9.2.20 We give a general construction of E-projective rtffr groups. Let S be an E-ring, let $I \doteq S$ be a projective *nonprincipal* ideal, and let N be an S-linear module. Then

$$G = I \oplus N$$

is an E-projective group that is generated by two elements as a left $\text{End}(G)$-module.

Proof: Since S is an E-ring I is S-linear and since N is S-linear

$$
\begin{aligned}
I &= \text{Hom}_S(S, I) = \text{Hom}(S, I) \text{ and} \\
N &= \text{Hom}_S(S, N) = \text{Hom}(S, N).
\end{aligned}
$$

The following chain of equations shows that $G = I \oplus N$ is S-linear.

$$
\begin{aligned}
G &= \text{Hom}_S(S, G) \\
&\cong \text{Hom}_S(S, I) \oplus \text{Hom}_S(S, N) \\
&\cong \text{Hom}(S, I) \oplus \text{Hom}(S, N) \\
&\cong \text{Hom}(S, I \oplus N) \\
&\cong \text{Hom}(S, G).
\end{aligned}
$$

The next series of equations shows us that G is direct summand of $\text{End}(G)^{(2)}$. Since $I \doteq S$ is a projective ideal Lemma 9.2.19 states that $I \oplus I \cong S \oplus K$ for some S-module K. Then because G is S-linear,

$$
\begin{aligned}
\text{End}(G) \oplus \text{End}(G) &\cong \text{Hom}(I \oplus I, G) \oplus \text{Hom}(N \oplus N, G) \\
&\cong \text{Hom}(S \oplus K, G) \oplus \text{Hom}(N \oplus N, G) \\
&\cong \text{Hom}(S, G) \oplus \text{Hom}(K \oplus N \oplus N, G) \\
&\cong G \oplus \text{Hom}(K \oplus N \oplus N, G)
\end{aligned}
$$

as left $\text{End}_S(G) = \text{End}(G)$-modules. Thus G is a projective left $\text{End}(G)$-module that is generated by at most two elements as a left $\text{End}(G)$-module.

EXAMPLE 9.2.21 Let S be an rtffr E-ring and let N be an rtffr S-linear module. By the above example $G = S \oplus N$ is an E-cyclic E-projective rtffr group.

Our work here will show that the above examples are the *only* kind of E-projective rtffr group. To motivate our discussion of E-cyclic E-projective rtffr groups, notice that by Lemmas 9.2.14 and 9.2.15 every E-projective rtffr group is quasi-isomorphic to an E-cyclic E-projective rtffr group.

THEOREM 9.2.22 [63, G.P. Niedzwecki, J.D. Reid] *Let G be an rtffr group, let $E = \text{End}(G)$, and let $S = \text{center}(E)$. Then G is an E-cyclic E-projective group iff $G \cong S \oplus N$ for some S-linear module N. In this case S is an rtffr E-ring.*

Proof: Suppose that $G = S \oplus N$ where $S = \text{center}(E)$ and where N is an S-linear module. We have shown in Example 9.2.21 that G is E-cyclic and E-projective.

Conversely let $E = \text{End}(G)$ and assume that G is an E-cyclic E-projective rtffr group. There is a split surjection

$$\pi : E \to G$$

of left E-modules so there is an $e^2 = e \in E$ such that $G \cong Ee$. Thus

$$S = \text{End}_E(G) = \text{End}_E(Ee) \cong eEe$$

by Lemma 1.2.4. If we let $N = (1 - e)G$ then

$$G = eG \oplus (1 - e)G \cong eEe \oplus N \cong S \oplus N$$

where N is an S-module. On the other hand because $G \cong S \oplus N$ each group homomorphism $f : S \to S$ commutes with each element of S so that

$$S \cong \text{Hom}_S(S, S) = \text{Hom}(S, S)$$

is an E-ring. Similarly

$$\text{Hom}_S(S, N) = \text{Hom}(S, N)$$

so that N is an S-linear module. This completes the proof of the theorem.

We have more to say about E-projective groups once we have studied E-generator groups. See Theorem 9.2.28.

It is known that P is a finitely generated projective right E-module iff P is a generator as a left $\mathrm{End}_E(P)$-module. This observation connects E-projective groups with the generator property as follows.

LEMMA 9.2.23   *Let G be an rtffr group and let $S = \mathrm{center}(\mathrm{End}(G))$. Then G is an E-projective group iff G is a generator in the category of right S-modules.*

Proof: Observe that $S = \mathrm{End}_{\mathrm{End}(G)}(G)$. Then by Lemma 9.2.15, G is E-projective iff G is E-finitely generated E-projective iff G is a generator as a right S-module.

The module U is called *quasi-projective* if U is projective relative to each short exact sequence

$$0 \longrightarrow K \longrightarrow U \longrightarrow V \longrightarrow 0$$

of left E-modules. The group G is *E-quasi-projective* if G is a quasi-projective left $\mathrm{End}(G)$-module. The reader can prove as an exercise that if E is an rtffr E-ring and if $E \doteq I$ is a projective ideal then E and I are locally isomorphic as groups and as E-modules.

PROPOSITION 9.2.24 [84, C. Vinsonhaler, W. Wickless] *Let G be an rtffr group. Then G is E-quasi-projective iff G is an E-finitely generated E-projective group.*

Proof: Let G be a quasi-projective left E-module. Because G is an rtffr group there is an integer $m \neq 0$ such that $\cap_{k>0} m^k G = 0$. Since G/mG is a finite group there is a finitely generated E-submodule $F \subset G$ such that $(F+mG)/mG = G/mG = M$. The usual properties of relative projectivity (see [6, Proposition 16.12]) show us that the quasi-projective left E-module G is projective relative to each short exact sequence

$$0 \to K \longrightarrow F \longrightarrow G/mG \to 0$$

of left E-modules. Thus the natural projection $G \to G/mG$ lifts to a map $f : G \to F$ such that $\ker f \subset mG$. Since G is torsion-free $\ker f$ is pure in G and so $\ker f = m(\ker f) = 0$ by our choice of m. Thus f is a regular endomorphism of G, hence $f(G) \subset F \subset G$, whence G/F is finite. (Prove that one, reader.) Then G is a finitely generated quasi-projective left E-module.

Conversely, suppose that G is E-projective and is generated by n elements. Because G has finite rank there is a finite set $\{x_1, \ldots, x_r\} \subset G$ such that

$$\text{ann}_E(x_1, \ldots, x_r) = 0.$$

Then $x = x_1 \oplus \cdots \oplus x_r \in G^{(r)}$ satisfies $\text{ann}_E(x) = 0$ so that $E \cong Ex \subset G^{(r)}$. That same theory of relative projectivity in [6] shows us that the quasi-projective E-module G is projective relative to each short exact sequence

$$0 \to K \longrightarrow E^{(n)} \longrightarrow G \to 0.$$

Then the identity 1_G lifts to a map $f : G \to E^{(n)}$. That is, our sequence is split exact, and therefore G is a projective left E-module. This completes the proof.

9.2.4 E-Generator Groups

A left E-module U is a generator over E iff U is finitely generated projective as a right $\text{End}_E(U)$-module. Thus we are motivated to study the *E-generator groups*. That is, those groups G that are generators in the category of left $\text{End}(G)$-modules. Recall that G is called an *E-progenerator* if G is a finitely generated projective generator in the category of left $\text{End}(G)$-modules.

THEOREM 9.2.25 [T.G. Faticoni] *Let G be an rtffr group, let $E = \text{End}(G)$, and let $S = \text{center}(E)$. The following are equivalent.*

1. G *is an E-progenerator group.*

2. G *is an E-generator group.*

3. G *is a finitely generated projective S-module.*

4. $G \cong U_1 \oplus \cdots \oplus U_s$ *for some projective ideals $U_1, \ldots, U_s \subset S$.*

Proof: $1 \Rightarrow 2$ is clear and $2 \Rightarrow 3$ follows from the introductory motivational statement.

$3 \Rightarrow 4$ Since G is a finitely generated projective module over the commutative rtffr ring S, Theorem 4.2.11 states that $G = U_1 \oplus \cdots \oplus U_s$ for some projective ideals $U_1, \ldots, U_s \subset S$.

$4 \Rightarrow 1$ Since G is a finitely generated projective S-module G is a generator for $\text{End}_S(G)$. The reader will recall that $E = \text{End}_S(G)$ so that G is a generator as a left E-module.

On the other hand we will show that G is a generator over S. Let Δ be the trace ideal in S for the projective S-module G. Since the trace is an idempotent ideal in S, Lemma 4.2.6 states that $\Delta = eS$ for some $e^2 = e \in S$. Inasmuch as $\text{ann}_S(G) = 0$, and since $eG = \Delta G = G$, $e = 1_G$. That is G generates S. Then G is a finitely generated projective left module over $\text{End}_S(G) = E$. This proves 1 which completes the logical cycle.

COROLLARY 9.2.26 *Let G be an rtffr group and let $S = \text{center}(\text{End}(G))$. Then G is an E-generator iff $G = U \oplus U'$ where U is an invertible ideal of S and U' is some projective S-module.*

Proof: Let $G = U \oplus U'$ for some invertible ideal U over S and some S-module U'. Then G is finitely generated projective over S, which implies that G is E-projective.

Conversely, suppose that G is an E-generator group. Since $S = S_1 \times \cdots \times S_t$ is a finite product of indecomposable commutative rtffr rings $S_1, \ldots, S_t$ there is no loss of generality if we prove the result for G under the assumption that S is indecomposable. So, Theorem 9.2.25 states that

$$G = U_1 \oplus \cdots \oplus U_s$$

for some indecomposable finitely generated projective ideals $U_1, \ldots, U_s$ of S. Let $\Delta = \text{trace}_S(U_1)$. By Lemma 4.2.6, there is an $e^2 = e \in S$ such that $\Delta = eS$. Since S is indecomposable $e = 1_G$ and hence U_1 generates S. In other words U_1 is an invertible ideal in S. This completes the proof.

The group G is *E-self-generating* if G generates each $\text{End}(G)$-submodule of $G^{(n)}$ for each $n > 0$. The following result shows that the notions of an E-generating and E-self-generating rtffr are interchangeable.

LEMMA 9.2.27 *Let G be an rtffr group. Then G is an E-self-generating group iff G is an E-generator group.*

Proof: Let $E = \text{End}(G)$. Let $x_1, \ldots, x_n \in G$ be a maximal linearly independent subset of G. Then $f(x_1, \ldots, x_n) = 0$ iff $f(G) = 0$ iff $f = 0$. Thus

$$E \cong E(x_1 \oplus \cdots \oplus x_n) \subset G^{(n)}$$

as left E-modules. Then G generates E so that G is a generator as a left E-module. The converse is clear, so the proof is complete.

9.2.5 E-Projective Rtffr Groups Characterized

Lemma 9.2.15 leaves open the question of just how many generators an E-projective rtffr group requires. That issue is addressed in the famous paper [16] written at the University of Connecticut during its Special Year in Algebra 1981. The main theorem of [16] is our next result. Let us agree to call the module M a *local direct summand of N* if given an integer $n \neq 0$ there is an integer m and module maps $f_n : N \to M$ and $g_n : M \to N$ such that $f_n g_n = m1_M$.

THEOREM 9.2.28 [UConn '81 Theorem] *Let G be an rtffr group, let $E = \text{End}(G)$, and let $S = \text{center}(E)$. The following are equivalent.*

1. *G is an E-projective group and generated by two elements.*

2. *G is an E-projective group.*

3. *G is an E-quasi-projective group.*

4. *G is a local direct summand of E.*

5. *G is E-finitely generated and G_p is a cyclic projective left E_p-module for each prime $p \in \mathbb{Z}$.*

6. *There is an rtffr E-ring S, an invertible ideal I in S, and an S-linear module N such that $G = I \oplus N$.*

Proof: We will prove the logical circuit $6 \Rightarrow 5 \Rightarrow 4 \Rightarrow 1 \Rightarrow 2 \Rightarrow 3 \Rightarrow 6$.

$6 \Rightarrow 5$ Suppose that $G \cong I \oplus N$ as in part 6. Since I is invertible, $\mathbb{Q}I$ is an invertible ideal in the Artinian commutative ring $\mathbb{Q}S$, so that $\mathbb{Q}I = \mathbb{Q}S$. Hence $I \doteq S$. Because I is a generator over S, there is an integer n and an S-module K such that $I^{(n)} \cong S \oplus K$. Subsequently

$$\text{Hom}_S(I, G)^{(n)} \cong \text{Hom}_S(I^{(n)}, G) \cong \text{Hom}_S(S, G) \oplus \text{Hom}_S(K, G)$$

as left E-modules. Then $G \cong \text{Hom}_S(S, G)$ is a direct summand of

$$\text{Hom}_S(I, G)^{(n)} \oplus \text{Hom}_S(N, G)^{(n)} \cong \text{Hom}_S(I \oplus N, G)^{(n)} \cong E^{(n)}$$

as left E-modules, and so G is E-finitely generated.

Furthermore, let $p \in \mathbb{Z}$ be prime. Choose $n \in \mathbb{Z}$ large enough that S_n is a reduced group. Since $I \doteq S$ is an invertible ideal in S, I_n is an invertible ideal in S_n. Since S_n/nS_n is then generated by I_n/nI_n and since the finite commutative ring S_n/nS_n is a product of local rings, there is a map $f : I_n \to S_n$ such that $f(I_n) + nS_n = S_n$. By Lemma 1.2.1 and our choice of n, $n \in \mathcal{J}(S_n)$, and hence $f(I_n) = S_n$. Since $S_n \doteq I_n$ have finite rank, $I_n \cong S_n$. Because S and I are finitely presented S-modules we can apply the Change of Rings Theorem 1.5.2 to show that

$$\mathrm{Hom}_S(I, G)_n \cong \mathrm{Hom}_{S_n}(I_n, G_n) \cong \mathrm{Hom}_{S_n}(S_n, G_n) \cong \mathrm{Hom}_S(S, G)_n.$$

Thus $G_n = \mathrm{Hom}_S(S, G)_n$ is a direct summand of $\mathrm{Hom}_S(I \oplus N, G)_n \cong E_n$ as a left E_n-module. This proves part 5.

5 $\Rightarrow$ 4 is proved along the lines of Lemma 2.5.2.

4 $\Rightarrow$ 1 Suppose that G is a local summand of E. We claim that G is E-projective and generated as a left E-module by two elements. Let $n \neq 0 \in \mathbb{Z}$. There is an integer m and E-module maps $f_n : E \to G$ and $g_n : G \to E$ such that $\gcd(m, n) = 1$ and $f_n g_n = m 1_G$. By part 4, there is an integer k and maps $f_m : E \to G$ and $g_m : G \to E$ such that $f_m g_m = k 1_G$. Since $\gcd(k, m) = 1$ there are integers a, b such that $ak + bm = 1$. Then the maps

$$
\begin{aligned}
f &: & E \oplus E \longrightarrow G &: & (x, y) \mapsto a f_m(x) + b f_n(y) \\
g &: & G \longrightarrow E \oplus E &: & z \mapsto g_m(z) \oplus g_n(z)
\end{aligned}
$$

satisfy

$$fg(z) = a f_m g_m(z) + b f_n g_n(z) = (ak + bm)z = z.$$

Thus $E \oplus E \cong G \oplus \ker f$ and hence G is projective and generated by two elements. This proves part 1.

1 $\Rightarrow$ 2 $\Rightarrow$ 3 are clear.

3 $\Rightarrow$ 6 By Proposition 9.2.24 the quasi-projective right E-module is a finitely generated projective left E-module, and by Lemma 9.2.23, G is a generator over S. We will proceed in a pair of lemmas. The first might be interesting in its own right.

LEMMA 9.2.29 *Let* G *be an* E-*finitely generated rtffr group and let* $S = \mathrm{center}(\mathrm{End}(G))$. *Then* S *is a Noetherian semi-prime commutative ring.*

Proof: Since G is E-finitely generated Theorem 9.2.12 states that there are integers $t, n_1, \ldots, n_t > 0$ and strongly indecomposable Dedekind E-rings $R_1, \ldots, R_t$ such that

$$G \doteq R_1^{(n_1)} \oplus \cdots \oplus R_t^{(n_t)} \oplus N \tag{9.1}$$

where N is generated by

$$R = R_1 \oplus \cdots \oplus R_t.$$

We will assume that $R_i \cong R_j \Rightarrow i = j$. Observe that $\mathrm{S}_{R_i}(R_j) = I_{ij}$ is an ideal in R_j. Since R_j is a Dedekind domain $(I_{ij})^{-1}I_{ij} = R_j$ for nonzero ideals I in R_j. Consequently if there is a nonzero map $R_i \to R_j$ then we can assume that R_j is a direct summand of N. Thus we can asssume that in (9.1), $\mathrm{Hom}(R_i, R_j) = 0$ for each $i \neq j$, and hence by Theorem 9.2.5, R is an E-ring. In particular R is a semi-prime ring.

Since R is an S-submodule of G, and since R generates a subgroup of finite index in G, $xR \neq 0$ for each $x \in S$. Thus

$$S \subset \mathrm{End}(R, +) \cong R$$

and hence S is a Noetherian semi-prime commutative rtffr ring. This proves the lemma.

LEMMA 9.2.30 *Under the hypotheses of Theorem 9.2.28, there is a map $f : G \to S$ such that $f(G) \doteq S$ is a progenerator S-module.*

Proof: Since S is a semi-prime rtffr ring Lemma 1.6.3 implies that there is an integer $m \neq 0$ large enough so that S_m is a reduced rtffr ring. Let $U = G/N_*$ and let $M_1, \ldots, M_s$ be a complete list of the maximal ideals of S that contain $m\overline{S}$. Since U is a generator over S there are maps $f_i : U \to S$ such that $f_i(U) \not\subset M_i$. By the Chinese Remainder Theorem 1.5.3 there is a map $f : U \to S$ such that

$$f(U) \not\subset M_i \quad \text{for any } i = 1, \ldots, s.$$

Then $f(U) + mS$ is an ideal in S that is not included in any of the maximal ideals that contain mS, whence $f(U) + mS = S$. Furthermore $f(U)_m + mS_m = S_m$ and by Lemma 1.2.1, $m \in \mathcal{J}(S_m)$, so that $f(U)_m = S_m \supset S$. Thus $f(U) \doteq S$ and therefore by Lemma 4.2.10, $f(U)$ is a progenerator ideal in S. It follows that G maps onto the progenerator $f(U)$, which proves the lemma.

We continue with the proof of Theorem 9.2.28. By the above lemma there is an S-module surjection $G \to G/N_* \to S$ whose image I is a progenerator over S. Evidently $I \doteq S$ is an invertible ideal of S. Hence $G \cong I \oplus N$ for some S-module N. Since S is then a quasi-summand of G, $\mathrm{Hom}(S, G) = \mathrm{Hom}_S(S, G)$ so that I and N are S-linear modules. This proves part 6 and completes the logical cycle.

9.2.6 Noetherian Endomorphism Modules

The group G is *E-Noetherian* if G is a Noetherian left End(G)-module. In this section we consider the quasi-direct sum decomposition of the E-Noetherian rtffr groups. In the process we classify Noetherian modules over rtffr rings.

LEMMA 9.2.31 *Let E be an rtffr ring. Then E is a right Noetherian ring iff for each integer $k > 0$, $\mathcal{N}(E)^k/\mathcal{N}(E)^{k+1}$ is a finitely generated right E-module.*

Proof: Suppose that for each integer $k > 0$, $\mathcal{N}(E)^k/\mathcal{N}(E)^{k+1}$ is a finitely generated right E-module. Then $\mathcal{N}(E)/\mathcal{N}(E)^2$ is finitely generated by the Noetherian ring $E/\mathcal{N}(E)$ so that $\mathcal{N}(E)/\mathcal{N}(E)^2$ and $E/\mathcal{N}(E)$ are Noetherian right E-modules. Hence $E/\mathcal{N}(E)^2$ is a Noetherian right E-module. Because E is rtffr $\mathcal{N}(E)^k = 0$ for some integer $k > 0$. Continuing inductively on k we eventually show that $E/\mathcal{N}(E)^k = E$ is right Noetherian. The converse is clear so the proof is complete.

The following result can be viewed as a generalization of R.S. Pierce's classification (Theorem 9.2.3) of the additive structure of a prime rtffr ring.

THEOREM 9.2.32 *Let E be an rtffr ring and let M be an rtffr left E-module. Then M is a Noetherian left E-module iff*

$$M \doteq R_1^{(n_1)} \oplus \cdots \oplus R_t^{(n_t)}$$

for some Dedekind E-rings $R_1, \ldots, R_t$.

Proof: Suppose that M is a Noetherian left rtffr E-module. Using the Beaumont-Pierce-Wedderburn Theorem 3.1.6 write $E \doteq T \oplus \mathcal{N}(E)$ for some semi-prime subring T of E. By Theorem 9.1.2

$$T \doteq R_1^{(m_1)} \oplus \cdots \oplus R_t^{(m_t)}$$

for some integers $m_1, \ldots, m_t$ and some distinct Dedekind E-rings $R_1, \ldots, R_t$.

Because M is a Noetherian E-module each E-submodule of each factor of M is Noetherian. In particular the following $E/\mathcal{N}(E)$-modules are Noetherian.

$$M_1 = \frac{M}{\mathcal{N}(E)M_*}, \quad M_2 = \frac{\mathcal{N}(E)M_*}{\mathcal{N}(E)^2 M_*}, \quad \cdots$$

Since $\mathcal{N}(E)$ is nilpotent there is an integer k such that $M_{k+1} = 0$. By our choice of T the modules $M_1, M_2, \ldots, M_k$ are Noetherian left T-modules. Thus an iterated application of Lemma 9.2.9 shows us that

$$M \doteq M_1 \oplus \cdots \oplus M_k$$

as left T-modules.

There is a finitely generated free T-module P and a surjection $\pi : P \to M$ of T-modules. There is a left T-submodule $P' \subset P$ such that $\ker \pi \oplus P' \doteq P$. Then $P' \doteq M$ and since $P \cong T^{(m)}$ for some integer m Jónsson's Theorem 2.1.10 states that

$$M \doteq R_1^{(n_1)} \oplus \cdots \oplus R_t^{(n_t)}$$

for some integers $n_1, \ldots, n_t \geq 0$.

The converse is an exercise for the reader. This completes the proof.

The proof of the main result of this section is an application of Lemma 9.2.32.

THEOREM 9.2.33 [64, A. Paras]. *Let E be an rtffr ring. The following are equivalent.*

1. *E is a right Noetherian ring.*

2. *E is a left Noetherian ring.*

3. *$E \doteq R_1^{(n_1)} \oplus \cdots \oplus R_t^{(n_t)}$ for some integers $n_1, \ldots, n_t$ and some distinct Dedekind E-rings.*

In the process of proving Theorem 9.2.32 we inadvertently proved the following result.

COROLLARY 9.2.34 *Let M be a Noetherian rtffr module over the rtffr ring E. Then*

$$M \doteq \frac{M}{\mathcal{N}(E)M_*} \oplus \frac{\mathcal{N}(E)M_*}{\mathcal{N}(E)^2 M_*} \oplus \cdots \oplus \mathcal{N}(E)^k M$$

as left $E/\mathcal{N}(E)$-modules for some integer $k > 0$.

We are now in a position to characterize the *E-Noetherian rtffr groups*. The E-Noetherian rtffr groups are characterized up to quasi-isomorphism by A. Paras [64] where she proves the following theorem.

THEOREM 9.2.35 [64, A. Paras]. *Let G be an rtffr group and let $E = \text{End}(G)$. Then G is an E-Noetherian rtffr group iff*

$$G \doteq R_1^{(n_1)} \oplus \cdots \oplus R_t^{(n_t)} \tag{9.2}$$

where $t, n_1, \ldots, n_t > 0$ are integers and where $R_1, \ldots, R_t$ are Dedekind E-rings. In this case $\text{End}(G)$ is a left Noetherian rtffr ring.

Proof: If G can be written as in (9.2) then G is E-Noetherian by Theorem 9.2.32.

Conversely, suppose that G is E-Noetherian. Since G has finite rank, there is an integer n and an embedding

$$E \longrightarrow G^{(n)}$$

of left E-modules. Then G, $G^{(n)}$, and E are left Noetherian E-modules. That is, E is a Noetherian ring. By Theorem 9.2.33 there are Dedekind E-rings $R_1, \ldots, R_t$ and integers $t, k_1, \ldots, k_t > 0$ such that

$$E \stackrel{.}{\cong} R_1^{(k_1)} \oplus \cdots \oplus R_t^{(k_t)}$$

and such that (9.2) is true. This completes the proof.

The E-Noetherian property provides us with another example of overlapping E-*properties* for rtffr groups. Of course an E-Noetherian group is an E-finitely generated group.

THEOREM 9.2.36 *Let G be an rtffr group and suppose that $E = \text{End}(G)$ is semi-prime. The following are equivalent for G.*

1. *G is an E-Noetherian group.*

2. *G is an E-finitely generated group.*

3. *G is quasi-isomorphic to an E-projective group.*

4. *G is quasi-isomorphic to an E-generator group.*

5. *G is quasi-isomorphic to an E-cyclic E-progenerator group.*

Proof: Let $E = \text{End}(G)$ be semi-prime.
$5 \Rightarrow 4$ and 3 is clear.
4 or $3 \Rightarrow 2$ follows from Corollaries 9.2.15 and 9.2.16.
$2 \Rightarrow 1$ Suppose that G is E-finitely generated. Since E is semi-prime, $\mathbb{Q}E$ is a semi-simple ring, and since $\mathbb{Q}G$ is a projective left $\mathbb{Q}E$-module

there is an integer $n > 0$ and a split embedding $f : \mathbb{Q}G \to \mathbb{Q}E^{(n)}$ of left $\mathbb{Q}E$-modules. Since G is finitely generated as a left E-module we may assume that $f(G)$ is a quasi-summand of $E^{(n)}$. By Theorem 9.2.3 there are integers $t, n_1, \ldots, n_t > 0$ and Dedekind E-rings $R_1, \ldots, R_t$ such that

$$E \overset{\cdot}{\cong} R^{(n_1)} \oplus \cdots \oplus R_t^{(n_t)}$$

so by Jónsson's Theorem 2.1.10, G has a quasi-direct sum decomposition (9.2) for some integers $m_1, \ldots, m_t \geq 0$. Then G is E-Noetherian by Theorem 9.2.35.

$1 \Rightarrow 5$ Say G is E-Noetherian. Then G has a quasi-direct sum decomposition

$$\overline{G} = R_1^{(m_1)} \oplus \cdots \oplus R_t^{(m_t)}$$

for some integers $t, m_1, \ldots, m_t > 0$ and Dedekind E-rings $R_1, \ldots, R_t$. We can assume without loss of generality that $R_i \overset{\cdot}{\cong} R_j \Rightarrow i = j$ for each $1 \leq i, j \leq t$. Then the Lemma 4.1.4 characterizing nilpotent sets and the semi-prime hypothesis imply that $\operatorname{Hom}(R_i, R_j) \subset \mathcal{N}(E) = 0$ for each $i \neq j$. Because $\operatorname{End}(R_i) = R_i$ for each $i = 1, \ldots, t$ we have

$$\operatorname{End}(\overline{G}) = \operatorname{Mat}_{n_1}(R_1) \times \cdots \times \operatorname{Mat}_{n_t}(R_t).$$

As a left $\operatorname{End}(\overline{G})$-module we have

$$\overline{G} = \begin{pmatrix} R_1 \\ \vdots \\ R_1 \end{pmatrix}_{n_1} \oplus \cdots \oplus \begin{pmatrix} R_t \\ \vdots \\ R_t \end{pmatrix}_{n_t}.$$

The reader will show as an exercise that $\overline{G}$ is a cyclic progenerator as a left $\operatorname{End}(\overline{G})$-module.

9.3   Homological Dimensions

Theorem 2.3.4 provides us with an effective method for constructing groups G whose structure as a left $\operatorname{End}(G)$-module has certain properties. This kind of example is an indispensable tool for developing intuition in studying properties of rtffr groups. We will use Theorem 2.3.4 to construct groups with prescribed homological dimension.

9.3.1 E-Projective Dimensions

Let E be an rtffr ring and let M be a left E-module. The *projective dimension of M*

$$\mathrm{pd}_E(M)$$

is the least integer $k \geq 0$ for which there is a long exact sequence

$$0 \longrightarrow P_k \longrightarrow P_{k-1} \longrightarrow \cdots \longrightarrow P_0 \longrightarrow M \longrightarrow 0 \qquad (9.3)$$

whose terms P_j are projective left E-modules for each $j = 0, \ldots, k$. Equivalently $\mathrm{pd}_E(M) = k$ if k is the least positive integer such that

$$\mathrm{Ext}_E^{k+j}(M, \cdot) = 0 \text{ for each integer } j > 0.$$

Then M is a projective left E-module iff $\mathrm{pd}_E(M) = 0$ iff $\mathrm{Ext}_E^1(M, \cdot) = 0$. (See [72].)

One of the first examples of an rtffr group with prescribed projective dimension is found in H.M.K Angad-Gaur's Thesis [7]. In it he shows that for each integer $n > 0$ there is an rtffr group G such that $\mathrm{pd}_E(G) = n$ where $E = \mathrm{End}(G)$. C. Vinsonhaler and W. Wickless [81, Corollary 6] show that for each integer $n > 0$ there is a *completely decomposable* rtffr group G such that $\mathrm{rank}(G) = 2n + 1$ and $\mathrm{pd}_E(G) = n$. We will use the construction in Theorem 2.3.4 to show that $\mathrm{pd}_{\mathrm{End}(G)}(G)$ can be almost any integer less than the left global dimension of $\mathrm{End}(G)$.

LEMMA 9.3.1 *Let E be an rtffr ring and let M be an rtffr left E-module. There is a short exact sequence*

$$0 \to M \oplus E \longrightarrow G \longrightarrow \mathbb{Q}E \oplus \mathbb{Q}E \to 0 \qquad (9.4)$$

of rtffr left E-modules such that $E \cong \mathrm{End}(G)$. Then

1. $\mathrm{pd}_E(G) = \mathrm{pd}_E(M)$ *if* $\mathrm{pd}_E(M) \geq 1$ *and*

2. $\mathrm{pd}_E(G) = 1$ *if* $\mathrm{pd}_E(M) = 0$.

Proof: 1 and 2. Let M be an rtffr left E-module, let $\mathrm{pd}_E(M) = k \geq 1$, and let N be any left E-module. By Theorem 2.3.4 there is a short exact sequence (9.4) such that $E \cong \mathrm{End}(G)$. Inasmuch as (9.4) is nonsplit,

$\operatorname{Ext}^1_E(\mathbb{Q}E, \cdot) \neq 0$. Furthermore think of $\operatorname{Ext}^k_E(\mathbb{Q}E, \cdot)$ as $\ker \delta_2/\text{image } \delta_3$ where

$$\operatorname{Hom}_E(\mathbb{Q}E, E^{(c_3)}) \xrightarrow{\delta_3} \operatorname{Hom}_E(\mathbb{Q}E, E^{(c_2)}) \xrightarrow{\delta_2} \operatorname{Hom}_E(\mathbb{Q}E, E^{(c_1)})$$

for some cardinals c_1, c_2, c_3. Inasmuch as $\operatorname{Hom}(\mathbb{Q}E, E^{(c)}) = 0$, $\operatorname{Ext}^k_E(\mathbb{Q}E, \cdot) = 0$ for each integer $k \geq 2$. Thus $\operatorname{pd}_E(\mathbb{Q}E) = 1$.

Given a fixed integer $k \geq 0$ an application of the contravariant functor $\operatorname{Hom}_E(\,, \star)$ to (9.4) yields the long exact sequence

$$\operatorname{Ext}^{k+j}_E(\mathbb{Q}E \oplus \mathbb{Q}E, \star) \rightarrow \operatorname{Ext}^{k+j}_E(G, \star)$$

$$\rightarrow \operatorname{Ext}^{k+j}_E(M, \star) \rightarrow \operatorname{Ext}^{k+j+1}_E(\mathbb{Q}E \oplus \mathbb{Q}E, \star)$$

for each integer $j > 0$. Since $\operatorname{pd}_E(\mathbb{Q}E) = 1 \leq k$ we have

$$\operatorname{Ext}^{k+j}_E(\mathbb{Q}E \oplus \mathbb{Q}E, \star) = \operatorname{Ext}^{k+j+1}_E(\mathbb{Q}E \oplus \mathbb{Q}E, \star) = 0$$

for each integer $j > 0$ and so

$$\operatorname{Ext}^{k+j}_E(G, \star) \cong \operatorname{Ext}^{k+j}_E(M, \star) \tag{9.5}$$

for integers $j > 0$.

In particular, if $\operatorname{pd}_E(M) = \infty$ then $\operatorname{Ext}^k_E(M, \star) \neq 0$ for all integers $k > 0$ so that by (9.5), $\operatorname{pd}_E(G) = \infty$.

If $1 \leq \operatorname{pd}_E(M) = k < \infty$ then $\operatorname{Ext}^{k+j}_E(M, \star) = 0$ for all integers $j > 0$ so that by (9.5), $\operatorname{pd}_E(G) \leq k$. Since $k \geq 1$, $\operatorname{Ext}^k_E(M, \star) \neq 0$ and so $\operatorname{pd}_E(G) = k$ by (9.5).

If $\operatorname{pd}_E(M) = 0$ then $\operatorname{Ext}^j_E(G, \star) \cong \operatorname{Ext}^j_E(M, \star) = 0$ for each $j > 0$ implies that $\operatorname{pd}_E(G) \leq 1$. Assume to the contrary that $\operatorname{pd}_E(G) = 0$. Because G is then an E-projective rtffr group Corollary 9.2.15 states that G is E-finitely generated. Hence $\mathbb{Q}E$ is finitely generated by E, a contradiction. Thus $\operatorname{pd}_E(G) = 1$ which completes the proof.

With the aid of the above lemma we will construct groups whose projective dimensions over their endomorphism rings are prescribed values. Recall that

$$\operatorname{gd}(E) = 1 + \{\operatorname{pd}_E(I) \mid I \subset E \text{ is a left ideal } \}.$$

THEOREM 9.3.2 Let E be an rtffr ring and suppose that there is a left ideal $I \subset E$ such that

$$\operatorname{pd}_E(I) = n \leq \infty.$$

There is an rtffr group G such that $\operatorname{End}(G) \cong E$ and $\operatorname{pd}_E(G) = n$.

Proof: Suppose that I is a *left ideal* of E such that $\mathrm{pd}_E(I) = n > 0$ and let $M = I \oplus E$. Then $\mathrm{pd}_E(M) = \mathrm{pd}_E(I) = n$ and by Theorem 2.3.4 there is a short exact sequence (9.4) of left E-modules such that $E \cong \mathrm{End}(G)$. Lemma 9.3.1 states that $\mathrm{pd}_E(G) = \mathrm{pd}_E(M) = n$.

THEOREM 9.3.3 *Let E be a Noetherian rtffr ring such that $\mathrm{gd}(E) = n + 1 < \infty$. For each integer $0 < k \leq n$ there is an rtffr group G_k such that $E \cong \mathrm{End}(G_k)$ and $\mathrm{pd}_E(G_k) = k$.*

Proof: Let $0 < k \leq n$ be an integer. There is a left ideal $I \subset E$ such that $\mathrm{pd}_E(I) = n$, so there is a long exact sequence

$$0 \to P_n \xrightarrow{\delta_n} \cdots \xrightarrow{\delta_2} P_1 \xrightarrow{\delta_1} P_0 \longrightarrow I \to 0$$

of left E-modules in which each P_i is a projective. Since E is Noetherian I is finitely presented so we can choose such a sequence in which each P_i is finitely generated projective. Let

$$M_k = \text{image } \delta_k.$$

The reader will note that the induced long exact sequence

$$0 \to P_k \xrightarrow{\delta_k} \cdots \xrightarrow{\delta_k} M_k \to 0$$

has exactly $n - k$ projective terms so that $\mathrm{pd}_E(M_k) = n - k$. Using $M = M_{n-k}$ we can use Theorem 2.3.4 to construct a short exact sequence (9.4) of left E-modules such that $E \cong \mathrm{End}(G)$. By Lemma 9.3.1, $\mathrm{pd}_E(G) = \mathrm{pd}_E(M_{n-k}) = k$.

We will state but will not prove a stronger result for countable reduced torsion-free groups whose ranks may not be finite.

THEOREM 9.3.4 [36, T.G. Faticoni] *Let E be a Corner ring and suppose that $\mathrm{gd}(E) = n + 1$. Then for each integer $0 < k \leq n$ there is a Corner group G_k such that $E \cong \mathrm{End}(G_k)$ and $\mathrm{pd}_E(G_k)$.*

EXAMPLE 9.3.5 Let $E = \mathbb{Z}[x]/(x^2)$, let $I = (x)/(x^2) \cong \mathbb{Z}$, and let $M = I \oplus E$. By Theorem 2.3.4 there is a short exact sequence (9.4) such that $E \cong \mathrm{End}(G)$. By Lemma 9.3.1, $\mathrm{pd}_E(G) = \mathrm{pd}_E(M)$, and it is readily verified that

$$\cdots \xrightarrow{x} E \xrightarrow{x} E \xrightarrow{x} I \to 0 \tag{9.6}$$

is a projective resolution of I such that $\ker x = I \subset E$ is not a direct summand of E. Thus $\mathrm{pd}_E(M) = \mathrm{pd}_E(I) = \infty$. By Lemma 9.3.1 there is an rtffr group G such that $E \cong \mathrm{End}(G)$ and $\mathrm{pd}_E(G) = \mathrm{pd}_E(I) = \infty$.

EXAMPLE 9.3.6 Let E be an rtffr ring and let G be a ring constructed according to Corner's Theorem such that $E \cong \operatorname{End}(G)$. Then G fits into a short exact sequence

$$0 \to E \longrightarrow G \longrightarrow \mathbb{Q}E \to 0$$

of left E-modules. By Lemma 9.3.1, $\operatorname{pd}_E(G) = 1$.

9.3.2 E-Flat Dimensions

The *flat dimension of M*

$$\boxed{\operatorname{fd}_E(M)}$$

is the least integer $k > 0$ for which there is a long exact sequence (9.3) in which each P_i is a flat left E-module for each $i = 0, \ldots, k$. Equivalently, $\operatorname{fd}_E(M) = k$ iff k is the least positive integer such that

$$\operatorname{Tor}_E^{k+j}(\star, M) = 0 \text{ for each integer } j > 0.$$

Then M is a *flat* left E-module iff $\operatorname{fd}_E(M) = 0$ iff $\operatorname{Tor}_E^1(\star, M) = 0$. Moreover M is a flat left E-module iff given a short exact sequence

$$0 \to K \longrightarrow N \longrightarrow L \to 0$$

of right E-modules then the induced sequence

$$0 \to K \otimes_E M \longrightarrow N \otimes_E M \longrightarrow L \otimes_E M \to 0$$

of groups is exact.

C. Vinsonhaler and W. Wickless [81, Theorem 5] show that for each integer $n > 0$ there is a *completely decomposable* rtffr group G such that $\operatorname{rank}(G) = 2n + 1$ and $\operatorname{fd}_E(G) = n$. We address the question of which integers are the flat dimensions of rtffr groups. The lemma indicates how we will construct rtffr groups of prescribed flat dimension. Recall that $\mathcal{O}(M) = \{q \in \mathbb{Q}E \mid qM \subset M\}$.

LEMMA 9.3.7 Let E be an rtffr ring and let M be a left rtffr E-module such that $E = \mathcal{O}(M)$. If there is a short exact sequence

$$0 \to M \longrightarrow G \longrightarrow \mathbb{Q}E \oplus \mathbb{Q}E \to 0 \tag{9.7}$$

of left E-modules then $\operatorname{fd}_E(G) = \operatorname{fd}_E(M)$.

Proof: Given $\mathrm{fd}_E(M) = k$ an application of $\star \otimes_E \star$ to (9.7) yields the long exact sequence

$$\mathrm{Tor}^{k+j-1}(\star, \mathbb{Q}E \oplus \mathbb{Q}E) \to \mathrm{Tor}^{k+j}(\star, M)$$
$$\to \mathrm{Tor}^{k+j}(\star, G) \to \mathrm{Tor}^{k+j}(\star, \mathbb{Q}E \oplus \mathbb{Q}E)$$

for each integer $j > 0$. Because $\mathbb{Q}E$ is a flat left E-module

$$\mathrm{Tor}^{k+j-1}(\star, \mathbb{Q}E \oplus \mathbb{Q}E) = \mathrm{Tor}^{k+j}(\star, \mathbb{Q}E \oplus \mathbb{Q}E) = 0$$

for each $k + j - 1 > 0$. Thus

$$\mathrm{Tor}^{k+j}(\star, M) \cong \mathrm{Tor}^{k+j}(\star, G) \tag{9.8}$$

for each $j > 0$.

If $\mathrm{fd}_E(M) = \infty$ then by (9.8), $\mathrm{fd}_E(G) = \infty$.

If $1 \leq \mathrm{fd}_E(M) = k < \infty$ then $\mathrm{Tor}^k(\star, M) \neq 0$ and $\mathrm{Tor}^{k+j}(\star, M) = 0$ for each integer $j > 0$. Then by (9.8), $\mathrm{fd}_E(G) = k$.

If $\mathrm{fd}_E(M) = 0$ then M and $\mathbb{Q}E$ are flat left E-modules. It follows from the exactness of

$$0 = \mathrm{Tor}^1(\star, M) \to \mathrm{Tor}^1(\star, G) \to \mathrm{Tor}^1(\star, \mathbb{Q}E \oplus \mathbb{Q}E) = 0$$

that $\mathrm{fd}_E(G) = 0$. This completes the proof.

We can apply Lemma 9.3.7 to the next result in the same manner that we used Lemma 9.3.1 to prove Theorem 9.3.2. In constructing (9.7) in the next proof choose $M = I \oplus E$ and then apply Theorem 2.3.4.

THEOREM 9.3.8 [T.G. Faticoni] *Let E be an rtffr ring and let I be a left ideal in E such that $\mathrm{fd}_E(I) = n$. There is an rtffr group G such that $E \cong \mathrm{End}(G)$ and $\mathrm{fd}_E(G) = n$.*

The following result is proved in the same manner that we proved Theorem 9.3.3, but use $\mathrm{Tor}(\star, \star)$ instead of $\mathrm{Ext}(\star, \star)$.

THEOREM 9.3.9 *Let E be a semi-prime rtffr ring and suppose that there is a left ideal I of E such that $\mathrm{fd}_E(I) = n$. Then for each integer $0 \leq k \leq n$ there is an rtffr group G_k such that $E \cong \mathrm{End}(G_k)$ and $\mathrm{fd}_E(G_k) = k$.*

EXAMPLE 9.3.10 Let $E = \mathbb{Z}[x]/(x^2)$, let $I = (x)/(x^2)$, and let $M = I \oplus E$. Then the group in the center term of (9.4) has $\mathrm{fd}_E(G) = \mathrm{fd}_E(I) = \infty$ as there is an infinite sequence (9.6) of flat modules E such that no kernel $\ker x$ is a direct summand of E.

EXAMPLE 9.3.11 Let E be an rtffr ring and let G be a group constructed according to Corner's Theorem 2.3.3 such that $E \cong \text{End}(G)$. There is a short exact sequence

$$0 \to E \longrightarrow G \longrightarrow \mathbb{Q}E \to 0$$

of left E-modules so G is a flat left E-module. That is, $\text{fd}_E(G) = 0$. Compare to Example 9.3.6 where we proved that $\text{pd}_E(G) = 1$.

9.3.3 E-Injective Dimensions

Define the *injective dimension of M*

$$\boxed{\text{id}_E(M)}$$

to be the least integer $k \geq 0$ for which there exists an exact sequence

$$0 \longrightarrow M \longrightarrow E_0 \longrightarrow E_1 \cdots \longrightarrow E_k \longrightarrow 0 \tag{9.9}$$

where each E_i is an injective left E-module. Equivalently $\text{id}_E(M) = k$ iff k is the least positive integer such that

$$\text{Ext}^{k+j}(\star, M) = 0 \text{ for each integer } j > 0.$$

The constructions illustrating the injective dimensions over an rtffr ring E turn out to be harder than the examples we gave about the projective and flat dimensions. For instance, while our techniques give us good results when we consider the injective dimension of $\mathbb{Q}G$, they say very little about the injective dimension of G. Thus our results are incomplete.

LEMMA 9.3.12 *Let E be an rtffr ring. There is a short exact sequence*

$$0 \to E \longrightarrow G \longrightarrow \mathbb{Q}E \to 0 \tag{9.10}$$

of left E-modules such that $E = \text{End}(G)$ and $\text{id}_E(\mathbb{Q}G) = \text{id}_E(\mathbb{Q}E)$.

Proof: Let $\text{id}_E(\mathbb{Q}E) = k$. Applying $\cdot \otimes \mathbb{Q}$ to 9.10 we have a split exact sequence

$$0 \to \mathbb{Q}E \longrightarrow \mathbb{Q}G \longrightarrow \mathbb{Q}E \to 0$$

Then $\text{id}_E(\mathbb{Q}E) = \text{id}_E(\mathbb{Q}G)$.

THEOREM 9.3.13 [T.G. Faticoni] *Let E be an rtffr ring and let G be the group constructed for E according to Corner's Theorem 2.3.3. If $\mathrm{id}_E(\mathbb{Q}E) = n$ then $\mathrm{id}_E(\mathbb{Q}G) = n$.*

Proof: By Corner's Theorem there is a short exact sequence (9.10) of left E-modules such that $E \cong \mathrm{End}(G)$. Then by Lemma 9.3.12, $n = \mathrm{id}_E(\mathbb{Q}E) = \mathrm{id}_E(\mathbb{Q}G)$.

Let E be an rtffr ring and let G be the group constructed according to Corner's Theorem 2.3.3. Examples 9.3.6 and 9.3.11 show us that

$$\mathrm{pd}_E(G) = 1, \mathrm{fd}_E(G) = 0, \text{ and } \mathrm{id}_E(\mathbb{Q}G) = \mathrm{id}_E(\mathbb{Q}E).$$

Of course $\mathrm{id}_E(\mathbb{Q}E) = 0$ if E is a semi-prime ring. So in the by now familiar manner the general condition $\mathrm{id}_E(\mathbb{Q}E) = 0$ leads us to the following theorem.

THEOREM 9.3.14 [T.G. Faticoni] *Let E be an rtffr ring such that $\mathrm{id}_E(\mathbb{Q}E) = 0$, let M be an rtffr left E-module such that $E = \mathcal{O}(M)$. There is a short exact sequence*

$$0 \to M \longrightarrow G \longrightarrow \mathbb{Q}E \oplus \mathbb{Q}E \to 0 \tag{9.11}$$

of left E-modules such that $E \cong \mathrm{End}(G)$ and such that $\mathrm{id}_E(G) = \mathrm{id}_E(M)$.

Proof: Use Theorem 2.3.4 to construct the short exact sequence (9.11) in which $E \cong \mathrm{End}(G)$. The usual argument shows us that $\mathrm{id}_E(G) = \mathrm{id}_E(M)$.

For example, if E is a semi-prime rtffr ring then $\mathrm{id}_E(G) = \mathrm{id}_E(E)$ for any group G constructed according to Corner's Theorem 2.3.3.

EXAMPLE 9.3.15 For each integer n there is a torsion-free group G of rank n such that $\mathrm{pd}_E(G) = \mathrm{fd}_E(G) = \infty$ and $\mathrm{id}_E(\mathbb{Q}G) = 0$ where $E = \mathrm{End}(G)$.

Proof: Let $E = \mathbb{Z}[x]/(x^n)$. Then E is a torsion-free ring of rank n whose additive structure is a free group. Butler's Construction in Theorem 2.3.1 states that $E = \mathrm{End}(G)$ for some group G such that $E \subset G \subset \mathbb{Q}E$ as left E-modules. The reader can show that $\mathbb{Q}E = \mathbb{Q}[x]/(x^n)$ is a self-injective ring so that $\mathrm{id}_E(\mathbb{Q}G) = \mathrm{id}_E(\mathbb{Q}E) = 0$. The rest follows from Lemmas 9.3.1 and 9.3.7.

EXAMPLE 9.3.16 Let A be a finite dimensional $\mathbb{Q}$-algebra and suppose that M is a finitely generated left A-module. If $\mathrm{ann}_A(M) = 0$ then there is a full subring $E \subset A$ and an rtffr group G such that $E = \mathrm{End}(G)$ and $\mathrm{id}_E(\mathbb{Q}G) = \min\{\mathrm{id}_E(\mathbb{Q}E), \mathrm{id}_E(M)\}$.

Proof: Let F be a full free subgroup of M and let

$$E = \mathcal{O}(F) = \{q \in A \,|\, qF \subset F\}.$$

It is an easy exercise to show that for each $q \in A$ there is an $n \neq 0 \in \mathbb{Z}$ such that $nq(F) \subset F$. Thus $\mathbb{Q}E = A$. Theorem 2.3.4 constructs a short exact sequence

$$0 \to F \longrightarrow G \longrightarrow \mathbb{Q}E \oplus \mathbb{Q}E \to 0$$

of left E-modules such that $E \cong \mathrm{End}(G)$. So

$$\mathrm{id}_E(\mathbb{Q}G) = \max\{\mathrm{id}_E(\mathbb{Q}E), \mathrm{id}_E(\mathbb{Q}M)\}.$$

Let us agree to call the rtffr group G an E-*injective* group if $\mathbb{Q}G$ is an injective left $\mathrm{End}(G)$-module. C. Vinsonhaler and W. Wickless [83] characterize the injective hull of a torsion-free group of finite rank while C. Vinsonhaler [79] characterizes certain E-injective rtffr groups.

EXAMPLE 9.3.17 Let $E = \begin{pmatrix} \mathbb{Z} & 0 \\ \mathbb{Z} & \mathbb{Z} \end{pmatrix}$ and let $I = \begin{pmatrix} \mathbb{Z} & 0 \\ \mathbb{Z} & 0 \end{pmatrix}$. Then I is a projective ideal of E such that $\mathrm{ann}_E(I) = 0$ and $\mathbb{Q}I$ is an injective left E-module. The reader can verify that $E = \{q \in \mathbb{Q}E \,|\, qI \subset I\}$. Theorem 2.3.4 produces an exact sequence

$$0 \longrightarrow I \longrightarrow G \longrightarrow \mathbb{Q}E \oplus \mathbb{Q}E \longrightarrow 0$$

of left E-modules such that $E \cong \mathrm{End}(G)$. Since $\mathbb{Q}E$ is a hereditary ring, $\mathrm{id}_E(M) \leq 1$ for each left E-module M. It is an interesting exercise to show that $\mathrm{Mat}_2(\mathbb{Q})$ is the injective hull of $\mathbb{Q}E$ so that $\mathrm{id}_E(\mathbb{Q}E) = 1$. Thus G and the group constructed according to Corner's Theorem 2.3.3 satisfy $\mathrm{pd}_E(G) = 1$, $\mathrm{fd}_E(G) = 0$, and $\mathrm{id}_E(G) = 1$.

The above constructions show us that any classification of the E-homological dimension of an rtffr group G will require an approach that differs significantly from the techniques used in this book. See [32] for a characterization of homological dimensions of modules over their endomorphism rings.

9.4 Self-Injective Rings

The literature contains a fairly complete characterization of those rtffr groups for which $\text{End}(G)$ is a left or right hereditary ring. See [4, 10, 12, 41, 35, 43, 53]. Let us consider the rtffr groups G such that $\mathbb{Q}\text{End}(G)$ is a *left self-injective* ring. Because $\mathbb{Q}\text{End}(G)$ is Artinian it is known (see [30]) that $\mathbb{Q}\text{End}(G)$ is right self-injective iff $\mathbb{Q}\text{End}(G)$ is left self-injective.

We will need the following ideas. G-*monomorphisms* are introduced in [37]. A G-monomorphism is a group map $\hat{\jmath} : H \to H'$ such that $\hat{\jmath} = \text{T}_G(\imath)$ for some injection $\imath : M \to M'$ of right $\text{End}(G)$-modules. The group H is G-*presented* if $H = G^{(n)}/K$ for some integer $n > 0$ and some G-generated subgroup $K \subset G^{(n)}$. It is easy to see that $\hat{\jmath} : K \to G$ is a monomorphism in *the category G-**Pre** of G-presented groups* iff $\text{S}_G(\ker \hat{\jmath}) = 0$. By [32, Lemma 5.1.1] a group map $\hat{\jmath} : K \to G$ is a G-monomorphism iff $\hat{\jmath} = \text{T}_G(\imath)$ for some injection $\imath : I \longrightarrow \text{H}_G(G)$ of right $\text{End}(G)$-modules iff $\hat{\jmath}$ is a *monomorphism in* the category G-**Pre**. Our characterization of rtffr groups such that $\mathbb{Q}\text{End}(G)$ is a right self-injective ring is in terms of a lifting property for G-monomorphisms.

THEOREM 9.4.1 [T.G. Faticoni] *Let G be an rtffr group and let $E = \text{End}(G)$. The following are equivalent.*

1. $\mathbb{Q}E$ *is self-injective.*

2. *If $\hat{\jmath} : K \to G$ is a monomorphism in G-**Pre** and if $f : K \to G$ is a group map then there is a map $g : G \to G$ and an integer n such that $nf = g\hat{\jmath}$.*

Proof: Assume part 1. Then $\mathbb{Q}E$ is right self-injective. Let $\hat{\jmath} : K \to G$ be a monomorphism in G-**Pre**, and let $h : K \to G$ be a group map. By [32, Lemma 5.1.1], $\hat{\jmath}$ is a G-monomorphism, so $\hat{\jmath} = \text{T}_G(\imath)$ for some injection $\imath : I \to E$ of right E-modules. Specifically $\text{T}_G(I) \cong K$. By the adjoint isomorphism

$$\text{Hom}_E(I, E) \cong \text{Hom}_E(I, \text{H}_G(G)) \cong \text{Hom}(\text{T}_G(I), G) = \text{Hom}(K, G)$$

and the reader can verify that this isomorphism sends each map $\phi : I \to E$ to $\text{T}_G(\phi)$. Thus $h = \text{T}_G(\phi)$ for some $\phi : I \to E$. Since $\mathbb{Q}E$ is self-injective there is a map $\gamma : \mathbb{Q}E \to \mathbb{Q}E$ such that $\phi = \gamma\imath$, and there is an integer n such that $n\gamma(E) \subset E$. Then

$$nh = n\text{T}_G(\phi) = n\text{T}_G(\gamma\imath) = (ng)\hat{\jmath}.$$

This proves part 2.

Conversely assume part 2, let $\imath : I \to \mathbb{Q}E$ be an inclusion of right $\mathbb{Q}E$-modules, and let $f : I \to \mathbb{Q}E$ be a $\mathbb{Q}E$-module map. Write $I = \mathbb{Q}J$ for some finitely generated right ideal $J \subset E$ so that there is an integer $n \neq 0$ such that $n\imath(J) \subset E$. By definition $\mathrm{T}_G(n\imath) : \mathrm{T}_G(J) \to \mathrm{T}_G(E) = G$ is a G-monomorphism. Choose an integer $m \neq 0$ such that $mf(J) \subset E$ so that $\mathrm{T}_G(mf) : \mathrm{T}_G(J) \to \mathrm{T}_G(E)$. By part 2 there is an integer $m \neq 0$ and a group map $g : \mathrm{T}_G(E) \to \mathrm{T}_G(E)$ such that

$$\mathrm{T}_G(mf) = g\mathrm{T}_G(n\imath).$$

By the Arnold-Lady-Murley Theorem 2.4.1 we can write $g = \mathrm{T}_G(h)$ for some $h : E \to E$. Then

$$\mathrm{T}_G(mf) = \mathrm{T}_G(nh\imath)$$

or in other words

$$\mathrm{T}_G(mf - nh\imath) = 0 : \mathrm{T}_G(J) \longrightarrow \mathrm{T}_G(E).$$

Since $J \subset E$ the adjoint isomorphism implies that $mf - nh\imath = 0$ whence $f = (\frac{n}{m}h)\imath$. This completes the proof.

COROLLARY 9.4.2 *Let G be an rtffr group such that $\mathbb{Q}E$ is self-injective. If $\hat{\jmath} : K \to G$ is a monomorphism in G-**Pre** (for instance, if $\hat{\jmath}$ is an injection) and if $f : K \to G$ is a group map then there is a map $\psi : G \to G$ and an integer n such that $nf = \psi\hat{\jmath}$.*

9.5   Exercises

Let E be an rtffr ring, let G, H be rtffr groups.

1. Let $E = \mathrm{End}(G)$ and $S = \mathrm{center}(E)$. Prove that

    (a) $E = \mathrm{End}_S(G)$.

    (b) $S = \mathrm{End}_E(G)$.

    (c) If $G = A \oplus B$ then A is an S-submodule of G.

2. Let M be a finitely generated right E-module whose additive structure is a torsion group. Show that $mM = 0$ for some integer $m \neq 0$.

3. A hard problem. Find a finitely generated right ideal in an rtffr ring that is not finitely presented.

4. Show that the finite modules over an rtffr ring E are finitely presented.

5. Let R be any ring and let N be a left R-module. Then N is generated by R as a *group*. Hint: For fixed $x \in N$ consider the map $\lambda_x : R \to N$ such that $\lambda_x(r) = rx$ for each $r \in R$.

6. Show that a finitely presented left E-module M possesses a projective resolution $\cdots P_2 \to P_1 \to P_0 \to M \to 0$ such that the projective modules $P_0, P_1, P_2, \ldots$ are finitely generated.

7. Let E be an rtffr ring and let I be a maximal right ideal of E. If $\mathrm{pd}_E(I) = n$ then for each integer $0 < k \leq n$ there is a *finite rank* group G_k such that $E = \mathrm{End}(G_k)$ and $\mathrm{pd}_E(G_k) = k$.

8. Let G be an E-finitely generated rtffr group $\mathcal{N} = \mathcal{N}(\mathrm{End}(G))$. Show that if G' is a left $\mathrm{End}(G)$-submodule of G such that $G \doteq G' + \mathcal{N}G$ then $G \doteq G'$.

9. Let G be an rtffr group and let $\mathcal{N}(\mathrm{End}(G)) = \mathcal{N}$. Construct rtffr G and G' such that $G' \subset G$ is a left $\mathrm{End}(G)$-submodule, $G \doteq G' + \mathcal{N}G$, but G' is not quasi-isomorphic to G.

10. If E is an rtffr semi-prime ring and if $P \subset E^{(m)}$ is a right E-submodule then any E-module surjection $\pi : E^{(n)} \to P$ is quasi-split. That is, there is a map $\sigma : P \to E^{(n)}$ and an integer k such that $\pi\sigma = k1_P$.

11. Let E be a semi-prime rtffr ring and let M be an E-submodule of $E^{(n)}$ for some integer $n \neq 0$. Then M is an E-Noetherian rtffr group.

12. Let E be an rtffr E-ring and let $E \doteq I$ be a projective ideal of E.

    (a) E is locally isomorphic to I as groups.

    (b) E is E-locally isomorphic to I as an E-module.

13. Let R be an rtffr integral domain. Prove that if $I \subset R$ is a nonzero ideal then R/I is finite.

14. Let R be an rtffr integral domain and let M be a finitely generated R-module. Then the R-torsion submodule T of M is finite and $M/T \stackrel{.}{\cong} R^{(m)}$ for some integer $m > 0$.

15. Suppose that G is an rtffr group and that $A \oplus B$ is a subgroup of finite index in G. Show that there is a *quasi-idempotent* $e \in \text{End}(G)$ corresponding to A. That is, there is an $e \in \text{End}(G)$ such that $e^2 = ne$ for some integer $n \neq 0$, $e(x) = nx$ for each $x \in A$, and $e(B) = 0$.

16. Show that if E is a ring then $E^{(n)}$ is a cyclic projective left $\text{Mat}_n(E)$-module.

17. Let R be an E-ring. Then $R^{(n)}$ is a cyclic projective left $\text{End}(R^{(n)})$-module.

18. Let E be an rtffr ring.

    (a) Find an rtffr left E-module M such that $\mathbb{Q}M$ is the injective hull of the left E-module E.

    (b) Find an rtffr group G such that $E \cong \text{End}(G)$ and $\text{id}_E(\mathbb{Q}G) = 0$.

    (c) Let E be an rtffr ring and let M be an rtffr left E-module. Let $\text{length}_E(M)$ denote the composition length of $\mathbb{Q}M$ as a left $\mathbb{Q}E$-module. Show that for each integer n there is an rtffr group G such that $\text{length}_{\text{End}(E)}(G) = n$.

19. Let E be a semi-prime rtffr ring and let M, N be left rtffr E-modules. Show that $\text{Ext}_E^n(M, N)$ is a bounded group for each integer $n \geq 1$.

20. Let A be a self-injective finite dimensional $\mathbb{Q}$-algebra and let M be any free *subgroup* of A such that $\mathbb{Q}M = A$.

    (a) Prove that there is a group G such that $M \subset G \subset A$ and such that $\mathbb{Q}\text{End}(G) = A$.

    (b) Prove that G is an E-injective rtffr group and $\mathbb{Q}\text{End}(G)$ is a self-injective ring.

    (c) By varying M produce a variety of properties on G. For instance, Show that under one choice of M, G has E-property while under another choice G does not have E-*property*. Be creative.

    (d) Is it possible that G is strongly indecomposable for one choice of M, and decomposable for another?

21. Find a better way of investigating the injective dimension of G as a left E-module and not of $\mathbb{Q}G$.

9.6 Questions for Future Research

Let G be an rtffr group, let E be an rtffr ring, and let $S = \text{center}(E)$ be the center of E. Let τ be the conductor of G. There is no loss of generality in assuming that G is strongly indecomposable.

1. We have characterized the finitely generated rtffr E-modules M in terms of direct sums of E-rings. Characterize larger classes of E-modules. For instance, consider the injective E-modules. Dual the definition of finitely generated E-modules to arrive at the finitely cogenerated E-modules. See [6].

2. Study The Beaumont-Pierce-Wedderburn Theorem for a ring E. Give a canonical or natural value for the semi-prime ring $T \subset E$.

3. Prove the converse of Theorem 9.2.11.

4. This question is due to R. Pierce. Let *property* be a module theoretic property, and say that G is an *E-property* group if G satisfies *property*. Pick *property* and characterize the *E-property* groups.

5. Fix *property*. Characterize all of the rtffr rings E such that each $G \in \Omega(E)$ satisfies *property*.

6. Fix *property*. Construct examples of indecomposable rtffr *E-property* groups.

7. Use Galois Theory to characterize the strongly indecomposable *E*-rings. See [39, 65, 66].

8. See [32]. Find an internal characterization of rtffr groups G that possess infinite flat, projective, or injective dimension over $\text{End}(G)$.

9. Find an internal characterization of rtffr groups G that possess finite flat, projective, or injective dimension over $\text{End}(G)$.

10. Characterize the rtffr groups G such that $\mathbb{Q}G$ is an injective left $\text{End}(G)$-module.

11. Characterize the rtffr groups G such that $\mathbb{Q}\text{End}(G)$ is an injective left or right $\text{End}(G)$-module.

Appendix A

Pathological Direct Sums

We have sectioned these examples into the appendices because of the different techniques used. Reread the first few results of the chapters. You will see the use of ring and module theory, and of functors and functorial methods. These ideas were not widely used in abelian group theory during the period of the 1950s and 1960s. A.L.S. Corner [23] broke that mold with the publication of his famous result now called Corner's Theorem F.1.1 wherein he proved that each countable reduced torsion-free ring is the group endomorphism ring of a torsion-free group. His technique was pure number theory and linear algebra over $\mathbb{Z}$. Compare this to the Arnold-Lady Murley Theorem 2.4.1 that is pure category theory. That is why we have these appendices. The results of the chapters use modern techniques while the techniques in the appendices are all number theory. They represent the techniques used in the early years of abelian groups.

A.1 Nonunique Direct Sums

We have already claimed that direct sums of the general rtffr group are not unique in any sense. In this appendix we will give examples to justify our claim. We begin with the most elementary of examples. These examples come from [46, Sections 90].

EXAMPLE A.1.1 Let $n \geq 2$ be an integer. There exists an rtffr group G such that
$$G = A_1 \oplus \cdots \oplus A_{n-1} \oplus B = C \oplus D$$
where the groups A_i, B, C, D are indecomposable, $\mathrm{rank}(A_i) = 1$, $\mathrm{rank}(B)$ $= \mathrm{rank}(C) = \mathrm{rank}(D) = n - 1$.

Proof: Let $n \geq 3$ be an integer, let $p, q, p_1, \cdots, p_{n-1}$ be different primes, and let

$$\{a_1, \ldots, a_{n-1}\} \text{ and } \{b_1, \ldots, b_{n-1}\}$$

be bases for the $\mathbb{Q}$-vector space V. Define

$$
\begin{aligned}
A_j &= \langle p_j^{-\infty} a_j \rangle \text{ for each } j = 1, \cdots, n-1 \\
B &= \langle p_1^{-\infty} b_1, \ldots, p_{n-1}^{-\infty} b_{n-1}, \\
&\quad p^{-1} q^{-1}(b_1 + b_2), \ldots, p^{-1} q^{-1}(b_1 + b_{n-2}) \rangle
\end{aligned}
$$

and let

$$G = A_1 \oplus \cdots \oplus A_{n-1} \oplus B.$$

LEMMA A.1.2 B *is indecomposable.*

Proof: Suppose that $B = U \oplus W$ as abelian groups. We have a quasi-isomorphism

$$U \oplus W \doteq \langle p_1^{-\infty} b_1, \ldots, p_{n-1}^{-\infty} b_{n-1} \rangle = B_o$$

and the subgroups $\langle p_j^{-\infty} b_j \rangle$ are fully inavariant in B_o because the p_j are different primes. Since these subgroups have rank one, there is a complete set of central indecomposable idempotents

$$e_1, \ldots, e_{n-1} \in \operatorname{End}_{\mathbb{Z}}(B_o)$$

such that

$$e_j(x) = x \text{ for each } x \in \langle p_j^{-\infty} b_j \rangle.$$

Thus each idempotent in $\operatorname{End}_{\mathbb{Z}}(B_o)$ is a sum of a subset of $\{e_j \mid i = 1, \ldots, n-1\}$. For example, given the canonical idempotent map

$$e : B_o \longrightarrow U$$

there are disjoint subsets σ, ψ of $\{1, \ldots, n-1\}$ such that

$$e = \oplus_{j \in \sigma} e_j.$$

Suppose for the sake of contradiction that there are $1 \in \psi$ and $j \in \sigma$. Then

$$e(p^{-1} q^{-1}(b_1 \oplus b_j)) = p^{-1} q^{-1}(b_j).$$

This is contrary to the fact that b_j has p-height 0 in $\langle p_j^{-\infty} b_j \rangle$. Thus $\sigma = \emptyset$, hence $U = 0$, whence B is indecomposable.

Proof of Example A.1.1: We choose an ordered base

$$a_1, \ldots, a_{n-1}, b_1, \ldots, b_{n-1}$$

for V and primes $p \neq q$. Choose integeres s and t such that $ps - qt = 1$ and then choose

$$c_j = pa_j + tb_j \text{ and } d_j = qa_j + sb_j$$

for each $j = 1, \ldots, n - 1$. Let

$$
\begin{aligned}
C &= \langle p_1^{-\infty} c_1, \ldots, p_{n-1}^{-\infty} c_{n-1}, p^{-1}(c_1 + c_2), \ldots, p^{-1}(c_1 + c_{n-2}) \rangle \\
D &= \langle p_1^{-\infty} d_1, \ldots, p_{n-1}^{-\infty} d_{n-1}, q^{-1}(d_1 + d_2), \ldots, q^{-1}(d_1 + d_{n-2}) \rangle.
\end{aligned}
$$

Then $C \cap D = 0$, $C, D \subset G$, and by the above lemma C and D are indecomposable. By our choices of c_j, d_j, s, t we have

$$
\begin{aligned}
a_j &= sc_j - td_j \\
b_j &= pd_j - qc_j \\
b_1 + b_j &= p(d_1 + d_j) - q(c_1 + c_j).
\end{aligned}
$$

Thus $G = C + D$ and therefore $G = C \oplus D$.

Because $A_1 \ncong C, D$ in the above example, it follows that a direct sum decomposition of an rtffr group G into indecomposables is not necessarily unique.

Appendix B

ACD Groups

B.1   Example by Corner

The next example shows that rtffr groups can have many direct sum decompositions into indecomposables.

EXAMPLE B.1.1 [Corner] (See [46].)   Let $n \geq k \geq 1$ be integers. There is a group $G = G(n)$ of rank n such that for each partition $r_1 + \cdots + r_k$ of n into k positive summands r_j there is an indecomposable direct sum decomposition

$$G = G_1 \oplus \cdots \oplus G_k$$

such that $r_j = \mathrm{rank}(G_j)$ for each $j = 1, \ldots, k$.

Proof: Let $p, p_1, \ldots, p_{n-k}, q_1, \ldots, q_{n-k}$ be a list of different primes, let V be a $\mathbb{Q}$-vector space, let

$$u_1, \ldots, u_k, x_1, \ldots, x_{n-k}$$

be a basis for V, and let

$$
\begin{aligned}
G \ = \ & \langle p^\infty u_1, \ldots, p^\infty u_k, \\
& p_1^\infty x_1, \ldots, p_{n-k}^\infty x_{n-k}, \\
& q_1^{-1}(u_1 + x_1), \ldots, q_{n-k}^{-1}(u_1 + x_{n-k}) \rangle.
\end{aligned}
$$

We will show that G has the indicated property.

Let $n = r_1 + \cdots + r_k$ be a partition of n with $r_j \geq 1$ for each $j = 1, \ldots, k$.

Suppose there are $s_1, \ldots, s_k \in \mathbb{Z}$ such that

$$s_1 + \cdots + s_k = 1$$

and unknowns $v_1, \ldots, v_k$. The system of linear equations

$$\begin{cases} s_1 v_1 & + & s_2 v_2 & + & \cdots & + & s_k v_k & = & u_1 \\ -v_1 & + & v_2 & & & & & = & u_2 \\ \vdots & & & & & & & & \vdots \\ -v_1 & & & & & + & v_k & = & u_k \end{cases}$$

has determinant 1. Hence the system has a solution $v_1, \ldots, v_k \in \mathbb{Z}$ such that

$$\langle v_1, \ldots, v_k \rangle = \langle u_1, \ldots, u_k \rangle.$$

Let

$$G_j = \langle p^{\infty} v_j, p_i^{\infty} x_i, q_i^{-1}(v_j + x_i) \mid i = t_j + 1, \ldots, t_{j+1} \rangle$$

where $t_1 = 0$ and $t_j = (r_1 - 1) + \cdots + (r_{j-1} - 1)$. Observe that

$$u_1 + x_i =$$
$$(v_j + x_i) + s_1 v_1 + \cdots + s_{j-1} v_{j-1} + (s_j - 1) v_j + \cdots + s_k v_k \ .$$

Then in choosing the $s_1, \ldots, s_k$ require that

$$s_j = \begin{cases} 1 \ (\mathrm{mod}\ )q_i & \text{for } i = t_j + 1, \cdots, t_{j+1} \\ 0 \ (\mathrm{mod}\ )q_j & \text{otherwise} \end{cases}.$$

[Comment: For example, let

$$Q_j = q_1 \cdots q_{j-1} q_{j+1} \cdots q_{n-k},$$

choose integers m_ℓ such that $\sum_\ell m_\ell Q_\ell = 1$, and then choose

$$s_j = m_{t_i+1} Q_{t_i+1} + \cdots + m_{t_{i+1}} Q_{t_{i+1}}.]$$

Subsequently $G_j \subset G$ for each $j = 1, \ldots, k$, and

$$\begin{aligned} & p^{\infty} u_1, \ldots, p^{\infty} u_k, \\ & p_1^{\infty} x_1, \ldots, p_{n-k}^{\infty} x_{n-k}, \\ & q_1^{-1}(u_1 + x_1), \ldots, q_{n-k}^{-1}(u_1 + x_{n-k}) \\ & \in G_1 + \cdots + G_k. \end{aligned}$$

Thus

$$G = G_1 \oplus \cdots \oplus G_k$$

where by construction $\mathrm{rank}(G_j) = r_j$ and by Lemma A.1.2, G_j is indecomposable. This completes the construction.

Appendix C

Power Cancellation

The group G satisfies *power cancellation* if given a group H and an integer $n \geq 1$ such that $G^{(n)} \cong H^{(n)}$ then $G \cong H$.

C.1   Failure of Power Cancellation

The next example shows that rtffr groups fail to satisfy the *power cancellation property*.

EXAMPLE C.1.1 Let $p \geq 3$ be a prime. There are nonisomorphic rtffr groups A and B of the same rank such that

$$A^{(n)} \cong B^{(n)} \text{ iff } p \text{ divides } n.$$

Proof: Let $\mathbf{k}$ be an algebraic number field of class number h such that p is a prime divisor of h. Let $\mathcal{O} = \mathcal{O}(\mathbf{k})$ denote the ring of algebraic integers in $\mathbf{k}$, and let Γ denote the ideal class group of $\mathcal{O}$. There is a subgroup $\Gamma(p)$ of the finite abelian group Γ of order p. Since $p \geq 3$ there are at least two isomorphism classes $(I) \neq (J)$ in $\Gamma(p)$ of order p. Then for integers $n > 0$

$$I^n \cong J^n \text{ iff } p \text{ divides } n.$$

Because $\mathcal{O}$ is a Dedekind domain, Steinitz' Theorem [69, Theorem 4.13] states that $I^{(n)} \cong J^{(n)}$ iff $I^n \cong J^n$. So $I^{(n)} \cong J^{(n)}$ iff p divides n.

Using Corner's Theorem F.1.3 there is an rtffr group G such that $\mathrm{End}_{\mathbb{Z}}(G) \cong \mathcal{O}$. By the Arnold-Lady-Murley Theorem 2.4.1 there are rtffr groups $A, B \in \mathbf{P}_o(G)$ such that $\mathrm{H}_G(A) = I$ and $\mathrm{H}_G(B) = J$. Then

$$\mathrm{H}_G(A^{(n)}) = I^{(n)} \cong J^{(n)} = \mathrm{H}_G(B^{(n)})$$

iff p divides n. By Theorem 2.4.1, $A^{(n)} \cong B^{(n)}$ iff p divides n. This completes the proof.

Fuchs and Loonstra give an example of nonisomorphic rtffr groups G and H such that $G^n \cong H^n$ for some integer $n > 1$.

EXAMPLE C.1.2 [L. Fuchs and F. Loonstra] (See [46, Theorem 90.3].) Let $m \geq 2$ be an integer. There are two torsion-free indecomposable rank 2 groups G and H such that

$$G^{(n)} \cong H^{(n)} \text{ iff } m \text{ divides } n.$$

Appendix D

Cancellation

The group G satisfies the *cancellation property* if for any groups K, L

$$G \oplus K \cong G \oplus L \text{ implies } K \cong L.$$

D.1   Failure of Cancellation

The next example shows that the category of rtffr groups does not enjoy the cancellation property.

THEOREM D.1.1 [L. Fuchs and F. Loonstra] *(See [46, Theorem 90.4].) Let $m \geq 1$ be an integer. There are groups G, K, and pairwise nonisomorphic indecomposable groups $H_1, \cdots, H_m$ such that $\mathrm{rank}(B) = 1$, $\mathrm{rank}(H_j) = 2$, and*

$$G \cong B \oplus H_j$$

for each $j = 1, \ldots, m$.

Proof: Begin with two infinite disjoint sets of primes P and Q, and a prime p_o not in $P \cup Q$. Let $b, x, y \neq 0 \in \mathbb{Q}$. Construct rank one groups

$$B = \langle p^{-1}b \,\big|\, p \in P \rangle, \quad X = \langle p^{-1}x \,\big|\, p \in P \rangle, \quad Y = \langle q^{-1}y \,\big|\, q \in Q \rangle.$$

By Lemma A.1.2,

$$H_1 = \langle X \oplus Y, p_o^{-1}(x \oplus y) \rangle$$

is an indecomposable rank 2 group. Define

$$G = B \oplus H_1.$$

Choose integers q_i, r_i, s_i, t_i such that $q_i t_i - r_i s_i = 1$ and let

$$b_i = q_i b + s_i x \quad x_i = r_i b + t_i x$$

for each $i = 2, \ldots, m$. We let

$$B_i = \langle p^{-1}b_i \,\big|\, p \in P \rangle_* \text{ and } X_i = \langle p_i^{-1}x_i \,\big|\, p \in P \rangle_*$$

be isomorphic rank one subgroups of $B \oplus X$. Then

$$B \oplus X = B_i \oplus X_i.$$

We will refine our choices of b_i and x_i so that for some integers $1 < k_i < p_o$,

$$G = B_i \oplus H_i \text{ with } H_i = \langle p_o^{-1}(k_i x_i + y), x_i \in X_i, y \in Y \rangle. \quad \text{(D.1)}$$

Notice that by Lemma A.1.2, in any choice of k_i, H_i is indecomposable. From our choices of q_i, r_i, s_i, t_i, and x_i we see that

$$k_i x_i + y = k_i r_i b + (k_i t_i - 1)x + (x + y)$$

is divisible by p. Hence

$$k_i r_i \equiv 0 \bmod p \quad \text{and} \quad k_i t_i \equiv 1 \bmod p. \quad \text{(D.2)}$$

Since $1 < k_1 < p$ we can choose $r_i = p$, $q_i = k_i$, and s_i, t_i such that $q_i t_i - r_i s_i = 1$. Then (D.2) is satisfied and hence (D.1) holds. It remains to prove the lemma.

LEMMA D.1.2 Let $H_1, \ldots, H_m$ be groups as defined in (D.1). Then the H_i are quasi-isomorphic rank 2 acd groups such that

$$H_i \not\cong H_j$$

for each $1 \le i \ne j \le m$.

Proof: Let $\phi : H_i \longrightarrow H_j$ be an isomorphism. Then $\phi(X) = X$ and $\phi(Y) = Y$ and $\phi(x_i) = \pm x_i$ and $\phi(y) = \pm y$. Thus

$$\phi(k_i x_i + y) = \pm(k_j x_j \pm y).$$

Preservation of divisibility implies that

$$k_j \equiv \pm k_i \bmod p.$$

Consequently with

$$k_1 = 1, k_2 = 2, \cdots, k_m = m$$

and $p > 2m - 1$ then $k_j \not\equiv \pm k_i \bmod p$. Thus no two of the $H_1, \dots, H_m$ are isomorphic. This proves the lemma and completes the proof of the example.

The above example shows that for each $m \geq 1$ there exists an rtffr of rank 3 G with m inequivalent direct sum decompositions. It also shows that cancellation fails m different times in G. It is interesting to note that E.L. Lady [10, Theorem 11.11(a)] has shown that in any group direct sum decomposition $G = B \oplus H \cong B \oplus K$ there are at most finitely many isomorphism classes of K.

Appendix E

Corner Rings and Modules

E.1   Topological Preliminaries

Fix a *Corner ring E*. That is, the additive group $(E, +)$ is a countable reduced torsion-free abelian group. A left E-module M is a *Corner E-module* if $(M, +)$ is a countable reduced torsion-free abelian group. A Corner E-module is always a left E-module. Fix a Corner E-module M and assume that $\operatorname{ann}_E(M) = 0$. Given a left E-module M

$$\mathcal{P}_o(M) \text{ is the set of finite subsets of } M.$$

We have a linear topology

$$\Gamma(\mathbb{Q}E, \mathbb{Q}M) = \Gamma(\mathbb{Q}M)$$

on $\mathbb{Q}E$ whose neighborhoods of 0 are $\operatorname{ann}_{\mathbb{Q}E}(F)$ over all finite subsets $F \subset \mathbb{Q}M$. This Haussdorf topology on $\mathbb{Q}E$ is called the $\mathbb{Q}M$-*topology*.

$$\widehat{\mathbb{Q}E} \text{ is the completion of } \mathbb{Q}E \text{ under the } \Gamma(\mathbb{Q}M)\text{-topology.}$$

Let

$$\hat{\mathcal{O}}(M) = \{q \in \widehat{\mathbb{Q}E} \mid qM \subset M\}.$$

Then M is a left $\widehat{\mathcal{O}}(M)$-module.

$\widehat{\Gamma}(M)$ is the linear topology whose basis of open neighborhoods of zero is $\{\operatorname{ann}_{\widehat{\mathcal{O}}(M)}(F) \mid F$ is a finite subset of $\mathbb{Q}M\}$.
This linear topology is called the M-topology on $\widehat{\mathcal{O}}(M)$.

We will prove the following.

THEOREM E.1.1 [36, T.G. Faticoni]. *Let E be a Corner ring and let M be a Corner E-module such that $\operatorname{ann}_E(M) = 0$. There is an exact sequence*

$$0 \longrightarrow M \longrightarrow G \longrightarrow \mathbb{Q}C \longrightarrow 0 \qquad (\text{E.1})$$

of left $\widehat{\mathcal{O}}(M)$-modules such that

1. *C is a direct sum of the cyclic E-submodules of $M^{(\aleph_o)}$.*

2. *There is a topological isomorphism $\widehat{\mathcal{O}}(M) \cong \operatorname{End}_{\mathbb{Z}}(G)$ where $\widehat{\mathcal{O}}(M)$ is endowed with the $\widehat{\Gamma}(M)$-topology and $\operatorname{End}_{\mathbb{Z}}(G)$ is endowed with the finite topology.*

Proof:

LEMMA E.1.2 $\mathbb{Q}M$ *is a left $\widehat{\mathbb{Q}E}$-module.*

Proof: Let $x \in \mathbb{Q}M$ and let $r \in \widehat{\mathbb{Q}E}$. Then r is the limit of a net $\{r_F \mid F\} \subset \mathbb{Q}E$ where ranges r_F ranges over the finite subsets of $\mathbb{Q}M$. Choose a finite set $F \subset \mathbb{Q}M$ such that $x \in F$ and $r_{F'} - r_F \in \operatorname{ann}_E(x)$ for each finite set $x \in F' \subset \mathbb{Q}M$. Then $(r_{F'} - r_F)x = 0$ implies that $r_F x = r_{F'}x$. Define $rx = r_F x$. This makes $\mathbb{Q}M$ a left $\widehat{\mathbb{Q}E}$-module.

LEMMA E.1.3 $\widehat{\mathcal{O}}(M)$ *is complete in the M-topology.*

Proof: A Cauchy net $\{r_F\}_F$ in $\widehat{\mathcal{O}}(M)$ is also a Cauchy net in $\widehat{\mathbb{Q}E}$ so that it converges to an element $\widehat{r} \in \widehat{\mathbb{Q}E}$ such that $\widehat{r}x = r_F x \in M$ for the subnet $\{r_F \mid x \in F\}$. Then $\widehat{r}M \subset M$ so that $\widehat{r} \in \widehat{\mathcal{O}}(M)$. This completes the proof.

E.2   The Construction of G

E.2.1 In what follows let $\widehat{M}$ denote *the $\mathbb{Z}$-adic completion of the Corner $\widehat{\mathcal{O}}(M)$-module M.* Let $\mathbf{P} \subset \widehat{\mathbb{Z}} \subset \widehat{E}$ be an uncountable domain whose elements are rational multiples of units in $\mathbf{P}$. (See [23, Section 2].) Let $\Pi \subset \mathbf{P}$ be a countable integral domain that satisfies the following lemma.

LEMMA E.2.2 *Let $\{\lambda_i \mid i \in \mathbb{N}\} = \mathcal{L} \subset \mathbf{P}$ be a Π-linearly independent set. Let $\{x_i \mid i \in \mathbb{N}\}$ be an ordered subset of $\mathbb{Q}M$.*

1. *If $\sum_{i \in \mathbb{N}} \lambda_i x_i = 0$ then $x_i = 0$ for each $i \in \mathbb{N}$.*

2. $\operatorname{ann}_{\widehat{\mathcal{O}}(M)}(\sum_{i \in \mathbb{N}} \lambda_i x_i) = \operatorname{ann}_{\widehat{\mathcal{O}}(M)}(\{x_i \mid i \in \mathbb{N}\})$.

LEMMA E.2.3 *There is a Corner submodule $N \subset \widehat{M}$ such that*

1. $M \cap N = 0$.

2. *For each finite set $F \subset M$ there is a $u_F \in N$ such that $\operatorname{ann}_E(u_F) = \operatorname{ann}_E(F)$.*

3. *N is a direct sum of cyclic submodules of $M^{(\aleph_o)}$.*

Proof: Because M is countable, $\mathcal{P}_o(M)$ is countable, and because $\mathbf{P}$ is uncountable, there is a countable set

$$\mathcal{L} = \{1, \lambda_{Fx} \mid F \in \mathcal{P}_o(M) \text{ and } x \in F\} \subset \mathbf{P}$$

that is Π-linearly independent. Given $F \in \mathcal{P}_o(M)$ let

$$u_F = \sum_{x \in F} \lambda_{Fx} x \tag{E.2}$$

and let

$$N = \sum_{\text{finite } F \subset \mathbb{Q}M} E u_F. \tag{E.3}$$

The value of N in (E.3) does not change in this appendix.

Because $\mathcal{L}$ is Π-linearly independent, Lemma E.2.2(1) implies that $\sum_{\text{finite } F \subset \mathbb{Q}M} E u_F$ is a direct sum and that $M \cap N = 0$. Thus N satisfies Lemma E.2.3(1).

Fix a finite subset $F \subset \mathbb{Q}M$, and form the element u_F in (E.2). Because $\mathcal{L}$ is Π-linearly independent, Lemma E.2.2(2) implies that $\operatorname{ann}_E(u_F) = \operatorname{ann}_E(F)$, so that Lemma E.2.3(2) is satisfied.

Finally, given a finite subset $F \subset \mathbb{Q}M$, Lemma E.2.3(2) implies that there is an injection $Eu_F \to M^{(F)}$ such that $u_F \to \oplus_{x \in F} x$ where $x \in M_x \cong M$. Thus N embeds in $M^{(\aleph_0)}$ as required by Lemma E.2.3(3).

Inasmuch as $M \oplus N$ is countable and $\mathbf{P}$ is uncountable there is a countable set of units

$$\mathcal{A} = \{ \epsilon_{mn} \mid m \in M, n \in N \} \subset \mathbf{P} \tag{E.4}$$

that is algebraically independent over Π. Let G be the pure subgroup of $\widehat{M}$ generated by $M \oplus N$ and ϵ_{mn} for each $m \oplus n \in M \oplus N$.

$$\begin{aligned} G &= \langle M, N, E\epsilon_{mn} \mid m \in M, n \in N \rangle_* \tag{E.5} \\ &= \widehat{M} \cap \mathbb{Q} \left(M \oplus N, \sum \{ E\epsilon_{mn} \mid m \in M, n \in N \} \right) \end{aligned}$$

Then G is a Corner E-submodule of $\widehat{M}$.

PROPOSITION E.2.4 *Let $\epsilon_{mn} \in \mathcal{A}$ be as in (E.4).*

1. *$\sum \{ E(m+n)\epsilon_{mn} \mid m \in M, n \in N \}$ is a direct sum of cyclic E-submodules of $M^{(\aleph_0)}$.*

2. *G is a left $\widehat{\mathcal{O}}(M)$-module.*

Proof: 1. The independence of the sum follows from Lemma E.2.2(1). Because ϵ_{mn} is a unit in $\mathbf{P}$ there is a natural embedding

$$E(m \oplus n)\epsilon_{mn} \cong E(m \oplus n) \subset M \oplus N \subset M \oplus M^{(\aleph_0)}.$$

This is part 1.

2. Observe that $\mathbb{Q}G$ is a left $\mathbb{Q}E$-module, so by Lemma E.1.2, $\mathbb{Q}G$ is a left $\widehat{\mathbb{Q}E}$-module. Thus $\mathbb{Q}G$ is a left $\widehat{\mathcal{O}}(M)$-module. Because M is a left $\widehat{\mathcal{O}}(M)$-module, $G = \widehat{M} \cap \mathbb{Q}G$ is a left $\widehat{\mathcal{O}}(M)$-module. This completes part 2.

E.3   Endomorphisms of G

The proof of the main result of this section is similar to the traditional proof of Corner's Theorem [46, Theorem 110.1]. Since [46, Theorem 110.1] is the only publication of the proof of Corner's Theorem in the last 30 years we will give a complete proof of that Theorem here.

LEMMA E.3.1 Let $\eta \in \text{End}_{\mathbb{Z}}(G)$. For each $m \in M$ and $n \in N$ there is an $r_{mn} \in E$ such that

$$\eta(m) = r_{mn}m \text{ and } \eta(n) = r_{mn}n.$$

Proof: Let $\eta \in \text{End}_{\mathbb{Z}}(G)$. Because G is pure in $\widehat{M}$, η lifts to a map

$$\widehat{\eta} : \widehat{M} \longrightarrow \widehat{M}$$

and to a $\mathbb{Q}$-linear homomorphism

$$\eta : \mathbb{Q}G \longrightarrow \mathbb{Q}G.$$

Let $m \in M$ and $n \in N$. By the definition of $\mathcal{A}$ in (E.4), there is a finite subset $F \subset M \oplus N$ such that

$$
\begin{aligned}
\eta((m \oplus n)\epsilon_{mn}) &= \widehat{\eta}(m \oplus n)\epsilon_{mn} & \text{(E.6)} \\
&= x + \sum_{m' \oplus n' \in F} r_{m'n'}(m' \oplus n')\epsilon_{m'n'} & \text{(E.7)} \\
\eta(m \oplus n) &= y + \sum_{m' \oplus n' \in F} s_{m'n'}(m' \oplus n')\epsilon_{m'n'} & \text{(E.8)}
\end{aligned}
$$

where $x, y \in \mathbb{Q}(M \oplus N)$ and $r_{m'n'}, s_{m'n'} \in E$. As in [46, Theorem 110.1], substitute (E.7) and (E.8) into (E.6). Then use Lemma E.2.2(1) to compare coefficients in equations (E.6) and (E.7) and show that for each $m' \oplus n' \in F$ with $m \oplus n \neq m' \oplus n'$ we have

$$
\begin{aligned}
x = r_{m'n'}(m' \oplus n') &= 0, \\
s_{m' \oplus n'}(m' \oplus n') &= 0, \\
y = r_{mn}(m \oplus n) &= \eta(m \oplus n). & \text{(E.9)}
\end{aligned}
$$

Finally, because $m \oplus n, 0 \oplus n \in M \oplus N$ there are $r_m, r_n \in E$ such that

$$\eta(m) = r_m m \text{ and } \eta(n) = r_n n.$$

Then

$$
\begin{aligned}
r_{mn}(m \oplus n) &= \eta(m \oplus n) \\
&= \eta(m) \oplus \eta(n) \\
&= r_m m \oplus r_n n. & \text{(E.10)}
\end{aligned}
$$

Using (E.9) and (E.10) we arrive at

$$r_m m = r_{mn}m \text{ and } r_n n = r_{mn}n.$$

This proves the lemma.

LEMMA E.3.2 *There is an isomorphism of rings*

$$\widehat{\mathcal{O}}(M) \cong \mathrm{End}_{\mathbb{Z}}(G).$$

Proof: By Proposition E.2.4, G is an $\widehat{\mathcal{O}}(M)$-submodule of $\widehat{M}$. Inasmuch as $rG = 0$ implies that $r\widehat{M} = 0$ implies that $r = 0$ for $r \in \widehat{\mathcal{O}}(M)$, there is an embedding of rings $\widehat{\mathcal{O}}(M) \longrightarrow \mathrm{End}_{\mathbb{Z}}(G)$. We claim that this embedding is an isomorphism.

Let $\eta \in \mathrm{End}_{\mathbb{Z}}(G)$ and let $F \subset M$ be a finite set. By Lemma E.2.3(2), there is a $u_F \in N$ such that

$$\mathrm{ann}_E(u_F) = \mathrm{ann}_E(F).$$

Lemma E.3.1 states that for each $m \in F$ there is an $r_{Fm} \in \mathbb{Q}E$ such that

$$\eta(m) = r_{Fm}m \text{ and } \eta(u_F) = r_{Fm}u_F \text{ for each } m \in F.$$

Let r_F be any one of the r_{Fx} with $x \in F$. Then

$$r_{Fm}u_F = \eta(u_F) = r_F u_F$$

for each $m \in F$, so that

$$r_F - r_{Fm} \in \mathrm{ann}_E(u_F) = \mathrm{ann}_E(F) \subset \mathrm{ann}_E(m).$$

Hence $r_F m = r_{Fm}m = \eta(m)$ for each $m \in F$.

Subsequently, the set $\{r_F \mid F \in \mathcal{P}_o(M)\}$ is a Cauchy net in $\widehat{\mathbb{Q}R}$ under the $\mathbb{Q}M$-topology. Let $\widehat{r}$ denote the limit in $\widehat{\mathbb{Q}R}$ of the Cauchy net. By (E.1.2),

$$\widehat{r}m = r_F m = \eta(m)$$

for each finite set $F \subset \mathbb{Q}M$ and each $m \in F$. Thus

$$[\eta - \widehat{r}](M) = 0.$$

Since M is pure and dense in $\widehat{M}$, $\eta = \widehat{r}$.

Therefore the embedding $\widehat{\mathcal{O}}(M) \longrightarrow \mathrm{End}_{\mathbb{Z}}(G)$ is an isomorphism.

LEMMA E.3.3 *The isomorphism of rings $\widehat{\mathcal{O}}(M) \cong \mathrm{End}_{\mathbb{Z}}(G)$ is a topological one, taking the M-topology on $\widehat{\mathcal{O}}(M)$ onto the finite topology of $\mathrm{End}_{\mathbb{Z}}(G)$.*

Proof: The proof, being natural, is left for the reader.

Proof of Theorem E.1.1: Given the Corner Module M construct N and G as in (E.3) and (E.5). Let

$$C \;=\; N \oplus \sum_{m \in M, n \in N} E(m \oplus n)\epsilon_{mn}.$$

Note that because M and G are pure and dense in $\widehat{M}$,

$$G/M \cong \mathbb{Q}C$$

as left $\widehat{\mathcal{O}}(M)$-modules. By Proposition E.2.4(1), C is a direct sum of cyclic submodules of $M^{(\aleph_o)}$. Thus Theorem E.1.1(1) is satisfied. Furthermore by Lemmas E.3.2 and E.3.3, $\widehat{\mathcal{O}}(M) \cong \mathrm{End}_{\mathbb{Z}}(G)$ topologically, where $\mathrm{End}_{\mathbb{Z}}(G)$ is endowed with the finite topology. This completes the proof of Theorem E.1.1.

Appendix F

Corner's Theorem

F.1 Countable Endomorphism Rings

THEOREM F.1.1 [23, A.L.S. Corner]. *Each countable reduced torsion-free ring E is the endomorphism ring of a countable reduced torsion-free abelian group G. Furthermore, there is a short exact sequence (E.1) of left E-modules in which C is a direct sum of cyclic submodules of $M^{(\aleph_0)}$.*

Proof: Corner's Theorem follows from Theorem E.1.1.

Let

$$\mathcal{O}(M) = \{q \in \mathbb{Q}E \mid qM \subset M\}.$$

THEOREM F.1.2 [43, T.G. Faticoni, H.P. Goeters]. *Let E be a reduced torsion-free finite rank ring and let M be a reduced torsion-free finite rank left E-module. Then $\mathcal{O}(M) \cong \operatorname{End}_{\mathbb{Z}}(G)$ for some reduced torsion-free finite rank group G. Furthermore, there is a short exact sequence (E.1) of left $\mathcal{O}(M)$-modules in which $C \cong \mathcal{O}(M) \oplus \mathcal{O}(M)$.*

Proof: When constructing G use a maximal linearly independent subset X of M instead of using each element $m \in M$.

THEOREM F.1.3 [23, A.L.S. Corner]. *If E is a reduced torsion-free finite rank ring then $E = \mathrm{End}_{\mathbb{Z}}(G)$ for some reduced torsion-free finite rank group G. Furthermore, there is a short exact sequence (E.1) of left E-modules in which C is a cyclic free left E-module.*

Appendix G

Torsion Torsion-Free Groups

G.1 E-Torsion Groups

Let E be an integral domain. A left E-module M is *torsion* if for each $x \in M$ there is an $r \neq 0 \in E$ such that $xr = 0$. We construct torsion-free groups G such that $\mathrm{End}_{\mathbb{Z}}(G)$ is an integral domain, and such that G is a torsion left $\mathrm{End}_{\mathbb{Z}}(G)$-module.

EXAMPLE G.1.1 There is a countable torsion-free group G such that $\mathrm{End}_{\mathbb{Z}}(G) = \mathbb{Z}[[x]]$ and such that G is a countable torsion left $\mathrm{End}_{\mathbb{Z}}(G)$-module.

Proof: Let $E = \mathbb{Z}[x]$ and let

$$M = \mathbb{Z}[x]/(x) \oplus \mathbb{Z}[x]/(x^2) \oplus \mathbb{Z}[x]/(x^3) \oplus \cdots.$$

Then $\Gamma(\mathbb{Q}M)$ is the x-adic topology on $\mathbb{Q}E$, $\widehat{\mathbb{Q}E} = \mathbb{Q}[[x]]$, and $\widehat{\mathcal{O}}(M) = \mathbb{Z}[[x]]$. Theorem E.1.1 states that there is a short exact sequence

$$0 \longrightarrow M \longrightarrow G \longrightarrow \mathbb{Q}C \longrightarrow 0$$

of left $\mathbb{Z}[[x]]$-modules and such that

1. $\mathrm{End}_{\mathbb{Z}}(G) = \mathbb{Z}[[x]]$, and

2. C is a direct sum of submodules of $M^{(\aleph_o)}$.

Then M, G, and $\mathbb{Q}C$ are torsion left $\mathbb{Z}[[x]]$-modules. Thus G is a torsion left $\mathrm{End}_{\mathbb{Z}}(G)$-module. This proves the Example.

G.2   Self-Small Corner Modules

The groups in the above constructions are torsion modules over their endomorphism rings. The next construction produces E-torsion-free groups.

The group G is *self-small* if for each cardinal c the natural mapping

$$\text{Hom}(G, G)^{(c)} \longrightarrow \text{Hom}(G, G^{(c)})$$

is an isomorphism of right $\text{End}_{\mathbb{Z}}(G)$-modules.

We will prove

THEOREM G.2.1 *Let E be a Corner ring and let M be a Corner E-module. There is a short exact sequence (E.1) in which*

1. *C is a free left $\mathcal{O}(M)$-module,*

2. *$\mathcal{O}(M) \cong \text{End}_{\mathbb{Z}}(G)$, and*

3. *G is a self-small group.*

We require a lemma.

LEMMA G.2.2 *As in (E.2.1) choose a countable integral domain $\Pi \subset \widehat{\mathbb{Z}}$ and an uncountable integral domain $\Pi \subset \mathbf{P} \subset \widehat{\mathbb{Z}}$. Let M be a Corner module such that $\text{ann}_E(M) = 0$. There is an element $u \in \widehat{M}$ such that $\text{ann}_E(u) = 0$.*

Proof: Write

$$M = \{m_1, m_2, m_3, \cdots\}$$

and since $\mathbf{P}$ is uncountable there is a countable set

$$\mathcal{L} = \{\lambda_1, \lambda_2, \lambda_3, \cdots\} \subset \mathbf{P}$$

that is algebraically independent over the integral domain Π. (See [23, Section 2].) There is a convergent sum

$$u = \sum_{k=1}^{\infty} m_k p^k$$

in the p-adic topology on $\widehat{M}$.

If $r \in E$ and $ru = 0$ then

$$\sum_{k=1}^{\infty} r m_k p^k = 0.$$

By Lemma E.2.2(2), $rm_k p^k = 0 = rm_k$ for each integer $k > 0$. Hence $rM = 0 = r$ by our hypothesis on M. Thus $u \in \widehat{M}$ is the element that we seek.

Proof of Theorem G.2.1: By Lemma G.2.2 there is an element $u \in \widehat{M}$ such that $\mathrm{ann}_E(u) = 0$. By taking λu for some $\lambda \in \mathbf{P}$ if necessary we can assume that
$$M \cap Eu = 0.$$
Then $M \oplus Eu \subset \widehat{M}$. Because $M \oplus Eu$ is countable there is an algebraically independent set
$$\mathcal{A} = \{\epsilon_m \mid m \in M\} \subset \mathbf{P}$$
over Π. Let
$$C = Eu + \sum_{m \in M} E(m \oplus u)\epsilon_m$$
and let
$$G = \mathbb{Q}(M \oplus C) \cap \widehat{M}. \tag{G.1}$$
Evidently G is a left E-submodule of $\widehat{M}$. Thus there is a short exact sequence
$$0 \longrightarrow M \longrightarrow G \longrightarrow \mathbb{Q}C \longrightarrow 0$$
of left E-modules. By our choice of u
$$E \cong Eu \cong E(m \oplus u)\epsilon_m$$
for each $m \in M$. Because $\mathcal{A} \cup \{1\}$ is algebraically independent over Π, C is a free module. (See Lemma E.2.3(2).) Thus Theorem G.2.1(1) is satisfied.

Because $\mathrm{ann}_{\mathcal{O}(M)}(G) = \mathrm{ann}_{\mathcal{O}(M)}(M) = 0$ there is an embedding
$$\mathcal{O}(M) \longrightarrow \mathrm{End}_{\mathbb{Z}}(G)$$
of rings. We claim that this is an isomorphism. Using Lemma 2.3.3, for each $m \in M$ there is an $r_m \in \mathbb{Q}E$ such that $\eta(m) = r_m m$ and $\eta(u) = r_n u$. Let r_o be one of the r_m. Then for each $m \in M$, $r_o u = r_m u$. Since $\mathrm{ann}_E(u) = 0$, $r_o = r_m$ for all $m \in M$. Hence $\eta(m) = r_o m$ for all $m \in M$. Thus $[\eta - r_o](M) = 0$. Then $[\eta - r_o](\widehat{M})$ is contained in the divisible subgroup of $\widehat{M}$, which in this case is 0. That is, $\eta = r_o$ and the proof of Theorem G.2.1(2) is complete.

Lastly, G is a countable group in which the finite topology is discrete ($u \in G$.) Then by [15, Proposition 2.1], G is self-small. This proves Theorem G.2.1(3) and completes the proof.

Appendix H

E-Flat Groups

Recall that G is *faithful* if $IG \neq G$ for each proper right ideal $I \subset \mathrm{End}_{\mathbb{Z}}(G)$, and that G is called *fully faithful* if $K \otimes_{\mathrm{End}_{\mathbb{Z}}(G)} G \neq 0$ for each nonzero right $\mathrm{End}_{\mathbb{Z}}(G)$-module K. G is E-flat if G is a flat left E-module. G is faithfully E-flat if G is a faithful group and a flat left E-module.

H.1   Ubiquity

THEOREM H.1.1 *The group constructed according to Corner's Theorem F.1.1 is E-flat.*

Proof: Let E be a Corner ring. By Corner's Theorem F.1.1 there is an exact sequence

$$0 \longrightarrow E \longrightarrow G \longrightarrow \mathbb{Q}C \longrightarrow 0$$

of left E-modules in which

1. $E \cong \mathrm{End}_{\mathbb{Z}}(G)$, and

2. C is a free left E-module.

Then there is an exact sequence of Tor groups

$$\mathrm{Tor}^1_E(\cdot, E) \longrightarrow \mathrm{Tor}^1_E(\cdot, G) \longrightarrow \mathrm{Tor}^1_E(\cdot, \mathbb{Q}C).$$

Since E is free, $\mathrm{Tor}^1_E(\cdot, E) = 0$, and since C is free, $\mathbb{Q}C$ is flat. Thus $\mathrm{Tor}^1_E(\cdot, \mathbb{Q}C) = 0$. It follows that $\mathrm{Tor}^1_E(\cdot, G) = 0$, whence G is a flat left E-module.

THEOREM H.1.2 *Let E be an rtffr ring, and let G be a group constructed according to Corner's Theorem F.1.3 such that $E = \text{End}_{\mathbb{Z}}(G)$. Then G is faithfully E-flat.*

Proof: By the previous theorem G is a flat left E-module.

To see that G is fully faithful (i.e., to see that $K \otimes_E G \neq 0$ for nonzero right E-modules K), it suffices to show that $IG \neq G$ for each *maximal right ideal* $I \subset E$. There is an exact sequence

$$\text{Tor}_E^1(E/I, \mathbb{Q}C) \longrightarrow E/I \otimes_E E \longrightarrow E/I \otimes_E G \longrightarrow E/I \otimes_E \mathbb{Q}C$$

of groups. Since E is rtffr, E/I is finite (Lemma 4.2.3). Thus $E/I \otimes_E \mathbb{Q}C = 0$. Furthermore since $\mathbb{Q}C$ is a flat left E-module $\text{Tor}_E^1(E/I, \mathbb{Q}C) = 0$. Hence

$$0 \neq E/I \cong G/IG$$

whence G is faithful. This completes the proof.

H.2   Unfaithful Groups

THEOREM H.2.1 *There is a countable torsion-free group G with endomorphism ring E such that*

1. *E is an integral domain,*

2. *$IG \neq G$ for each nonzero ideal $I \subset E$,*

3. *There is a nonzero E-module K such that $K \otimes_E G = 0$.*

That is, there is a countable torsion-free group G that is faithful but not fully faithful.

Proof: 1. Let $E = \mathbb{Z}[x]_{(x,2)}$ be the localization of $\mathbb{Z}[x]$ at the ideal generated by x and 2. Let

$$M = \bigoplus_{k=1}^{\infty} E/x^k E.$$

Then the reader shows that $\mathcal{O}(M) = E$. By Theorem E.1.1 there is a short exact sequence

$$0 \longrightarrow M \longrightarrow G \longrightarrow \mathbb{Q}C \longrightarrow 0$$

of left E-modules such that C is a direct sum of submodules of $M^{(\aleph_0)}$. Then G is a torsion E-module.

Choose the unique maximal right ideal I. Then $E/I \cong \mathbb{Z}/2\mathbb{Z}$ so that

$$E/I \otimes_E \mathbb{Q}C = 0 = \operatorname{Tor}_E^1(E/I, \mathbb{Q}C)$$

and so

$$0 \neq M/IM \cong G/IG.$$

This proves part 2.

For part 3 let $c_0, c_1, c_2, \cdots$ be elements such that

$$c_k 2 = 0 \text{ for all } k \geq 0$$
$$c_0 x = 0 \text{ and}$$
$$c_{k+1} x = c_k \text{ for all } k \geq 0.$$

Let

$$K = \sum_{k=0}^{\infty} c_k E.$$

Because $\mathbb{Q}C$ is divisible and K is an elementary 2-group $K \otimes_E \mathbb{Q}C = 0$. Because M is x-torsion while K is x-divisible, $K \otimes_E M = 0$. Then $K \otimes_E G = 0$. This proves part 3 and completes the proof.

THEOREM H.2.2 *There is a faithful group that is not E-flat.*

THEOREM H.2.3 *There is an E-flat group that is not faithful.*

Proof: Let $E = \begin{pmatrix} \mathbb{Z} & 0 \\ \mathbb{Z} & \mathbb{Z} \end{pmatrix}$ and let M be the projective left E-module $M = \begin{pmatrix} \mathbb{Z} \\ \mathbb{Z} \end{pmatrix}$. Evidently $\mathcal{O}(M) = \begin{pmatrix} \mathbb{Z} & 0 \\ \mathbb{Z} & \mathbb{Z} \end{pmatrix} = E$.

A little more effort (using matrix units) shows that the trace ideal of M in E is $I = \begin{pmatrix} \mathbb{Z} & 0 \\ \mathbb{Z} & 0 \end{pmatrix}$. Then $IM = M$ so that $(I + nE)M = M$ for each integer n. By Theorem E.1.1 there is a short exact sequence

$$0 \longrightarrow M \longrightarrow G \longrightarrow \mathbb{Q}C \longrightarrow 0$$

in which C is a finitely generated free left E-module. Since M is projective as a left E-module, G is flat as a left E-module. Then

$$E/(I + nE) \otimes_E M = E/(I + nE) \otimes_E \mathbb{Q}C = 0$$

for each integer $n > 0$. An application of $E/(I + nE) \otimes_E \cdot$ to the short exact sequence shows us that $(I + nE)G = G$. Thus G is E-flat but not faithful.

THEOREM H.2.4 *There is a countable torsion-free group that is not faithful and not E-flat.*

Proof: Let

$$E = \begin{pmatrix} \mathbb{Z} & 0 & 0 \\ \mathbb{Z} & \mathbb{Z} & 0 \\ \mathbb{Z} & \mathbb{Z} & \mathbb{Z} \end{pmatrix}$$

and let

$$M = \begin{pmatrix} \mathbb{Z} & 0 \\ \mathbb{Z} & \mathbb{Z} \\ \mathbb{Z} & \mathbb{Z} \end{pmatrix} \bigg/ \begin{pmatrix} 0 & 0 \\ 0 & 0 \\ 0 & \mathbb{Z} \end{pmatrix} = \begin{pmatrix} \mathbb{Z} & 0 \\ \mathbb{Z} & \mathbb{Z} \\ \mathbb{Z} & 0 \end{pmatrix}.$$

Since

$$\mathbb{Q}M = \begin{pmatrix} \mathbb{Q} & 0 \\ \mathbb{Q} & \mathbb{Q} \\ \mathbb{Q} & 0 \end{pmatrix}$$

is not a projective left $\mathbb{Q}E$-module, M is not flat as a left E-module. This is because the indecomposable projective $\mathbb{Q}E$-modules are the columns of

$$\mathbb{Q}E = \begin{pmatrix} \mathbb{Q} & 0 & 0 \\ \mathbb{Q} & \mathbb{Q} & 0 \\ \mathbb{Q} & \mathbb{Q} & \mathbb{Q} \end{pmatrix}.$$

Since $\begin{pmatrix} 0 & 0 & 0 \\ 0 & 0 & 0 \\ 0 & \mathbb{Z} & \mathbb{Z} \end{pmatrix}$ does not contain a nonzero ideal of E, $\operatorname{ann}_E(M) = 0$. Observe that $I = \begin{pmatrix} \mathbb{Z} & 0 & 0 \\ \mathbb{Z} & \mathbb{Z} & 0 \\ \mathbb{Z} & \mathbb{Z} & 0 \end{pmatrix}$ is a right ideal in E such that

$$IM = \begin{pmatrix} \mathbb{Z} & 0 & 0 \\ \mathbb{Z} & \mathbb{Z} & 0 \\ \mathbb{Z} & \mathbb{Z} & 0 \end{pmatrix} \begin{pmatrix} \mathbb{Z} & 0 \\ \mathbb{Z} & \mathbb{Z} \\ \mathbb{Z} & 0 \end{pmatrix} = \begin{pmatrix} \mathbb{Z} & 0 \\ \mathbb{Z} & \mathbb{Z} \\ \mathbb{Z} & 0 \end{pmatrix} = M.$$

Then $JM = M$ for each proper right ideal

$$J = \begin{pmatrix} \mathbb{Z} & 0 & 0 \\ \mathbb{Z} & \mathbb{Z} & 0 \\ \mathbb{Z} & \mathbb{Z} & n\mathbb{Z} \end{pmatrix}.$$

We note that $I \subset J \subset E$ and that E/J is finite.

Now use Theorem F.1.2 to construct a short exact sequence

$$0 \longrightarrow M \longrightarrow G \longrightarrow \mathbb{Q}C \longrightarrow 0$$

of left E-modules such that C is a free E-module and $E \cong \mathrm{End}_{\mathbb{Z}}(G)$. The by now familiar arguments show us that since $JM = M$, $JG = G$, and since M is not flat, G is not E-flat.

Appendix I

Zassenhaus and Butler

I.1 Statement

A is a finite dimensional $\mathbb{Q}$-algebra and $V = A\epsilon$ is a cyclic free left A-module with free basis ϵ. Identify A via left multiplication with the subring A_L of $\mathrm{End}_{\mathbb{Q}}(V)$. Let

$$Q = \text{the set of primes } p \in \mathbb{Z}.$$

For primes $p \in \mathbb{Z}$ and groups X we let

$$X_p = \mathbb{Z}_p X \subset \mathbb{Q}X.$$

Let

$$\mathcal{O}(X) = \mathcal{O}_A(X) = \{a \in A_L \,\big|\, aX \subset X\}.$$

Let $\epsilon \in M \subset V$ be a full *free abelian subgroup*. Then $\mathbb{Q}M = V$ and M_p is a free $\mathbb{Z}_p$-module for each $p \in Q$. Because M is a free abelian group we have the local-global relationships

$$M = \bigcap_{p \in Q} M_p,$$

$\mathcal{O}(M)_p = \mathcal{O}(M_p)$ for each $p \in Q$, and

$$\mathcal{O}(M) = \bigcap_{p \in Q} \mathcal{O}(M_p).$$

We will prove the following theorem.

THEOREM I.1.1 [T.G. Faticoni] *Let A, V, and M be as above. There is a subgroup $M \subset G \subset V$ such that*

1. *G is a locally free group,*

2. *$\mathcal{O}(M) \subset \mathrm{End}_{\mathbb{Z}}(G) \subset A$,*

3. *$\mathrm{End}_{\mathbb{Z}}(G)_p / \mathcal{O}(M)_p$ is a finite p-group for each prime $p \in \mathbb{Z}$.*

I.2   Proof

We construct G as the intersection $G = \bigcap_p K_p$ where the K_p are $\mathcal{O}(M_p) = \mathcal{O}(M)_p$-submodules of V. The K_p are chosen in a series of lemmas.

Since V has finite $\mathbb{Q}$-dimension we can enumerate the elements of

$$\mathrm{End}_{\mathbb{Q}}(V) \setminus A_L = \bigcup \{\phi_p \mid p \in Q\}. \tag{I.1}$$

LEMMA I.2.1 *Let $p \in Q$, let $\phi = \phi_p$, and suppose that*

$$\phi(M_p) \not\subset M_p.$$

Then

$$\text{choose } K_p = M_p. \tag{I.2}$$

In particular, K_p contains M_p, K_p is a left E_p-module, and K_p is a finitely generated free $\mathbb{Z}_p$-module.

LEMMA I.2.2 *Let $0 \neq \phi = \phi_p$ for some $p \in Q$. Suppose that ϕ satisfies*

1. *$\phi(M_p) \subset M_p$, and*

2. *$\phi(\epsilon) = 0$.*

There is a cyclic E_p-submodule $N_p \subset V$ such that $\phi(N_p) \not\subset N_p$.

Proof: Let $p \in Q$ and let $0 \neq \phi = \phi_p$ satisfy conditions 1 and 2 above. Let $E_p = \mathcal{O}(M_p)$ and let $R_p = E_p[\phi]$ denote the subring of $\mathrm{End}_{\mathbb{Q}}(V)$ generated by E_q and ϕ.

Assume, for the sake of contradiction that

$$\text{each cyclic } E_p\text{-submodule } N_p \subset V \text{ satisfies } \phi(N_p) \subset N_p. \tag{I.3}$$

The contradiction that we are looking for is $\phi = 0$.

Since each E_p-submodule of V is a sum of cyclic E_p-submodules of V, condition (I.3) implies that

$$\text{each } E_p\text{-submodule of } V \text{ is an } R_p\text{-module.} \tag{I.4}$$

Arbitrarily choose $x \in V$. Since ϵ is a free basis for V, $E_p\epsilon$ is a cyclic free left E_p-module with basis ϵ. There is then a short exact sequence

$$0 \longrightarrow L\epsilon \longrightarrow E_p\epsilon \xrightarrow{\ \rho\ } E_p x \longrightarrow 0$$

of left E_p-modules in which $\rho(s\epsilon) = sx$ for each $s \in E_p$.

Tensoring with the right E_p-module R_p yields the commutative diagram with exact rows

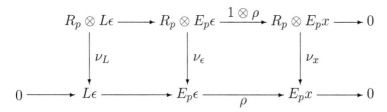

where the maps ν_Y are scalar multiplication maps

$$\nu_Y(r \otimes y) = ry.$$

By (I.4), $L\epsilon$, $E_p\epsilon$, and $E_p x$ are left R_p-modules, so ν_L, ν_ϵ, and ν_x are well defined R_p-module maps. Because ν_ϵ, $1 \otimes \rho$, and ν_x are surjections, the commutativity of the diagram shows us that for each $r \in R_p$ and $y = s\epsilon \in E_p\epsilon$

$$
\begin{aligned}
\rho(ry) &= \rho(\nu_\epsilon(r \otimes y)) \\
&= \rho(\nu_\epsilon(r \otimes s\epsilon)) \\
&= \nu_x(1 \otimes \rho)(r \otimes s\epsilon) \\
&= \nu_x(r \otimes sx) \\
&= r(sx) \\
&= r\rho(y)
\end{aligned}
$$

Hence ρ is an R_p-module map.

Recall the hypothesis that $\phi(\epsilon) = 0$. Then

$$\phi(x) = \phi(\rho(\epsilon)) = \rho(\phi(\epsilon)) = 0$$

so that $\phi(x) = 0$ for any $x \in V$. Thus $\phi(V) = 0$ which implies that $\phi = 0$. This contradicts our choice of $\phi \neq 0$ so condition (I.3) is false, that is, $\phi N_p \not\subset N_p$ some cyclic E_p-submodule of V. This proves the lemma.

COROLLARY I.2.3 Let $0 \neq \phi = \phi_p$ for some $p \in Q$, suppose that $\phi(M_p) \subset M_p$ and that $\phi(\epsilon) = 0$. There is an $\mathcal{O}(M_p)$-submodule $M_p \subset K_p \subset V$ such that K_p is a finitely generated free $\mathbb{Z}_p$-module, and $\phi(K_p) \not\subset K_p$.

Proof: Let $\mathcal{O}(M_p) = E_p$. By Lemma I.2.2, given ϕ such that $\phi(M_p) \subset M_p$ and $\phi(\epsilon) = 0$ there is a cyclic E_p-submodule $N_p \subset V$ such that $\phi(N_p) \not\subset N_p$. Then N_p is a torsion-free quotient of the finitely generated free $\mathbb{Z}_p$-module E_p, hence N_p is a finitely generated free $\mathbb{Z}_p$-module, whence $p^k M_p + N_p$ is a finitely generated free $\mathbb{Z}_p$-module for each integer $k > 0$.

Let $x \in N_p$ be such that $\phi(x) \notin N_p$. Since the q-adic topology on the free $\mathbb{Z}_p$-module $M_p + N_p$ is discrete there is an integer $k = k_p > 0$ such that $\phi(x) \notin p^k M_p + N_p$.

$$\text{Choose } K_p = M_p + p^{-k} N_p \tag{I.5}$$

and observe that $M_p \subset K_p$, that K_p is a finitely generated free $\mathbb{Z}_p$-module, and that $\phi(K_p) \not\subset K_p$.

There is one case left in our choices of K_p.

LEMMA I.2.4 Let $0 \neq \phi = \phi_p$ for some $p \in Q$, suppose that $\phi(M_p) \subset M_p$, and suppose that $\phi(\epsilon) \neq 0$. There is a finitely generated $\mathcal{O}(M_p)$-module K_p such that $M_p \subset K_p \subset V$ and $\phi(K_p) \not\subset K_p$.

Proof: Let $\mathcal{O}(M_p) = E_p$. Since $\phi(M_p) \subset M_p$ and since $\phi \notin A_L$,

$$0 \neq m\phi - r \notin E_p = \mathcal{O}(M_p) \tag{I.6}$$

for any integer $m \neq 0$ and element $r \in E_p$.

Because M_p and $E_p\epsilon$ are full subgroups of V we can write

$$0 \neq \phi(m\epsilon) = r\epsilon$$

for some integer $m \neq 0$ and some $r \in E_p$. Because $\phi(M_p) \subset M_p$ and $r \in E_p$ we have $(m\phi - r)(M_p) \subset M_p$ and $(m\phi - r)\epsilon = 0$. By (I.6), $m\phi - r \neq 0$ so we can apply Corollary I.2.3 to show that there is a left E_p-submodule $M_p \subset K_p \subset V$ such that K_p is a finitely generated free $\mathbb{Z}_p$-module, and such that $(m\phi - r)(K_p) \not\subset K_p$. Since $r(K_p) \subset K_p$, $m\phi(K_p) \not\subset K_p$, whence $\phi(K_p) \not\subset K_p$. This proves the lemma.

For $\phi = \phi_p \notin A_L$ such that $\phi(M_p) \subset M_p$,

$$\text{choose } M_p \subset K_p \subset V \tag{I.7}$$

as in Lemma I.2.4. This takes into account all of the cases for $\phi \in \text{End}_{\mathbb{Q}}(V) \setminus A_L$.

Proof of Theorem I.1.1: Let $E = \mathcal{O}(M)$, let K_p denote the left E_p-modules chosen in (I.2), (I.5), and (I.7) and let

$$G = \bigcap_{p \in Q} K_p. \tag{I.8}$$

Each K_p is an E_p-module, each K_p is a finitely generated free $\mathbb{Z}_p$-module, and $\epsilon \in G$ is a unimodular element. Since localization commutes with intersections, $G_p = K_p \doteq M_p$ for all primes $p \in \mathbb{Z}$. In each of our choices, K_p is a finitely generated free $\mathbb{Z}_p$-module. Hence G is a locally free full subgroup of V, as required by Theorem I.1.1(1).

Since $E = \bigcap_{p \in Q} E_p$, G is a left E-module such that $\psi(\epsilon) \neq 0$ for each $0 \neq \psi \in E \subset A_L$. Hence

$$E \subset \text{End}_{\mathbb{Z}}(G) \subset \text{End}_{\mathbb{Q}}(V).$$

By our choices of ϕ_p and K_p, for each $\phi \in \text{End}_{\mathbb{Q}}(V) \setminus A_L$ there is a $p \in Q$ such that $\phi = \phi_p$ and $\phi(K_p) \not\subset K_p$. However for $\psi \in \text{End}_{\mathbb{Z}}(G)$,

$$\psi(K_p) = \psi(G_p) \subset G_p = K_p$$

for each $p \in Q$. Thus

$$\text{End}_{\mathbb{Z}}(G) \bigcap (\text{End}_{\mathbb{Q}}(V) \setminus A_L) = \emptyset$$

and hence

$$\text{End}_{\mathbb{Z}}(G) \subset A_L.$$

This proves condition Theorem I.1.1(2).

Since $G_p = K_p$ for all primes $p \in \mathbb{Z}$ it follows that

$$\text{End}_{\mathbb{Z}}(G)_p \subset \mathcal{O}(K_p).$$

By our construction of K_p in (I.2), (I.5), and (I.7) we see that K_p/M_p is a finite p-group so that

$$\mathcal{O}(K_p)/\mathcal{O}(M_p)$$

is a finite p-group. Thus $\operatorname{End}_{\mathbb{Z}}(G)_p/\mathcal{O}(M_p)$ is a finite p-group for each $p \in Q$. This is what is required in Theorem I.1.1(3).

This completes the proof of the theorem.

THEOREM I.2.5 [90, H. Zassenhaus]. *Let E be a ring such that $(E,+)$ is a finitely generated free ring. There is a group $E \subset G \subset \mathbb{Q}E$ such that $E = \operatorname{End}_{\mathbb{Z}}(G)$ where the action of E on G is the natural one as an E-submodule of $\mathbb{Q}E$.*

THEOREM I.2.6 [21, M.C.R. Butler]. *Let E be a locally free rtffr ring. There is a group $E \subset G \subset \mathbb{Q}E$ such that $E = \operatorname{End}_{\mathbb{Z}}(G)$ where the action of E on G is the natural one as an E-submodule of $\mathbb{Q}E$.*

Appendix J

Countable E-Rings

J.1 Countable Torsion-Free E-Rings

An E-*ring* is a ring E for which the left representation embedding $\lambda :$ $E \longrightarrow \text{End}_{\mathbb{Z}}(E)$ is an isomorphism of rings. A *Dedekind E-ring* is an E-ring R that is a Dedekind domain.

THEOREM J.1.1 *An E-ring is a commutative ring.*

R. Bowshell and P. Schultz [20] characterize the rtffr E-rings.

THEOREM J.1.2 [20, R. Bowshell, P. Schultz]. *The rtffr ring E is an E-ring iff*
$$E \doteq R_1 \oplus \cdots \oplus R_t$$
for some integer $t > 0$ and some Dedekind E-rings $R_1, \ldots, R_t$ such that

$$\text{Hom}_{\mathbb{Z}}(R_i, R_j) = 0 \text{ for each } 1 \leq i \neq j \leq t.$$

This structure theorem seemed to indicate that the structure of E-rings was restricted as far as rtffr groups go. The following theorem shows that the countable torsion-free E-rings have a much more diverse additive structure. The group G is p-local if $pG \neq G$.

THEOREM J.1.3 [42, T.G. Faticoni]. *Let S be a countable reduced torsion-free commutative ring. There is a E-ring E and a pure and dense embedding of rings $S \longrightarrow E$.*

Before proving this theorem an example or two along with a comparison to Theorem J.1.2 will illustrate the difference between rtffr E-rings and countable torsion-free E-rings.

EXAMPLE J.1.4 Let $S = \mathbb{Z}[x_1, x_2, x_3, \ldots]/I$ where $x_1, x_2, x_3, \ldots$ are commuting indeterminants, and where I is an ideal in S. By Theorem J.1.3, S is a pure and dense subring of an E-ring E. We point out that every commutative ring can be realized in this way. Specifically $\mathbb{Z}[x]/(x^n)$ embeds in an E-ring E. This E-ring contains a nonzero nilpotent element x. Rtffr E-rings do not contain nonzero nilpotent elements.

Proof of Theorem J.1.3: As in the proof of Corner's Theorem 2.3.3 we require an uncountable integral domain. The proof is left as an exercise for the reader.

LEMMA J.1.5 Let K be a uncountable field extension of $\mathbb{Q}$, and let B be a commutative K-algebra. Let S be a countable $\mathbb{Q}$-subalgebra of B. There exists an uncountable set $U \subset K$ that is algebraically independent over S.

For groups G and primes $p \in \mathbb{Z}$, let us define

$$G_p = \frac{G \otimes \mathbb{Z}_p}{\operatorname{div}(G \otimes \mathbb{Z}_p)}.$$

Then G_p is a reduced p-local group, and G_p is countable if G is countable. To reduce to the p-local case, observe that if S is a countable reduced torsion-free commutative ring then S is a pure and dense subring of $\prod_p S_p$ where p ranges over the primes of $\mathbb{Z}$. Thus for each prime $p \in \mathbb{Z}$ we will embed each S_p as a pure and dense subring of an E-ring E_p. We can then embed S as a pure and dense subring of the E-ring $\prod_p E_p$.

Let
$$W \cup \{1\}$$
be a p-basis of S. Then $\widehat{S}$ is a free $\widehat{\mathbb{Z}}_p$-module with basis $W \cup \{1\}$. Since S is countable there is an algebraically independent subset $U \subset \widehat{\mathbb{Z}}_p$ of units such that

$$\mathrm{card}(U) = \mathrm{card}(W) + 1.$$

Write $U = \{d, u_w \mid w \in W\}$. We will show that

$$E = \langle S[u_w + dw \mid w \in W] \rangle_*$$

is the E-ring we seek, the purification taking place in $\widehat{S}$. Certainly

$$S \subset E \subset \widehat{S}$$

so that S is a pure and dense subring of E. It remains to prove that E is an E-ring.

LEMMA J.1.6 Let S, W, u_w, d, E be as above. Then

1. $E \cap dE = 0$.

2. Let $\phi : E \to \widehat{S}$ be a group homomorphism such that $d\phi(S) \cap \phi(S) = 0$. Then $f\phi = 0$ if $f\phi(1) = 0$.

Proof: 1. It suffices to show that $\{d, u_w + dw \mid w \in W\}$ is algebraically independent over $\mathbb{Q}S$.

Let

$$X = \{x_d, x_w \mid w \in W\}$$

be a set of indeterminants and consider the evaluation maps

$$\alpha : \mathbb{Q}S[X] \longrightarrow \mathbb{Q}S[U] \text{ and } \beta : \mathbb{Q}S[X] \longrightarrow \mathbb{Q}S[U]$$

such that

$$
\begin{aligned}
\alpha(x_d) = \beta(x_d) &= d \\
\alpha(x_w) &= u_w + dw \\
\beta(x_w) &= u_w.
\end{aligned}
$$

Since U is algebraically independent and since $W \subset S$, β is a ring isomorphism.

Let $P(X)$ be a polynomial of minimial x_d-degree such that $\alpha(P(X)) = 0$. We can write

$$P(X) = x_d Q(X) + R(X)$$

where the x_d-degree of $R(X)$ is 0. Then $\alpha(P(X)) = 0$ implies that

$$\alpha(R(X)) = -d\alpha(Q(X)).$$

Arbitrarily enumerate the terms of $R(X)$ and let $R_k(X)$ be the k-th term in $R(X)$. For each integer $k \geq 1$ and $w \in W$ let

$$e(w, k) = \text{ the } x_w\text{-degree of } R_k(X).$$

Then

$$\alpha(R_k(X)) = a_k \left(\prod_{w \in W} (u_w + dw)^{e(w,k)} \right) = a_k \left(\prod_{w \in W} u_w^{e(w,k)} \right) + dB_k$$

for some $a_k \in \mathbb{Q}S$ and some $B_k \in \mathbb{Q}S[U]$. Then

$$\prod_{w \in W} u_w^{e(w,k)} = \beta(R_k(X))$$

so that

$$\alpha(R_k(X)) = \beta(R_k(X)) + dB_k.$$

By setting $B = \sum_k B_k$ then we have

$$\alpha(R(X)) = \beta(R(X)) + bB = -d\alpha(Q(X)).$$

It follows that

$$R(X) = x_d \beta^{-1}[-B - \alpha(Q(X))].$$

Since $R(X)$ has x_d-degree 0 it must be that $R(X) = 0$. That is

$$P(X) = x_d Q(X).$$

Now d is a unit in $\widehat{Z}_p$ and α is $\widehat{Z}_p$-linear so

$$\alpha(Q(X)) = d^{-1}\alpha(x_d Q(X)) = d^{-1}\alpha(P(X)) = 0.$$

This contradiction to the minimality of the x_d-degree of $P(X)$ shows that $\alpha : \mathbb{Q}S[X] \longrightarrow \mathbb{Q}S[U]$ is an isomorphism. In particular $\{d, u_w + dw \,|\, dw \in W\}$ is algebraically independent over $\mathbb{Q}S$. This proves part 1.

2. Assume that $\phi : E \longrightarrow \widehat{S}$ is a function such that $\phi(1) = 0$ and $\phi(E) \cap d\phi(E) = 0$. Since ϕ lifts to a $\widehat{Z}_p$-module map $\phi : \widehat{E} \longrightarrow \widehat{S}$,

$$\phi(u) = u\phi(1) \text{ for each } u \in \widehat{Z}_p.$$

Thus, for each $w \in W$,

$$\phi(u_w + dw) = u_w\phi(1) + d\phi(w) = d\phi(w) \in \phi(E) \cap d\phi(E) = 0$$

by part 1. Therefore

$$0 = \phi(u_w + dw) = d\phi(w) = \phi(w) \text{ for all } w \in W.$$

Inasmuch as $\widehat{E} = \widehat{Z}_p \cdot W$ and since $\phi(1) = 0$ by hypothesis, $\phi = 0$. This proves part 2 and completes the proof of the lemma.

Proof of Theorem J.1.3: Let S be a countable reduced torsion-free commutative ring. We have already observed that $S \subset E$ is a pure and dense embedding of rings. Let $\phi : E \longrightarrow E$ be a group homomorphism, let $u = \phi(1)$, and consider $\phi - u$. By Lemma J.1.6(1), $E \cap dE = 0$, and by Lemma J.1.6(2), $\phi - u = 0$. I.e. $\phi = u \in E$, hence E is an E-ring. This concludes the proof of Theorem J.1.3.

Appendix K

Dedekind E-Rings

Our references in this section for commutative ring theory are [6, 49]. The number theory we use can be found in [58, 66]. The ring R is a *Dedekind E-ring* if the group $(R, +)$ is an E-ring and if the ring R is a Dedekind domain. We will show that the Dedekind domain R is an E-ring provided it is divisible by a certain set of primes in the ring of algebraic integers in algebraic number field $\mathbb{Q}R$.

K.1 Number Theoretic Preliminaries

At all times $p \in \mathbb{Z}$ is a prime number, E and F are algebraic number fields, $\mathcal{O}_E$ denotes the ring of algebraic integers in E, and $\mathcal{L}(E)$ denotes the lattice of subrings of E containing $\mathcal{O}_E$. Let $\mathrm{spec}(E)$ denote the set of maximal ideals in a ring $\mathcal{O}_E$. For more general commutative rings R, $\mathrm{spec}(R)$ denotes the set of maximal ideals in R.

Let $E \subset F$ be a subfield, let $P \in \mathrm{spec}(E)$ and $M \in \mathrm{spec}(F)$. We say that M *lies over* P if $P \subset M$. Equivalently, M lies over P if $P \subset M \cap E$. For unramified primes $P \in E$ it is known that

$$[F : E] \;=\; \sum \{ [\mathcal{O}_F/M : \mathcal{O}_E/P] \,|\, M \in \mathrm{spec}(F) \text{ and}$$
$$M \text{ lies over } P.\}$$

If there are at least two different primes in $\mathcal{O}_F$ lying over P then we say that P *splits* in F or in $\mathcal{O}_F$. If $[F : E]$ primes in $\mathcal{O}_F$ lie over P then we say that P *splits completely* in F. Thus P splits completely in F iff $\mathcal{O}_F/M \cong \mathcal{O}_E/P$ for each $M \in \mathrm{spec}(F)$ lying over P. Complete splitting occurs often enough as [58, Theorem 6] states that infinitely many primes $P \in \mathrm{spec}(E)$ split completely in F.

K.2   Integrally Closed Rings

Let $\mathcal{O}_F \subset R \subset F$ be a Dedekind domain. The *support of R in F* is the set

$$\sigma_F(R) = \{M \in \mathrm{spec}(F) \mid MR \neq R\}$$

and the *divisibility of R in F* is

$$\delta_F(R) = \{M \in \mathrm{spec}(F) \mid MR = R\}.$$

For a Dedekind domain $\mathcal{O}_F \subset R \subset F$, the sets $\sigma_F(R)$ and $\delta_F(R)$ form a partition of $\mathrm{spec}(F)$. Given a set $\sigma \subset \mathrm{spec}(F)$ then let

$$\mathcal{O}_\sigma = \cap\{(\mathcal{O}_F)_M \mid M \in \sigma\}$$

where $(\mathcal{O}_F)_M$ is the classic localization of $\mathcal{O}_F$ at the maximal ideal M in $\mathcal{O}_F$.

K.2.1 Let $\sigma \subset \mathrm{spec}(F)$. Then $\sigma_F(\mathcal{O}_\sigma) = \sigma$ and $\delta(\mathcal{O}_\sigma) = \sigma' = $ the complement of σ in $\mathrm{spec}(F)$. Moreover, if $\mathcal{C}(R) = \{c \in \mathcal{O}_F \mid cR = R\}$ then by [49, page 73, exercise 7]

$$R = \mathcal{O}_F[\mathcal{C}^{-1}].$$

Furthermore $\sigma_F(R) = \{M \in \mathrm{spec}(F) \mid M \cap \mathcal{C} = \emptyset\}$ and $\delta_F(R) = \{M \in \mathrm{spec}(F) \mid c \in M \text{ for some } c \in \mathcal{C}\}$.

Thus to study E-rings we need to study localizations $\mathcal{O}_\sigma$ for subsets $\sigma \subset \operatorname{spec}(F)$.

A classic result from commutative ring theory follows. Let $S \subset R$ be rings. Then R is *integral over* S if R is finitely generated by S.

LEMMA K.2.2 Let $S \subset R \subset F$ be subrings of F. Then R is integral over S iff $R = S\mathcal{O}_F$.

Proof: If $R = S\mathcal{O}_F$ then R is finitely generated by S because $\mathcal{O}_F$ is a finitely generated free abelian group. Hence R is integral over S.

Conversely if R is integral over S then R is a finitely generated S-module. Since R is integrally closed $S\mathcal{O}_F \subset R$ and since R is finitely generated over the integrally closed domain $S\mathcal{O}_F$, $S\mathcal{O}_F = R$.

THEOREM K.2.3 *[10, Theorem 14.3]* Suppose that $\mathcal{O}_F \subset R \subset F$ and that R is a Dedekind domain. The following are equivalent.

1. $(R, +)$ is strongly indecomposable.

2. F is the smallest subfield $E \subset F$ such that R is integral over $R \cap E$.

3. F is the smallest subfield $E \subset F$ such that R is finitely generated by $R \cap E$.

4. R is an E-ring.

Proof: The proof follows from the Lemma and from [10, Theorem 14.3].

THEOREM K.2.4 *[39, Proposition 2.1]* Let F be an algebraic number field, and let $\sigma \subset \operatorname{spec}(F)$. The following are equivalent.

1. $\mathcal{O}_\sigma$ is an E-ring.

2. For each proper subfield $E \subset F$ there are $M, M' \in \operatorname{spec}(F)$ such that $M \cap E = M' \cap E$ while $M \in \sigma$ and $M' \notin \sigma$.

Proof: Assume that $R = \mathcal{O}_\sigma$ is not an E-ring. By Theorem K.2.3, there is a proper subfield $E \subset F$ such that $R = S\mathcal{O}_E$ where $S = R \cap E$. Since S is a Dedekind domain with field of quotients E, $S = \mathcal{O}_E[\mathcal{C}^{-1}]$ for some multiplicatively closed subset $\mathcal{C} \subset \mathcal{O}_E$. Then $\mathcal{O}_F[\mathcal{C}^{-1}] = R$.

Fix $M' \in \mathrm{spec}(F)$ such that $M'R = R$. There is a $c \in C$ such that $1 \cdot c = c \in M'$. Thus $MR = R$ for each maximal ideal M lying over $M' \cap S$. This is the negation of part 2.

Conversely, suppose that $R = \mathcal{O}_\sigma$ is an E-ring, let $E \subset F$ be a proper subfield of F, and let $S = R \cap E$. Since R is an E-ring, R is not integral over S. Since the integral property is a local-global property in F there is a prime $P \in \mathrm{spec}(E)$ such that R_P is not integral over the discrete valuation domain S_P.

If $R_P = F$ then

$$P_P = P_P R_P \cap S_P = F \cap S_P = S_P,$$

contrary to the fact that the ring S_P is never equal to its Jacobson radical P_P. Thus $R_P \neq F$.

S and R are Dedekind domains and R is not finitely generated by S, so

$$S_P = (\mathcal{O}_E)_P \subset S_P \mathcal{O}_F = (\mathcal{O}_F)_P \neq R_P.$$

Since $R_P \neq F$ there is an ideal $M \in \mathrm{spec}(F)$ lying over P such that

$$(M_P)_M = M_M \neq R_M = (R_P)_M \neq F.$$

It follows that $MR_P \neq R_P$.

R_P is not integral over $(\mathcal{O}_F)_P$, so there is an ideal $M' \in \mathrm{spec}(F)$ lying over P such that $(R_P)_{M'} = R_{M'}$ is not finitely generated by the discrete valuation domain $((\mathcal{O}_F)_P)_{M'} = (\mathcal{O}_F)_{M'}$. Then $R_{M'} = F$, and hence $M'R_P = R_P$. Localizing at primes $Q \in \mathrm{spec}(E)$ other than P yields $M'R_Q = R_Q$ because there is an element $c \in M' \cap E \setminus Q$. By the Local-Global Theorem, $M'R = R$.

Since M and M' both lie over P we have proved part 2.

THEOREM K.2.5 *Let $p \in \mathbb{Z}$ be prime and let F be a minimal field extension of $\mathbb{Q}$. Then F is the field of fractions of a p-local E-ring iff p splits in F.*

Examples of minimal field extensions are found by examining the finite lattice of subfields of the algebraic number field F. Thus quadratic number fields and in general fields F such that $[F : \mathbb{Q}] = p$ a prime are minimal field extensions. Another method is to take a Galois extension $K/\mathbb{Q}$ and then take a maximal subgroup $G \subset \mathrm{Gal}(K/\mathbb{Q})$. The fixed field F of G is minimal by the Galois correspondence. The degree $[F : \mathbb{Q}]$ is the degree $[\mathrm{Gal}(K/\mathbb{Q}) : G]$.

Bibliography

[1] U. Albrecht, Endomorphism rings of faithfully flat abelian groups, Resultate der Mathematik, **17**, (1990), 179-201.

[2] U. Albrecht, Locally A-Projective abelian groups and generalizations, Pac. J. Math. **141**, No. 2, (1990), 209-228.

[3] U. Albrecht, Faithful abelian groups of infinite rank, Proc. Am. Math. Soc. **103**, (1), (1988), 21-26.

[4] U. Albrecht, Baer's Lemma and Fuch's Problem 84a, Trans. Am. Math. Soc. **293**, (1986), 565-582.

[5] U. Albrecht, H.P. Goeters, A dual to Baer's Lemma, Proc. Am. Math. Soc. **105**, (1989), 217-227.

[6] F.W. Anderson, K.R. Fuller, *Rings and Categories of Modules*, Graduate texts in Mathematics **13**, Springer-Verlag, New York and Berlin, (1974).

[7] H.M.K Angad-Gaur, The homological dimension of a torsion-free group of finite rank as a module over its ring of endomorphisms, Rend. Sem. Mat. Univ. Padova **57**, (1977), 299-309.

[8] D.M. Arnold, *Abelian Groups and Representations of Finite Par- tially Ordered Sets*, Canadian Mathematical Society: Books in Mathematics, Springer, New York, (2000).

[9] D.M. Arnold, A finite global Azumaya Theorem in additive categories, Proc. Am. Math. Soc. **91**, No. 1, May, (1984), 25-29.

[10] D.M. Arnold, *Finite Rank Abelian Groups and Rings*, Lecture Notes in Mathematics **931**, Springer-Verlag, New York, (1982).

[11] D.M. Arnold, Endomorphism rings and subgroups of finite rank torsion-free abelian groups, Rocky Mt. J. Math. **12**, No. 2, (1982), 241-256.

[12] D.M. Arnold, J. Hausen, Modules with the summand intersection property, Comm. Algebra **18**, (1990), 519-528.

[13] D.M. Arnold, R. Hunter, F. Richman, Global Azuamya Theorems in additive categories, Jr. Pure and Applied Alg. **16**, (1980), 223-242.

[14] D.M. Arnold, E.L. Lady, Endomorphism rings and direct sums of torsion-free abelian groups, Trans. Am. Math. Soc. **211**, (1975), 225-237.

[15] D.M. Arnold, C.E. Murley, Abelian groups A such that $\operatorname{Hom}(A, \cdot)$ preserves direct sums of copies of A, Pac. J. Math. **56**, (1), (1975), 7-20.

[16] D.M. Arnold, R.S. Pierce, J.D. Ried, C. Vinsonhaler, W. Wickless, Torsion-free abelian groups of finite rank projective as modules over their endomorphism rings, J. Algebra **71**, No. 1, July (1981), 1-10.

[17] M.F. Atiyah, I.G. Macdonald, Introduction to Commutative Algebra, Addison-Wesley Publishing Co., Reading, MA-Sydney, (1969).

[18] R.A. Beaumont, R.S. Pierce, Subrings of algebraic number fields, Acta Sci. Math. (Szeged), **22**, (1961), 202-216.

[19] R.A. Beaumont, R. S. Pierce, Torsion-free rings, Illinois J. Math., **5**, (1961), 61-98.

[20] R. Bowshell, P. Schultz, Unital rings whose additive endomorphisms commute, Math. Ann. **228**, (1977), 197-214.

[21] M.C.R. Butler, On locally free torsion-free rings of finite rank, J. London Math. Soc. **43**, (1968), 297-300.

[22] B. Charles, Sous-groupes fonctoriels et topologies, Studies on Abelian Groups, Paris, (1968), 75-92.

[23] A.L.S. Corner, Every countable torsion-free ring is an endomorphism ring, Proc. London Math. Soc. **13**, (1963), 23-33.

[24] A.L.S. Corner, Endomorphism Rings of Torsion-Free Abelian Groups, *Proceedings of the International Conference on the Theory of Groups*, edited by L.G. Kovacs and B.H. Neumann, Gordon and Breach, New York, London, and Paris, (1967), 59-70.

[25] A.L.S. Corner, R. Göbel, Prescribing endomorphism algebras, Proc. London Math. Soc. **50**, (3), (1985), 447-494.

[26] A. Dress, On the decomposition of modules, Bull. Am. Math. Soc. **75**, (1969), 984-986.

[27] M. Dugas, T.G. Faticoni, Cotorsion-free groups cotorsion as modules over their endomorphism rings, *Abelian Groups*, Lecture Notes in Pure and Applied Mathematics, Marcel Dekker, New York, (1993), 111-127.

[28] M. Dugas, A. Mader, C. Vinsonhaler, Large E-rings exist, J. Algebra, **108**, (1987), 88-101.

[29] C. Faith, *Algebra I: Algebra and Categories*, Springer-Verlag, New York and Berlin, (1974).

[30] C. Faith, *Algebra II: Ring Theory*, Springer-Verlag, New York, and Berlin, (1976).

[31] C. Faith, E.A. Walker, Direct sum representations of injective modules, J. Alg. **5**, 203-221, (1967).

[32] T.G. Faticoni, *Modules over Endomorphism Rings*, submitted, (2006).

[33] T.G. Faticoni, Azumaya theorems for classes of abelian groups, manuscript.

[34] T.G. Faticoni, Direct sums and refinement, Comm. Algebra **27**(1), (1999), 451-464.

[35] T.G. Faticoni, Modules over endomorphism rings as homotopy classes, Abelian Groups and Modules, A. Facchini, C. Menini, Kluwer Academic Publishers, Mathematics and its Applications, **343**, (1995), 163-183.

[36] T.G. Faticoni, Torsion-free groups torsion as modules over their endomorphism rings, Bull. Aust. Math. Soc. **50** (2), (1994), 177-195.

[37] T.G. Faticoni, *Categories of Modules Over Endomorphism Rings*, Memoirs of the Am. Math. Soc. **492**, (May, 1993).

[38] T.G. Faticoni, The endlich Baer splitting property, Pacific J. Math **157** (2), (1993), 225-240.

[39] T.G. Faticoni, *E*-rings as Localizations of Orders, Acta Math. Hung. **57**, (**3-4**), (1991), 265-274.

[40] T.G. Faticoni, Gabriel filters on the endomorphism ring of a torsion-free abelian group, Comm. Algebra **18**, (9), (1990), 2841-2883.

[41] T.G. Faticoni, On the lattice of right ideals of the endomorphism ring of an abelian group, Bull. Aust. Math. Soc. **38**, (2), (1988), 273-291.

[42] T.G. Faticoni, Each countable reduced torsion-free commutative ring is a pure subring of an *E*-ring, Comm. Algebra **15**, (12), (1987), 2545-2564.

[43] T.G. Faticoni, H.P. Goeters, Examples of torsion-free groups flat as modules over their endomorphism rings, Comm. Algebra **19**, (1), (1991), 1-28.

[44] T.G. Faticoni, H.P. Goeters, On torsion-free Ext, Comm. Algebra **16**, (9), (1988), 1853-1876.

[45] T.G. Faticoni, P. Schultz, Direct sum decompositions of acd groups with primary regulating quotient, Abelian Groups and Modules, Lecture Notes in Pure and Applied Mathematics, Marcell-Dekker, (1996).

[46] L. Fuchs, *Infinite Abelian Groups I, II*, Academic Press, New York and London, (1969, 1970).

[47] K. Fuller, Density and Equivalence, J. Alg. **29**, (1974), 528-550.

[48] M. Huber, R.B. Warfield, Jr., Homomorphisms between cartesian powers of an abelian group, *Abelian Group Theory*, Lecture Notes in Mathematics **874**, Springer-Verlag, New York and Berlin, (1981), 202-227.

[49] I. Kaplansky, *Commutative Rings*, Allyn and Bacon Inc., Boston-New York, (1970).

[50] I. Kaplansky, *Fields and Rings*, The University of Chicago Press, Chicago-London, (1969).

[51] I. Kaplansky, *Infinite Abelian Groups*, The University of Michigan Press, Ann Arbor, Michigan, (1954).

[52] S.M. Khuri, Modules with regular, perfect, Noetherian or Artinian endomorphism rings, *Non-commutative Ring Theory*, Lecture Notes in Mathematics **1448**, Springer-Verlag, New York and Berlin, (1989), 7-18.

[53] P.A. Krylov, Torsion-free abelian groups with hereditary rings of endomorphisms, Algebra and Logic **27**, (3), (1989), 184-190.

[54] P.A. Krylov, A.V. Mikhalev, A.A. Tuganbaev, *Endomorphism Rings of Abelian Groups*, Kluwer Academic Publishers, Boston and London, (2003).

[55] R. Kuebler, J.D. Reid, On a paper of Richman and Walker, Rocky Mt. J. Math. **5**, (4), (1975), 585-592.

[56] E.L. Lady, Nearly isomorphic torsion-free abelian groups, J. Algebra **35**, (1975), 235-238.

[57] E.L. Lady, *Torsion-Free Modules over Dedekind Domains*, preprint.

[58] S. Lang, *Algebraic Number Theory*, Addison-Wesley, New York, (1970).

[59] L.S. Levy, Direct-sum cancellation and genus class groups, Methods in Module Theory, G. Abrams, J. Haefner, K.M. Rangaswamy editors, *Methods in Module Theory*, Lecture Notes in Pure and Applied Mathematics, **140**, (1993), 203-218.

[60] A. Mader, *Almost Completely Decomposable Groups*, Gordon and Breach Science Publishers, France-Germany, (2000).

[61] A. Mader, C. Vinsonhaler, Torsion-free E-modules, J. Algebra, **108**, (1987), 88-101.

[62] C. Murley, The classification of certain classes of torsion-free abelian groups, Pac. J. Math. **40**, (1972), 647-665.

[63] G.P. Niedzwecki, J. Reid, Abelian groups projective over their endomorphism rings, J. Algebra **159**, (1993), 139-149.

[64] A.T. Paras, Abelian groups as Noetherian modules over their endomorphism rings, *Abelian Group Theory and Related Topics*, Contemporary Mathematics, **171**, Ed. R. Göbel, P. Hill, W. Lieber, Providence, (1993), 325-332.

[65] R.S. Pierce, Realizing Galois fields, CISM Courses and Lectures, **287**, *Abelian Groups and Modules*, Springer-Verlag, Wein and New York, (1984), 291-304.

[66] R.S. Pierce, C. Vinsonhaler, Realizing algebraic number fields, *Abelian Group Theory*, Lecture Notes in Mathematics **1006**, Springer-Verlag, New York and Berlin, (1982-3), 49-96.

[67] J.D. Reid, Abelian groups finitely generated over their endomorphism rings, *Abelian Group Theory*, Lecture Notes in Mathematics **874**, Springer-Verlag, New York and Berlin, (1981), 41-52.

[68] J.D. Reid, On quasi-decompositions of torsion-free abelian groups, Proc. Am. Math. Soc. **13**, (1961), 550-554.

[69] I. Reiner, *Maximal Orders*, Academic Press Inc., New York, (1975).

[70] F. Richman, An extension of the theory of completely decomposable torsion-free abelian groups, Trans. Am. Math. Soc **279**, (1), (1983), 175-185.

[71] F. Richman, E.A. Walker, Primary abelian groups as modules over their endomorphism rings, Math. Z. **89**, (1965), 77-81.

[72] J. Rotman, *An Introduction to Homological Algebra*, Pure and Applied Mathematics **85**, Academic Press, New York, San Francisco, and London, (1979).

[73] P. Schultz, The endomorphism rings of the additive group of a ring, J. Australian Math. Soc. **15**, (1973), 60-69.

[74] J. Seltzer, A cancellation criterion for finite rank torsion-free abelian groups, Proc. Amer. Math. Soc. **94**, (1985), 363-368.

[75] B. Stenstrom, *Rings of Quotients: An Introduction to Ring Theory*, Springer-Verlag, Die Grundlehren Band **217**, New York and Berlin, (1975).

[76] I. Stewart, D. Tall, *Algebraic Number Theory*, 2nd ed., Chapman and Hall publishers, London and New York, (1987).

[77] R.G. Swan, Projective modules over group rings and maximal orders, Ann. of Math. (2) **76**, (1962), 55-62.

[78] C. Vinsonhaler, Torsion-free abelian groups quasi-projective over their endomorphism rings II, Pac. J. Math. **74**, (1978), 261-265.

[79] C. Vinsonhaler, The divisible and *E*-injective hulls of a torsion-free group, *Abelian Groups and Modules, Proceedings of the Udine Conference*, Udine, April 1984, Ed. R. Göbel, C. Metelli, D. Orsatti, L. Salce, Springer-Verlag **287**, Wein and New York, (1984), 163-181.

[80] C. Vinsonhaler, K. O'Meara, Separative cancellation and multiple isomorphism in torsion-free abelian groups, J. of Algebra **221**, (1999), 536-550.

[81] C. Vinsonhaler, W. Wickless, Homological dimension of completely decomposable groups, *Abelian Groups*, Lecture Notes in Pure and Applied Mathematics **146**, Ed. L. Fuchs, R. Göbel, (1993), 247-258.

[82] C. Vinsonhaler, W. Wickless, Balanced projective and cobalanced injective torsion-free groups of finite rank, Acta Math. Hung. **46** (3-4), (1985), 217-225.

[83] C. Vinsonhaler, W. Wickless, Injective hulls of torsion-free abelian groups as modules over their endomorphism rings, J. of Algebra, **58** (1), (1979), 64-69.

[84] C. Vinsonhaler, W. Wickless, Torsion-free abelian groups quasi-projective over their endomorphism rings, Pac. J. Math. (2) **70**, (1977), 1-9.

[85] C.P. Walker, Properties of Ext and quasi-splitting of abelian groups, Acta Math. Acad. Sci. Hungary **1**, (1964), 157-160.

[86] E.A. Walker, Quotient categories and quasi-isomorphisms of abelian groups, Proc. Colloq. Abelian Groups, Budapest, (1964), 147-162.

[87] R.B. Warfield, Jr., Extensions of torsion-free abelian groups of finite rank, Arch. Math. **23**, (1972), 145-155.

[88] R.B. Warfield, Jr., Cancellation for modules and stable range in endomorphism rings, Pac. J. Math. **91**, (1980), 457-485.

[89] R.B. Warfield, Jr., Countably generated modules over commutative Artinian rings, Pac. J. Math. **60** (2), (1975), 289-302.

[90] H. Zassenhaus, Orders as endomorphism rings of modules of the same rank, J. London Math. Soc. **42**, (1967), 180-182.

[91] B. Zimmerman-Huisgen, Endomorphism rings of self-generators, Pac. J. Math. **61**, (1), (1975), 587-602.

Index

9 780367 389321